CALIFORNIA STATE UNIVERSITY, SACRAMENTO

This book is due on the last date stamped below.
Failure to return books on the date due will result in assessment of overdue fees.

The Ecology of Adaptive Radiation

Dolph Schluter
University of British Columbia, Canada

UNIVERSITY PRESS

OXFORD
UNIVERSITY PRESS

Great Clarendon Street, Oxford OX2 6DP

Oxford University Press is a department of the University of Oxford.
It furthers the University's objective of excellence in research, scholarship,
and education by publishing worldwide in

Oxford New York

Athens Auckland Bangkok Bogotá Buenos Aires Calcutta
Cape Town Chennai Dar es Salaam Delhi Florence Hong Kong Istanbul
Karachi Kuala Lumpur Madrid Melbourne Mexico City Mumbai
Nairobi Paris São Paulo Singapore Taipei Tokyo Toronto Warsaw
with associated companies in
Berlin Ibadan

Oxford is a registered trade mark of Oxford University Press
in the UK and in certain other countries

Published in the United States
by Oxford University Press Inc., New York

© Dolph Schluter, 2000

The moral rights of the authors have been asserted

Database right Oxford University Press (maker)

First published 2000

All rights reserved. No part of this publication may be reproduced,
stored in a retrieval system, or transmitted, in any form or by any means,
without the prior permission in writing of Oxford University Press,
or as expressly permitted by law, or under terms agreed with the appropriate
reprographics rights organization. Enquiries concerning reproduction
outside the scope of the above should be sent to the Rights Department,
Oxford University Press, at the address above

You must not circulate this book in any other binding or cover
and you must impose this same condition on any acquirer

British Library Cataloguing in Publication Data
Data available

Library of Congress Cataloging in Publication Data

1 3 5 7 9 10 8 6 4 2

ISBN 0 19 850523 X (hbk)
0 19 850522 1 (pbk)

Typeset by Newgen Imaging Systems (P) Ltd., Chennai, India
Printed in Great Britain
on acid-free paper by
Biddles Ltd, Guildford & King's Lynn

Contents

Preface		**vii**
1	**The origins of ecological diversity**	**1**
	1.1 Introduction	1
	1.2 The issues	2
	1.3 Fifty years later	6
	1.4 General implications	7
	1.5 The book	7
2	**Detecting adaptive radiation**	**10**
	2.1 Introduction	10
	2.2 Definitions	10
	2.3 Examples	22
	2.4 Adaptation vs adaptive radiation	31
	2.5 Discussion	32
3	**The progress of adaptive radiation**	**36**
	3.1 Introduction	36
	3.2 Generalist ancestors, specialized descendants?	37
	3.3 Repeatable rules of niche spread	49
	3.4 Replicate radiations	55
	3.5 Phenotypic evolution near the end of adaptive radiation	59
	3.6 Discussion	64
4	**The ecological theory of adaptive radiation**	**65**
	4.1 Introduction	65
	4.2 The ecological theory	65
	4.3 Extensions and alternatives	71
	4.4 Discussion	83
5	**Divergent natural selection between environments**	**84**
	5.1 Introduction	84
	5.2 Natural selection and the adaptive landscape	85
	5.3 Comparison with the neutral expectation	90

5.4	Reciprocal transplant experiments	95
5.5	Direct measurements of natural selection	104
5.6	Estimating adaptive landscapes from environments	111
5.7	How do adaptive peak shifts occur?	115
5.8	Discussion	121

6 Divergence and species interactions — 123

6.1	Introduction	123
6.2	Divergence between competitors	124
6.3	Observational evidence	129
6.4	Evidence from prediction	147
6.5	Evidence from field experiments	150
6.6	Other interactions promoting divergence	152
6.7	Discussion	161

7 Ecological opportunity — 163

7.1	Introduction	163
7.2	Ecological opportunity and morphological divergence	164
7.3	Ecological opportunity and speciation rate	175
7.4	Key evolutionary innovations	181
7.5	Discussion	186

8 The ecological basis of speciation — 188

8.1	Introduction	188
8.2	Models of ecological speciation	189
8.3	Tests of ecological speciation	197
8.4	Divergent sexual selection	207
8.5	Discussion	212

9 Divergence along genetic lines of least resistance — 215

9.1	Introduction	215
9.2	Quantitative genetic framework	216
9.3	Divergence along genetic lines of least resistance	224
9.4	Divergent natural selection in retrospect	231
9.5	Discussion	235

10 The ecology of adaptive radiation — 236

10.1	Finale	236
10.2	General features of adaptive radiation	236
10.3	Fate of the ecological theory	238

References — 244

Index — 285

Preface

Adaptive radiation is a spectacular feature of evolution. It is also widespread, more so than the list of familiar cases, including the Galápagos finches, the cichlid fishes of East African lakes, and the Hawaiian silversword alliance, alone would suggest. Much of life's diversity, perhaps even most of it, has arisen during similar episodes of speciation and phenotypic and ecological divergence. My main goal in this book is to assess how far we have come in understanding the causes of this remarkable process. Before I began the book I held the naive notion that my years of study on the Galápagos finches, where my fascination with adaptive radiation began, on African and American finches, and more recently on fishes of postglacial lakes, had taught me enough about adaptive radiation that my task would involve little more than writing down all I knew before I forgot it. As the book got underway the limits of my knowledge became distressingly apparent, and I now feel I learned most of its contents along the way. Here I aim to put the results of many studies together to ask whether they conform or contrast with the dominant 'ecological' theory of adaptive radiation that was formulated in the first half of the last century. This theory was compelling, indeed I believe that it represented the greatest single application of the then-new modern synthesis, but it was based on very little evidence. Now seemed an ideal time for a thorough evaluation. The time required to write the book was borrowed from other duties, and I have many to thank for making this possible. The start of the project was facilitated by a E.W.R. Steacie Memorial Fellowship from the Natural Sciences and Engineering Research Council (NSERC) of Canada. I continued the project while on sabbatical leave at James Cook University in Queensland, Australia. Howard Choat and Julian Caley of Marine Biology were my hosts there, and provided me with space and considerable resources. Ken MacCrimmon and the Peter Wall Institute of Advanced Studies at The University of British Columbia later gave me the facilities and time necessary to complete the book. Finally my family endured many lonely Sundays so that I might catch up on sleep.

Many people deserve thanks for reading and commenting on individual chapters, sometimes more than one draft. Foremost of these is Trevor Price, who read different versions of every chapter as many as three times, on each occasion densely covering the margins of its pages with often savage but always constructive criticisms. I probably would not have attempted to write the book in the first place had I not known

I could rely on his feedback. The Series editor, Paul Harvey, encouraged the project from the beginning and suggested many improvements. Andrea Schluter's suggestions improved the writing and organization immeasurably. John Endler dissected Chapter 5, and I hope the reconstruction measures up. The many others who provided significant input and expertise include Peter Abrams, Scott Armbruster, Mark Blows, Julian Caley, Micheal Doebeli, Peter Grant, Todd Hatfield, Steve Heard, Menna Jones, Jeff McKinnon, Arne Mooers, Sarah Otto, Beren Robinson, Simon Robson, Howard Rundle, Stewart Schultz, Steve Vamosi, and Michael Whitlock. I thank many individuals for sending their data and preprints of unpublished work, especially Peter Abrams, Bruce Baldwin, Louis Bernatchez, Neils Bouton, Jerry Coyne, Bernie Crespi, Dan Funk, Peter Grant, Jessica Gurevitch, Thomas Hansen, Jana Heilbuth, Duncan Irschick, Scott Hodges, Glen Johns, Jonathan Losos, Eric Nagy, Shyril O'Steen, Allen Orr, David Pfennig, Trevor Price, Bob Ricklefs, Beren Robinson, Stephen Rothstein, Doug Schemske, Michael Sanderson, and Olle Seehausen. Howard Rundle read the final version from cover to cover and repaired many awkward defects. Scott Hodges and Jonathan Losos sent me the photos that grace the front and back covers. Other credits too numerous to list precede the actual writing, but I must acknowledge my sister, Alice, for teaching me to read.

Dolph Schluter
Vancouver
9 July, 2000.

For Andrea and Magdalen
who helped the most

1

The origins of ecological diversity

Theoretically, at least, the whole of the diversity of life is explicable by these two not sharply distinct processes

—Simpson (1953)

1.1 Introduction

Adaptive radiation is the evolution of ecological diversity within a rapidly multiplying lineage. It is the differentiation of a single ancestor into an array of species that inhabit a variety of environments and that differ in traits used to exploit those environments. It includes the origin of new species and the evolution of ecological differences between them. It is regarded as the hallmark of adaptive evolution and may well be the most common syndrome in the origin and proliferation of taxa. The most dramatic living examples of adaptive radiation, such as the Galápagos finches (Fig. 1.1), the Hawaiian silverswords (Fig. 1.2), and the East African cichlid fishes (Fig. 1.3), rank with the dinosaur extinctions and the origin of our own species among the most celebrated events in the history of life.

This book addresses the causes of adaptive radiation. Its focus is a theory developed by naturalists working in the first half of the twentieth century, and whose origins trace back to Darwin (1859). The theory received its strongest and clearest formulation in the writings of Lack (1947), Dobzhansky (1951) and especially Simpson (1953), and it was widely accepted by the middle of the twentieth century. The theory holds that adaptive radiation—phenotypic divergence as well as speciation—is ultimately the outcome of divergent natural selection stemming from environments, resources, and resource competition.

This 'ecological theory' of adaptive radiation represents the last major synthesis of ideas addressing the origins of ecological diversity. My principal aim in the book is to re-evaluate this theory in light of all the evidence that has accumulated since its formulation. Much progress has been made in testing the most important elements of the theory, and many new factors have come to be recognized as also playing a role in adaptive radiation. A modern revision is long overdue.

My treatment proceeds on two fronts. On the first, I test the major elements of the theory as originally formulated, and in the process demonstrate the many environmental mechanisms of divergent natural selection. On the second, I evaluate the

2 • *The origins of ecological diversity*

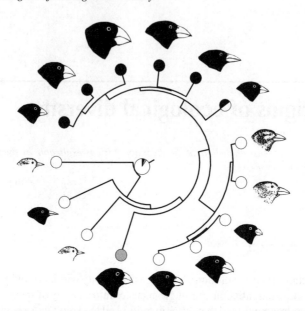

Fig. 1.1 Adaptive radiation in beak size and shape in the Darwin's finches. Phylogenetic relations are based on microsatellite distances (Petren *et al.*, 1999). Diets are indicated by shading: seeds (filled), insects (open), and vegetation (shaded). The pie at the centre provides an estimate of the diet of the last common ancestor of all extant species, with area of portions giving relative support for the three diet states (cf. Schluter *et al.*, 1997). Modified from Schluter (1996a) after Grant (1986), Bowman (1961) and Swarth (1931).

importance of other mechanisms not incorporated or poorly developed in the original formulation of the theory. My overarching goal is to make progress toward a richer and more complete theory of adaptive radiation.

1.2 The issues

1.2.1 *Adaptive and nonadaptive radiation*

My first goal is to identify those features that mark a *radiation* and that justify calling it *adaptive*. Almost everyone's definition of adaptive radiation would encompass the most famous cases (e.g. Figs 1.1–1.3). The task is to pull out the most salient general features to classify the rest. I adopt a straightforward definition close to the traditional one: Adaptive radiation is the evolution of ecological and phenotypic diversity within a rapidly multiplying lineage. It occurs when a single ancestor diverges into a host of species that use a variety of environments and that differ in traits used to exploit those environments. Comparative methods in evolutionary biology help to apply this definition, and a number of examples will be shown (Chapter 2).

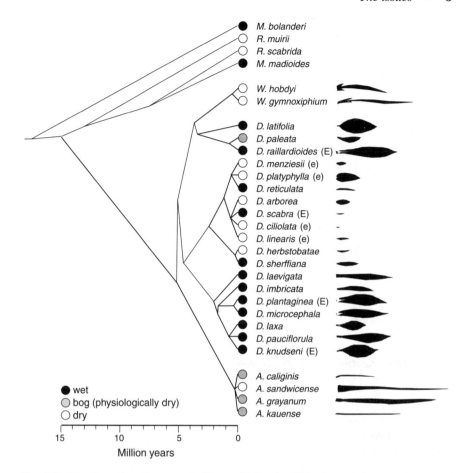

Fig. 1.2 The Hawaiian silversword alliance (*Dubautia*, *Wilkesia*, and *Argyroxiphium*) and related California tarweeds (*Madia* and *Raillardioides*). Circles at branch tips indicate habitat, taken from Robichaux *et al.* (1990). Leaf profiles are from Carr *et al.* (1989; reprinted with permission of Oxford University Press); they show a tendency for narrower leaves in dry environments. 'E' and 'e' indicate high and low leaf tissue elastic modulus, respectively, from Robichaux and Canfield (1985). A high modulus means a low capacity for maintaining turgor as tissue water content decreases, whereas a low modulus indicates a high capacity. The tree was modified from Baldwin and Sanderson (1998) by adding several species according to the topology and nucleotide differences in Baldwin and Robichaux (1995).

Satisfying the definition is the first step, paving the way for tests of the theory that divergent natural selection is the ultimate cause, against alternative hypotheses for rapid speciation and phenotypic divergence. Recognizing that an adaptive radiation has taken place does not by itself distinguish these theories.

4 • *The origins of ecological diversity*

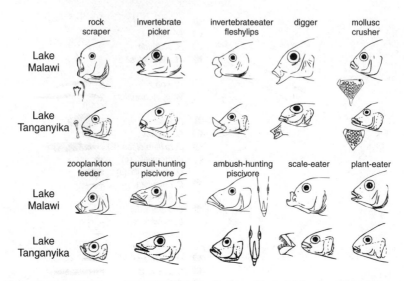

Fig. 1.3 Similar phenotypes in the cichlid fish radiations of Lake Malawi and Lake Tanganyika in East Africa. Images are from Fryer and Iles (1972; Malawi) and Liem (1991; Tanganyika) and are reprinted with kind permission of G. Fryer and Kluwer Academic Publishers, respectively.

1.2.2 Specialization, niche expansion, and other trends

A better understanding of adaptive radiation would result from accurate descriptions of how ecological and phenotypic diversity are modified throughout the process. What trends exist and how repeatable are they?

One trend widely discussed in many early writings is the tendency of species to become more specialized. Founders of adaptive radiations, particularly on remote archipelagoes, are often regarded as unspecialized immigrants whose descendants eventually made use of more narrow segments of the resource base (e.g. Lack, 1947; Simpson, 1953). The evidence for and against this assumption is explored in detail.

Other questions about sequences of niche differentiation are also investigated in the taxa to which they apply, in particular whether expansion to new resources and environments follows a characteristic ordering. For example, Grant (1949) posited that floral differentiation preceded vegetative differentiation in angiosperm lineages pollinated by specialized animal vectors, whereas this sequence is reversed in lineages pollinated by unspecialized animal vectors or abiotic agents such as wind. Other taxa have been identified as undergoing a characteristic order of differentiation along the major resource dimensions of habitat, food size, and food type. An alternative hypothesis is that divergence is idiosyncratic, lineage-specific, and never replicated. I also describe how phenotypic diversity changes during an adaptive radiation, whether it expands most rapidly early in a radiation and whether it falls when species diversity declines as a radiation nears its end.

1.2.3 *The three processes of the ecological theory*

The ecological theory incorporates three main processes, to be evaluated in turn. The first is *phenotypic differentiation between populations and species caused directly by differences in the environments they inhabit and the resources they consume*. According to this view, each environment subjects a species to unique selection pressures, owing to the advantages of specific combinations of traits for efficiently exploiting that environment. Different environments and resources provide the alternative fitness 'peaks' that bring about divergence in phenotype.

The second process is *divergence in phenotype resulting from resource competition*. Competition drives co-occurring populations and species to exploit new resource types and environments where they become subject to distinct selection pressures. This second process is like the first, but here each species alters environments in such a way that natural selection pressures on ecologically similar species are changed, favouring divergence.

The other side of competition between close relatives is *ecological opportunity*, the wealth of different resource types underutilized by species in other taxa. Divergence of a lineage proceeds apace when ecological opportunity is high, according to the theory, and then slows as niches became filled. Ecological opportunity often means being in the right place at the right time: colonizing a remote archipelago or surviving a mass extinction. Less intuitively, new niches may become available upon the acquisition of a novel trait or 'key innovation,' which then precipitates adaptive radiation.

The third process of the ecological theory is *ecological speciation*, whereby new species arise in adaptive radiation by the same processes that drive differentiation of phenotypes, namely divergent natural selection stemming from environment and resource competition. By speciation I mean the buildup of reproductive isolation, which in adaptive radiation evolves around the time that populations diverge in ecology and aspects of the phenotype.

The importance of each of the three processes has been debated since the theory's formulation. The first part, phenotypic divergence driven by divergent natural selection between environments, seems almost self-evident when obviously useful traits such as beak size or floral display are concerned. Yet, the ecological theory of adaptive radiation was formulated entirely without direct evidence that natural selection was truly divergent. Establishing that populations are separated by fitness valleys resulting from environment and resources is surprisingly difficult even now. The importance of resource competition in divergence has been even more controversial, and debates have lingered to the present time. The role of ecological opportunity was poorly quantified. The existence of key innovations was not demonstrated, nor was there evidence that they spurred adaptive radiation by increasing ecological opportunity. Finally, the prevalence of ecological speciation was doubted from the beginning. Even by 1950, most evolutionists realized that many nonecological processes also lead to speciation. The key issue is whether divergent natural selection plays the dominant role when speciation is rapid during adaptive radiations.

1.2.4 Genetics and the path of adaptive radiation

The fact that mendelian genetics underlies all evolutionary change was established before the ecological theory itself was developed. It was widely known that standing variation in populations had a partly genetic basis and that evolution could occur without the introduction of new mutations (e.g. Wright, 1945). Simpson (1953) appreciated that traits may be correlated genetically through pleiotropy or linkage disequilibrium and that selection for one trait may cause another to evolve even against the force of selection on it.

Lacking in these early studies of adaptive radiation is the use of genetics in a predictive sense. Can measures of genetic variation and covariation within populations help us to determine why phenotypic divergence proceeded in some directions and not others, or why divergence was accompanied by speciation in some instances and not others? The quantitative framework for answering these questions, evolutionary quantitative genetics, was undeveloped until recently and hence was not a part of the initial ecological theory. One of my goals is to establish whether the promise of quantitative prediction has been fulfilled, and whether an expanded theory of adaptive radiation can now more thoroughly incorporate the genetics of phenotypic divergence.

1.3 Fifty years later

A re-evaluation of the ecology of adaptive radiation is timely for a number of reasons. Foremost is the passing of nearly 50 years without such an overview, despite considerable advances in our abilities to study the process. When the ecological theory was developed, there had been no direct demonstrations of divergent natural selection in the wild. The experimental study of competition and other species interactions was in its infancy. No tests of key innovation hypotheses had even been attempted. The biological species concept of Mayr (1942) was young and no concrete link had been established between divergent natural selection and the origin of reproductive isolation. The grounds for believing the ecological theory were frail and indirect.

The contrast with the present could hardly be more striking. Studies estimating natural selection on wild populations have proliferated, giving us a view of the prevalence and force of divergent natural selection in nature and its environmental agents. We have an idea in a few instances of what 'landscapes' of selection look like in nature, and whether they have multiple peaks and valleys. Our understanding of competition and other interactions is vastly greater now than 50 years ago, and the number of alleged cases of competition-induced divergence has ballooned. Tests of ecological opportunity have made great strides, and several candidate key innovations have been identified. Experimental models of ecological speciation have proliferated, and knowledge of the role of divergent natural selection in natural speciation is advancing steadily. Sexual selection has been observed in many natural populations, and its consequences for speciation in adaptive radiation are potentially enormous. Phylogenetic trees of lineages undergoing adaptive radiation are now estimable, and statistical methods based on these trees have improved our ability to identify adaptive radiations. The trees are also providing insights into the timing and sequence of ecological changes throughout a radiation. Quantitative genetic theory has given us the tools to predict the direction and rate of divergence in complex suites of traits under natural selection, and even to estimate the history of net selection pressures that led to the diversity of modern forms.

The last 50 years have also witnessed advances in theory that challenge or extend the original synthesis. These alternatives include models of genetic drift combined with divergent natural selection and/or sexual selection; phenotypic divergence via interactions other than competition; critiques of the notion of ecological opportunity; and models of nonecological speciation. These ideas make plain that a diversity of processes may underlie adaptive radiation and that an expanded theory may be overdue.

1.4 General implications

A re-evaluation of the ecology of adaptive radiation has broader implications, because adaptive radiations greatly impact our perception of the causes of evolution in general. Simpson (1953) went so far as to claim that adaptive radiation could explain all of life's diversity.[1] Adaptive radiation is undoubtedly a major feature of evolution whether or not it explains everything. Knowledge of its ecological causes therefore advances our understanding of the origins of all biological diversity. This generality has always been one of the prime reasons for studying adaptive radiation.

1.5 The book

In the chapters to follow, I synthesize the observational, experimental, predictive, and historical evidence concerning the ecology of adaptive radiation. Studies of closely related populations and species predominate. In practice, I focus on patterns and processes occurring in taxa ranging from conspecific populations on the one hand to congeneric species on the other. This focus may at first seem restricted. After all, adaptive radiation is a macroevolutionary phenomenon, and most of our ideas about the causes of adaptive radiation have come from investigations of higher taxa.

Nevertheless, most recent progress in applying the concept of adaptive radiation and testing its mechanisms has come instead from comparative analysis using young

---

[1] Simpson (1953) distinguished between 'adaptive radiation,' a term he applied strictly to simultaneous bursts in the number of lineages descended from a common ancestor, and 'progressive occupation of adaptive zones,' in which the timing of speciation events within a lineage were staggered in time. However, Simpson regarded these two processes as 'not sharply distinct' and I do not distinguish them. In his discussion of factors contributing to life's diversity, he also added a third influence: the spatial separation of different ancestral lineages. This, he believed, allowed different taxa to diversify into the same adaptive zone in different regions of the globe.

clades and from experimental and observational study of populations and closely related species in nature. The logic of this focus on adaptive radiation at low taxonomic levels is clearest when addressing speciation. Understanding the mechanisms driving the evolution of reproductive isolation requires study of the process around the time it is underway. Conspecific populations and recently-diverged species are then also the most relevant taxonomic levels for understanding the causes of phenotypic divergence and its link to speciation. A microevolutionary focus amounts to the study of 'macroevolution in action.'

Throughout, I concentrate on mechanisms acting on wild populations. Laboratory studies have greatly assisted the development of theories of ecological diversification, but simulation is their principal contribution to evaluating these theories. For example, laboratory experiments have revealed that divergent natural selection can cause the buildup of reproductive isolation between populations. This result enhances the plausibility of the hypothesis of ecological speciation, but does not inform us whether any species in nature have formed by such a mechanism.

My strategy is to evaluate each question broadly, which usually requires that I consider evidence beyond that from cases fulfilling the strict definition of adaptive radiation. This is partly to include studies of taxa that show many of the signs of adaptive radiation but for which the evidence is still incomplete. It allows me to bring as much evidence as possible to bear on an issue. To compensate, I return frequently to the issue of whether a particular result really is one that characterizes true adaptive radiations.

In Chapter 2, I define adaptive radiation along with other terms used throughout the book. I explain the main criterion for concluding whether or not an adaptive radiation has occurred, and illustrate it with a selection of cases. These cases illustrate the kinds of traits that diverge during an adaptive radiation and the variety of environmental agents that may drive them. In Chapter 3, I use case studies to test for trends in the evolution of niche use during adaptive radiation. Of special interest is the alleged trend toward specialization in resource use, and other regularities in the sequence of steps in divergence of niche use. I also describe how morphological diversity may change from beginning to end of adaptive radiation.

Chapter 4 summarizes the ecological theory in detail and the types of patterns that stimulated its development. Alternative hypotheses for these patterns, and for adaptive radiation itself, are then presented. This establishes the framework for the next four chapters, which review evidence for each of the three processes important to the ecological theory of adaptive radiation. Chapter 5 presents the evidence that divergent natural selection stemming from resources and environment drives phenotypic differentiation in adaptive radiation. Chapter 6 summarizes models for divergence between competing species and the evidence from natural populations that competition is involved in phenotypic divergence. The chapter identifies other interactions between species that might be equally important in divergence but about which much less is known. Chapter 7 evaluates evidence addressing the hypothesis of ecological opportunity and its putative mechanisms. I compare rates of morphological evolution and speciation on islands and mainlands, proxies for high and low levels of ecological opportunity, and case studies of several possible key innovations. Chapter 8 summarizes the evidence for divergent natural selection between environments in the evolution of reproductive isolation. I include a discussion of sexual selection's role in speciation, and how environments may precipitate the evolution of divergent mate preferences.

Chapter 9 reviews the contributions of quantitative genetic theory to understanding the paths of divergence and speciation in adaptive radiation. Chapter 10 synthesizes the results of the previous chapters and summarizes the revised ecological theory of adaptive radiation. In it I also review the most pressing work that lies ahead.

2

Detecting adaptive radiation

Seeing this gradation and diversity of structure in one small, intimately related group of birds, one might really fancy that, from an original paucity of birds in this archipelago, one species had been taken and modified for different ends.
—C. Darwin (1842)

2.1 Introduction

What is adaptive radiation? A working definition and the means to apply it are crucial to successful study of the process. Modern usage derives from Simpson (1953, p. 223): 'adaptive radiation strictly speaking refers to more or less simultaneous divergence of numerous lines from much the same adaptive type into different, also diverging adaptive zones'. This definition is not fully operational by itself, but it incorporates two core processes that most researchers continue to regard as essential: a rise in the rate of appearance of new species and a concurrent increase in ecological and phenotypic diversity.

In this chapter, I define adaptive radiation as it will be used throughout the book, and give an overview of progress and problems in testing putative examples. To better explicate the concept, I describe in some detail four cases that satisfy the definition of adaptive radiation reasonably well. Two are of animals (Darwin's finches on the Galápagos Islands, and *Anolis* lizards on large Caribbean islands), and two are of plants (the Hawaiian silversword alliance, and mainland columbines *Aquilegia*). I comment on the distinction between phenotypic divergence in adaptive radiation and modern concepts of adaptation. I close by highlighting some unresolved issues.

2.2 Definitions

2.2.1 *Adaptive radiation*

Adaptive radiation is the evolution of ecological and phenotypic diversity within a rapidly multiplying lineage. It involves the differentiation of a single ancestor into an array of species that inhabit a variety of environments and that differ in the morphological and physiological traits used to exploit those environments. The process includes both speciation and phenotypic adaptation to divergent environments. This definition of adaptive radiation preserves the crucial elements of its predecessors (e.g. Simpson, 1953; Lack, 1947; Mayr, 1963; Carlquist, 1974; Grant, 1986; Futuyma, 1986; Skelton, 1993; Schluter, 1996a) and incorporates key features seen in the most impressive cases (e.g. Figs 1.1–1.3).

According to this definition, an adaptive radiation may be detected by four features that I will employ as criteria. The first is common ancestry of component species. The second is phenotype–environment correlation, the significant association between environments utilized and the morphological and physiological traits used to exploit those environments. The third is trait utility, the performance or fitness advantages of trait values in their corresponding environments. These second and third criteria establish that phenotypes of species 'fit' the divergent environments they exploit, and hence that species have 'been taken and modified for different ends' (Darwin, 1842). The fourth criterion is rapid speciation, the presence of one or more bursts in the emergence of new species around the time that ecological and phenotypic divergence is underway. I elaborate on these four features below.

Common ancestry—Common ancestry is straightforward to assess from phylogenetic trees of species relationships (Givnish, 1997). My focus on radiations of low taxonomic level assumes further that common ancestry is recent. Common ancestry is not the same as monophyly, however, which would demand that *all* descendants of the common ancestor be included in the radiation. No theory of adaptive radiation predicts monophyly, and indeed field studies are frequently carried out on pruned lineages—clades with uninteresting, obscure, or geographically distinct branches removed. I revisit this point in Section 2.5.

Phenotype–environment correlation—A fit between the diverse phenotypes of descendant species and their divergent environments fulfills the 'adaptive' criteria. Here I mean adaptive in the traditional sense of current utility (cf. Reeve and Sherman, 1993) rather than in the more restrictive sense of evolutionarily derived traits that retain their historical function (Gould and Vrba, 1982; Coddington, 1988; Baum and Larson, 1991; Harvey and Pagel, 1991). My focus on adaptive radiations at low taxonomic level minimizes the significance of the distinction between the two concepts of adaptation because traits in the youngest taxa are likely to experience current selection pressures similar to those initially favouring their appearance.

A significant phenotype–environment correlation is the first indication of such a fit. The evidence typically emerges from field observations that species' differences in phenotype (e.g. body size) are associated with use of different resources or other features of environments (e.g. prey size). Statistical testing of such correlations between phenotype and environment, controlling if necessary for nonindependence of species attributes that phylogeny may induce (the 'comparative method'; Harvey and Pagel, 1991), is the principal means of establishing whether such correlations are

12 • *Detecting adaptive radiation*

real (Box 2.1). Phenotypic differences between species must be genetically based, not just the outcome of plastic responses to divergent environmental signals during development. (Inducibility itself is a genetically based trait that may evolve to different levels between environments.)

Box 2.1 Phylogeny and tests of trait–environment association.

Statistical tests of association between phenotype and environment or performance confront the problem that under some models of phenotypic evolution, especially 'random walks' in time, population and species means are not independent if they are related to varing degrees by descent. This is because two species sharing a relatively recent common ancestor are expected to be more similar in their trait values than two species picked at random (Felsenstein, 1985; Harvey and Pagel, 1991). The solution to this problem powerfully advanced the study of adaptive radiation. Applying the solution, however, usually requires an accurate phylogeny, implying that adaptive radiation can scarcely be tested without one. For this reason, in Table 2.1 I have emphasized cases that incorporate phylogenetic information in tests, or cases for which sufficient information on phylogeny is available to indicate that its explicit incorporation is unnecessary.

The solution comprises several methods suited to different types of data. Under the null hypothesis of no association, all the methods assume that changes in traits, performance, and environment mimic independent 'random walks' through time. The alternate hypothesis is that of a correlated random walk. The method of independent contrasts (Felsenstein, 1985) tests associations between continuously varying traits and environmental states such as body size (X) and prey size (Y). For discrete states (e.g. presence/absence of predators vs presence/absence of defensive armour) Pagel's (1994) likelihood ratio test is recommended. Hansen (1997) has presented a method to deal with discrete environmental states and continuous phenotypic or performance traits. Martins and Hansen (1996) present a general framework for handling all data types.

Observation suggests that if the method of independent contrasts rejects the null hypothesis of no association between two continuous traits, then a conventional statistical test that ignores phylogeny is also likely to reject it (Ricklefs and Starck, 1996; Price, 1997). Apparently, phylogeny matters least when a radiation is indeed an adaptive radiation (this conclusion may not hold for discrete traits). In any case, phylogeny's unimportance cannot be assumed *a priori*. But the real possibility that phylogeny would not matter in the end may justify alternative testing approaches. For example, before doing anything else, test whether phylogenetic similarity indeed predicts similarity in phenotype, environment, and performance (Gittleman and Kot, 1990; Mooers *et al.*, 1999; Abouheif, 1999). If the hypothesis of 'no influence of phylogeny' cannot be rejected, and if it can be shown that this is not simply the outcome of low statistical power, then conventional statistical methods may be justified. Such an approach has been used in *Anolis* lizards (Irschick *et al.*, 1997).

However, the random walk is not the only possible model of trait evolution. Harvey and Rambaut (2000) have compared the performance of methods for testing correlated evolution under other models of continuous trait evolution. For example, in Price's (1997) model competition between species prevents any two of them from having closely similar values for trait and environment. Instead, species are assigned to vacancies in the ellipse

> of feasible values for trait and environment. A vacancy is most likely to be filled by a speciation event in the branch of the clade with the nearest living member. Simulation has revealed that conventional regression and correlation performs better than the method of independent contrasts under this model (Harvey and Rambaut, 2000) even though an influence of phylogeny is built in.
> The most appropriate test of correlated evolution therefore depends on the most relevant model of phenotypic evolution, which is not easy to determine *a priori*. A robust approach may be to report results from both phylogeny-based and conventional statistical methods.

Trait utility—Evidence that morphological and physiological traits of species are indeed useful where they are employed is the next requirement for establishing a fit between phenotypes and environments. The usual way is to test whether traits consistently associated with particular environments consistently enhance performance there. Experimental or theoretical methods (ideally, both) create this crucial link. Such tests are important because they illuminate the mechanisms underlying the phenotype–environment correlation, they expose and eliminate trivial associations between phenotype and environment (Wainwright, 1994), and they connect trait values with fitness (Arnold, 1983). Direct measurements of natural selection on traits are an additional way to test trait utility. I consider direct measurements of selection further when evaluating theories of adaptive radiation (Chapter 5).

Rapid speciation—Speciation refers to the evolution of reproductive isolation, defined as the complete absence of interbreeding between individuals from different populations (should they encounter one another), or the strong restriction of gene flow sufficient to prevent collapse of genetically distinct populations that continue to interbreed at a low rate. This is basically the biological species concept (Mayr, 1942) but accomodates the fact that a great many sexual species hybridize (e.g. V. Grant, 1981; Gill, 1989; Grant and Grant, 1992; Rieseberg and Wendel, 1993; Mallet, 1995), yet existing levels of assortative mating do not decay.

Not everyone's definition of adaptive radiation includes speciation (e.g. Givnish, 1997), but the process characterizes all the best-known cases. To Simpson (1953), adaptive radiation marked the birth of a higher taxon, and lineage splitting (usually accomplished in sexual species by the evolution of reproductive isolation) is the requisite first stage. Regardless of one's perspective on the necessity of speciation in adaptive radiation, the evolution of reproductive isolation is common and has significant consequences for phenotypic divergence. Speciation therefore represents one of the most significant problems for research on adaptive radiation. Speciation (however defined) is not a major issue in adaptive radiation of fully asexual taxa because lineage splitting occurs each generation. Nevertheless, individuals of many asexual taxa often undergo gene exchange (e.g. via conjugation, transformation, and transduction) and how and whether the rate of exchange declines with phenotypic and genetic differentiation is an interesting question for study (Vulić *et al.*, 2000).

Specifying *rapid* speciation recognizes that speciation rates are highly variable between contemporaneous clades and between time periods within a clade (e.g. Fig. 2.1A). If speciation is episodic then so is adaptive radiation, if speciation is an important part. By focussing on the bursts, we distinguish adaptive radiation from events that precede and follow it, or from events in other lineages existing at the same time.

'Rapid' speciation is not a well-defined concept but at least four operational standards might be used. The first three are statistical (Box 2.2) and detect episodes of branching in phylogenetic trees by identifying: (1) periods of time within a clade in which rates of branching exceed those before and/or later; (2) asymmetries between contemporaneous clades in their number of descendant species; (3) periods of time or lineages in which speciation events significantly exceed extinction events. A fourth standard should also be considered, one that identifies periods or lineages in which reproductive isolation itself evolves unusually rapidly. For example, hybrid sterility and/or inviability usually evolve quite slowly (Coyne and Orr, 1989; Mahmood *et al.*, 1998) and this period may be used as a benchmark against which to measure truly rapid speciation.

2.2.2 Adaptive radiation in a narrow sense

The focus of this book is adaptive radiation at low taxonomic levels. In practice, I concentrate on or near the level of closely related species within the same genus. The reason is that mechanisms of divergence are clearest at this level. Adaptive radiation is also a macroevolutionary concept applicable to high taxonomic levels (e.g. 'the angiosperms' or 'the passerine birds') but it is best to use it in this way only if the criteria for adaptive radiation are also met within its sublevels. Speciation neither preceded, accompanied, nor soon followed by ecological and phenotypic divergence is nonadaptive radiation (see below) whatever the pattern at high taxonomic levels.

This 'narrow-sense' perspective on adaptive radiation is consistent with Simpson (1953). He applied the concept primarily to higher taxa but believed that patterns seen there were the magnified outcome of a population process: 'Whether we represent the outcome as radiation of genera, orders, or phyla, depends, so to speak, on how far back we stand, how much of the whole picture we take in and how much we generalize on a pattern which is in detail no more and no less than constantly recurring radiation of populations and constantly occurring succession of adaptive changes in populations.' (p. 227-8).

2.2.3 Nonadaptive radiation

Nonadaptive radiation is rapid proliferation of species accompanied by negligible or infrequent ecological differentiation (Gittenberger, 1991), or by morphological and physiological differentiation unrelated to patterns of resource use and environment (Brooks *et al.*, 1985). The 75 species of *Albinaria* land snails of Greece and its

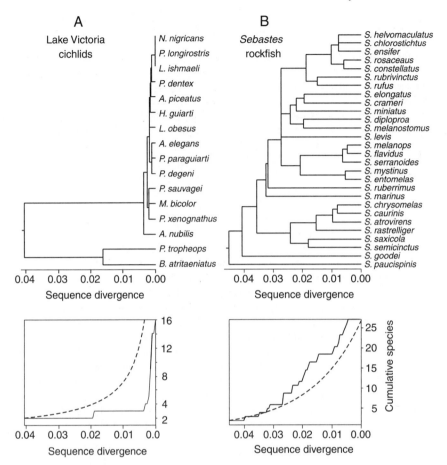

Fig. 2.1 Partial phylogenies of two fish radiations (upper panels) and the cumulative number of lineages (lower panels). The smooth dotted curve in each lower panel is the number of species expected over time under a random birth/death process with constant speciation and extinction rates, estimated using maximum likelihood (Nee et al., 1994). The random expectation is preliminary because the method used to generate them assumes complete trees. A. The cichlid fishes of Lake Victoria, East Africa show too few splitting events early in the tree and unduly many branching events in the very recent past compared with the random expectation. B. The *Sebastes* rockfishes show a lesser deviation from the random expectation but exhibit an excess of branching events in the middle part of the tree and no speciation events at the very tip. The phylogenies are maximum likelihood estimates from mtDNA sequences (Meyer et al., 1990; Johns and Avise, 1998). The cichlid tree used the nonendemic *Astateoreochromis allauadi* from Lake Victoria as the outgroup. The *Sebastes* tree was rooted with *Sebastopolus alascanus*. Modified from Johns and Avise (1998), with permission of the Society for the Study of Evolution.

Box 2.2   Detecting rapid speciation.

There is no standard criterion for 'rapid' speciation. The concept is relative and the outcome of a test may depend on the measure used and the benchmark employed. Nevertheless, the main goal is clear: to establish whether speciation rates are episodic and, if so, to locate the peak periods. Below I summarize three criteria that might be used as the basis for detecting speciation episodes. All three are based on analyses of branching rates of phylogenetic trees. This rate depends not only on the rate of speciation but also on the lineage extinction rate, and it can be difficult to distinguish these. A fourth criterion not discussed here examines instead the rate at which reproductive isolation evolves when a branching event occurs.

(1) Temporal variation in speciation and extinction rates—Nee *et al.* (1992) introduced the cumulative lineages-through-time plot as a way of visualizing the gradual rise in the number of species in a clade from the origin of the lineage to the present time. Data from such plots were then used to test deviations between observed cumulative frequencies and trends expected from simple stochastic 'birth–death' models in which speciation and extinction rates do not change through time. Two such lineages-through-time plots are shown in the lower panel of Fig. 2.1. One is from a sample of Lake Victoria cichlid fishes (16 of about 500 described species) and the other represents a sample of marine rockfishes *Sebastes* (28 of about 60 species). Within each panel, a jump in the height of the curve as one moves from the past to the present time corresponds to a branching (speciation) event in the tree immediately above. The observed lineages-through-time are compared with that under the null hypothesis that speciation rates b ('birth rate') and extinction rates d ('death') are constant through time (calculated from eqn (17) in Nee *et al.*, 1994). These examples are illustrative only, since the method assumes complete trees. The two taxa have sharply different trends. The cichlids show a late burst whereas cumulative numbers of *Sebastes* conform reasonably well with the random expectation.

Nee *et al.* (1994) suggested a goodness-of-fit test to determine whether observed lineages-through-time plots are different from that expected under the null model of constant rates. Under the null hypothesis the time intervals T_i between the ith and $(i+1)$th speciation event (corresponding to the time interval between consecutive jumps in the cumulative lineages-through-time plot) have an approximately exponential distribution with parameter $i(b-d)$. The observed values of $i \times T_i$ should have an approximately exponential distribution with parameter $(b-d)$, the difference between the true per lineage speciation and extinction rates. The exponential distribution is rejected in the cichlid example in Fig. 2.1A (KS-test using Stephens' approximation (implemented in S-Plus 2000; Mathsoft, 1999); $D = 0.54$, $P = 0.0003$), suggesting that $b-d$ has not been constant. The exponential distribution is not rejected in the case of *Sebastes* (Fig. 2.1B, $D = 0.16$, $P > 0.38$). Wollenberg *et al.* (1996) presented an alternate method, but its null distribution of lineages-through-time plots included only the surviving lineages from a large simulation of a birth and death process in which $b = d$. In contrast, the Nee *et al.* method fits any birth–death process in which $b > d$.

The above methods assume that we have complete phylogenetic trees, whereas most data sets are incomplete (corrections are presumably possible but have not been developed). A coalescent approach not requiring complete trees was used by Price *et al.* (1998), who suggested that bursts of branching could be detected by comparing the observed lengths of branches leading to tip species with the length of branches deeper in the tree, i.e. those connecting nodes. Comparatively long branches leading to the tips indicate a slowing down of speciation (or an increasing extinction rate) toward the present. The null hypothesis assumes that the number of species has remained constant through time since the origin of the clade.

(2) Differences in the numbers of descendants—The second approach to testing rapid speciation compares two or more ancestral lineages of equal age by the number of their contemporary descendants. If one lineage left many more descendants than the other, beyond that expected under the null hypothesis of equal speciation and extinction rates, then a significant episode of fast speciation (or reduced extinction) is indicated.

Slowinski and Guyer (1989) proposed a simple statistical procedure for sister taxa (two taxa sharing a common ancestor and therefore of equal age). If n is the total number of descendant species in the two sister taxa, then all partitions of n into n_1 species and n_2 species (with $n_1 + n_2 = n$) are equiprobable under the null hypothesis that sister taxa have equal speciation rates and equal extinction rates. The P value for a test of equal rates is then the probability of a partition at least as extreme as that observed,

$$P = \begin{cases} 2n_1(n-1)^{-1}, & \text{if } n_1 < n_2; \\ 1, & \text{if } n_1 = n_2. \end{cases}$$

If $P < 0.05$ then the null hypothesis of equal rates is rejected. This test is not statistically powerful. For example, if we set $n_1 = 1$ (i.e. one of the sister taxa never branched) then n_2 must be at least 40 to reject the null hypothesis.

A potentially more powerful approach contrasts the fit of two models to the data. In the null model speciation rates are equal between sister taxa, whereas the rates are unequal in the alternate model. Define $F(n_i, \lambda_i)$ to be the probability of observing n_i species at the present time in one of two sister taxa. Under a pure birth process (no extinctions),

$$F(n_i, \lambda_i) = e^{-\lambda_i}(1 - e^{-\lambda_i})^{n_i - 1}$$

(Sanderson and Donoghue, 1994). λ is the speciation rate multiplied by the time interval over which speciation is measured, here set to 1. Under the null model of equal speciation rates the maximum likelihood (ML) estimate of $\lambda_1 = \lambda_2$ is $\hat{\lambda} = \ln((n_1+n_2)/2)$. Under the alternate model of unequal rates the two ML estimates are $\hat{\lambda}_1 = \ln(n_1)$ and $\hat{\lambda}_2 = \ln(n_2)$. Under the null hypothesis of equal rates the following likelihood ratio test statistic

$$L = -2\ln\left(\frac{F(n_1, \hat{\lambda})F(n_2, \hat{\lambda})}{F(n_1, \hat{\lambda}_1)F(n_2, \hat{\lambda}_2)}\right)$$

has an approximate χ^2 distribution with 1 df.

Nee *et al.* (1996) generalize Slowinski and Guyer's (1989) formula to multiple taxa. They ask whether a given taxon has an unusually large number of descendants compared with other comparable contemporaneous taxa. Consider k lineages at some fixed time in the past that gave rise to a combined total of n contemporary species today. Under the

> null hypothesis that all lineages have equal speciation rates (and equal extinction rates) the probability that one of these lineages i had n_i descendants or more is
>
> $$P = 1 - \sum_v (-1)^v \binom{k}{v} \binom{n - v(n_i - 1) - 1}{k - 1} \bigg/ \binom{n - 1}{k - 1},$$
>
> where $v = 0, 1, \ldots, (n - k)/(n_i - 1)$ and $n/k \leq n_i \leq n - k + 1$. For example, $k = 3$ *Sebastes* lineages existed at the time corresponding to a sequence divergence of 0.04 (Fig. 2.1B). One of these lineages gave rise to $n_1 = 26$ of the $n = 28$ contemporary species in the sample, a result inconsistent with the hypothesis of equal rates ($P = 0.0085$).
>
> *(3) Speciation in excess of extinction*—The above methods are relative, identifying radiations by the presence of significant variation in speciation rates or extinction rates in time or between contemporaneous lineages. The problem with a relative measure is that a given clade may appear to be a radiation when compared against some lineages but not when compared against others. An absolute standard may therefore be desired. For example, one may wish to test whether speciation rates *substantially* exceed extinction rates over a lineages history. The *Sebastes* phylogeny (Fig. 2.1B) is best fit by a birth–death process in which the probability of extinction is zero. If statistically supportable, this would indeed represent a sustained excess of speciation over extinction. The formulas of Nee *et al.* (1994) would allow one to put likelihood bounds on the degree of excess. Applying these reveals that the speciation rate of *Sebastes* is at least an order of magnitude greater than its extinction rate.

surroundings have been put forth as a possible example. They exhibit a wide variety of shell forms but are apparently indistinguishable by environment (Gittenberger, 1991). Of course, the possibility exists in such cases that researchers have overlooked important features of environment, and that further study would uncover a full-blown adaptive radiation. Nevertheless, adaptive radiation cannot be assumed *a priori*, and nonadaptive radiation is a logical null hypothesis.

Radiations in which phenotypic divergence is limited to secondary sexual traits and preferences for these traits also qualify as nonadaptive radiations if the evolution of preferences is not guided by environment. Carson and Kaneshiro (Table 2.1) offered the Hawaiian *Drosophila* as a possible example of the latter. Their interpretation leaves unexplained the broad diversification of larval ecology and female reproductive traits in this group (Kambysellis and Craddock, 1997). The possible interaction between sexual selection and ecological divergence is discussed in Chapters 4 and 8.

Species proliferation in which ecological differentiation is accompanied only by divergence in resource preference or host recognition should also be considered nonadaptive radiation. Behavioural differentiation of this sort only implies that the species utilize different environments but not that selection in these environments have improved them. 'Adaptive radiation' is reserved for cases in which this ecological divergence is accompanied also by morphological and physiological changes that elevate the capacity to utilize those environments.

2.2.4 Other terms

Hereafter, I use *environment* as a general term including all extrinsic factors that potentially influence population dynamics and natural selection. 'Different environments' may refer to spatially distinct areas of the globe having contrasting biotic or abiotic regimes, or to different resources exploited by species inhabiting the same geographic region (e.g. habitat, host plant species, or food type). The main emphasis of this book is on *resources*, by which I mean any utilizable and potentially depletable aspect of environment, physical or biological, including raw nutrients, food, pollinators, and 'enemy-free space' (Jeffries and Lawton, 1984). Resources may also refer to the medium through which the animal must move to acquire food, and the devices that must be overcome to render food edible.

I use the terms *phenotype* and *phenotypic trait* to refer to aspects of morphology, physiology, and most behaviours. I focus on traits such as flower structure, bill size, sensory systems for finding prey, digestive chemistry, and predator evasion tactics, i.e. traits that affect the ability of individual organisms to exploit particular classes of resources. However, divergence between environments may include life history traits and mating systems (e.g. Stebbins, 1970, 1971). Most of the literature on adaptive radiation is concerned with morphological traits

The *adaptive zone* is a key component of the ecological theory of adaptive radiation (see Chapter 4). Simpson (1953) defined it essentially as a collection of empty niches that may be exploited by a set of species varying in phenotype but descended from a common ancestor. It refers to the set of ecological niches that do not require major discontinuities in way of life.

Niche has several meanings in the ecological literature (e.g. Schoener, 1989; Leibold, 1995) but evolutionary biologists largely retain a version of Grinnell's definition (Grinnell, 1917, 1924) and I do the same here. Grinnell's niche represents the set of environmental features utilizable by a single species within a particular region and abiotic setting. It is more a property of the environment than of the species '... in the figurative sense of a "recess" or place in the community' (Schoener, 1989). This metaphor of a 'recess' implies that Grinnell's niche is somehow discrete. This discreteness does not arise from attributes of resources alone, which may in principle be used in infinite combinations. It arises instead from the fact that morphology and physiology constrain the efficiencies with which sets of available resources can be utilized, and consequently that a given resource base will accomodate some phenotypes better than others. Grinnell's niche may therefore correspond to a 'peak' in the adaptive landscape of potential organism designs (Schluter, 1988*a*). This equating of niche and adaptive peak is implicit in Simpson (1944, 1953).

Table 2.1 Progress in testing the 'fit' of species to their diverse environments (correlated evolution of phenotype and environment, and performance consequences) in cases of apparent adaptive radiation

Taxon	Region	Phenotype–environment correlation	Performance consequences	Sources
Vertebrates				
Geospiza	Galápagos	Bill size and shape–seed size and hardness	Handling time; breaking stress; crushing force	Bowman (1961); Abbott *et al.* (1977); Schluter (1982); Schluter and Grant (1984); Grant (1981, 1986)
Loxia	N. America	Bill size–cone strength, stage	Handling time	Benkman (1993); Benkman and Lindholm (1991)
Phylloscopus	Asia	Body size–prey size; beak/body shape habitat and foraging; plumage–light regime		Price (1991); Richman and Price (1992); Marchetti (1993)
Parus	N. and W. Europe	Body size–feeding substrate; bill shape–habitat; limb length and musculature–substrate	Foraging ability; hanging and perching ability	Suhonen *et al.* (1994); Partridge (1976); Moreno and Carrascal (1993)
Vidua	Africa	Gape colouration–host species		Payne (1977); Klein and Payne (1998)
Anolis	Greater Antilles	Body size and hindlimb length–perch diameter and height	Sprint speed; jump distance; running stability	Losos (1990*a*, *b*, 1992); Losos and Irschick (1996); Irschick *et al.* (1997)
Ctenotus	Australia	Body size–habitat		Pianka (1969, 1986); Garland and Losos (1994)
Centrarchus	N. America	Body size and gape–prey size; pharyngeal jaw neuro/musculature–snails	Handling time; crushing force	Werner (1977); Lauder (1983); Mittelbach (1981, 1984); Wainwright and Lauder (1992)
Coregonus	N. America	Gill rakers–habitat		Bodaly (1979); Bernatchez *et al.* (1996)

Taxon	Region	Traits	Measured variables	References
Gasterosteus	N. America	Body size–prey size; body shape–habitat	Foraging success; growth rate	Lavin and McPhail (1987); Schluter and McPhail (1992); Schluter (1993, 1995); Walker (1997)
Neochromis	Lake Victoria	Premaxilla angle and length, hyoid position–diet	Biting and suction forces	Bouton et al. (1999)
Invertebrates				
Cicindela	N. America	Mandible length–prey size		Pearson and Mury (1979); Vogler and Goldstein (1997)
Cancer	Global	Body size–habitat		Harrison and Crespi 1999
Plants				
Dubautia	Hawaii	Tissue elastic modulus–habitat	Turgor pressure	Robichaux (1984); Robichaux and Canfield (1985); Carr et al. (1989)
Dalechampia	C. and S. America	Reward system–pollinator type; flower dimensions–pollinator species; mating system–habitat		Armbruster (1988, 1990, 1993)
Aquilegia	N. America	Flower orientation–pollinators; spur length–pollinators; flower colour–pollinators	Pollinator visits; pollen removal	Chase and Raven (1975); Miller (1981); Hodges (1997); Fulton and Hodges (1999)
Encelia	N. America	Pubescence–temperature, moisture regime	Transpiration, photosynthesis	Ehleringer and Clark (1988)
Brocchinia	S. America	Nutrient uptake, trichomes–habitat		Givnish et al. (1997)
Platanthera	N. hemisphere	Flower colour–pollinator; spur length–pollinator mouthpart length	Pollinia removal	Hapeman and Inouye (1997); Nilsson (1988)
Schiedea	Hawaii	Breeding system, pollination vector–habitat		Weller et al. (1990); Sakai et al. (1997)

2.3 Examples

It would not be possible to review or even list all cases of adaptive radiation that have been proposed. Instead I illustrate the concept by summarizing four compelling cases. I do not review our full state of knowledge about each of the four radiations, but rather focus on how each of them passes the four tests of recent common ancestry, phenotype–environment correlation, trait utility, and rapid speciation. I then comment on progress in fulfilling the criteria in a wider set of putative cases (Table 2.1). Throughout, I assume that phenotypic differences between species have a genetic basis, although this has not been confirmed in all cases.

2.3.1 *Galápagos finches*

The Darwin's finches need little introduction. They are the best-studied adaptive radiation from an ecological perspective. Lack's (1947) perceptive analysis of the Galápagos species guided not only the course of future work on this group (they should be called Lack's finches, not Darwin's) but also the genesis of the ecological theory. The long-term field studies by 'el grupo Grant' have tested numerous aspects of the theory (e.g. Boag and Grant, 1981; Ratcliffe and Grant, 1983; Price *et al.*, 1984*a*; Schluter and Grant, 1984; Grant, 1986; Gibbs and Grant, 1987*a*; Grant and Grant, 1989), and this work will accordingly resurface at points in later chapters. Here I focus on those parts of the work which illustrate the concept of adaptive radiation.

Recent common ancestry—The recent suggestion of at least two cryptic allopatric species of warbler finch, *Certhidea*, by Petren *et al.* (1999) brings the known number of species in the group to 15 (Fig. 1.1). Fourteen species occur on the Galápagos Islands whereas the last one inhabits Cocos Island off Central America and is probably descended from an earlier Galápagos ancestor (Petren *et al.*, 1999). The 15 species form a clearly monophyletic clade with affinities to mainland Emberizinae. The nearest mainland relative may be the dull-coloured grassquit from South America (Sato *et al.*, 1999). The islands were colonized about three million years ago (Grant and Grant, 1996). The ancestor probably ate seeds, but the most recent common ancestor of the present-day finch species was probably an insectivore (Fig. 1.1). This implies that the early seed eaters died out and that granivory re-evolved more recently from a warbler-like finch.

Phenotype–environment correlation—The diets and morphology of the six ground finches *Geospiza* are best quantified. These six form a monophyletic group within the larger finch clade (Fig. 1.1). Their most conspicuous phenotypic differences are in bill size and shape, especially bill length, depth, width, and curvature. Bill depth shows a striking correlation with the hardness and size of seeds consumed, a relationship that holds among species, among populations within a species, and within populations (Fig. 2.2; Abbott *et al.*, 1977; Schluter and Grant, 1984; Grant, 1986;

Price, 1987). Bill depth is correlated with the other dimensions of the bill and with body size. Larger bills are also associated with increased curvature of the culmen (top edge of the upper mandible), a higher ratio of bill depth to bill length, increased mass of mandible musculature, and increased length and curvature of bones in the skull (Bowman, 1961). Seed hardness is only one of several ecological dimensions differentiating the *Geospiza* and the 15 species as a whole, more of which are summarized in Grant (1986).

Trait utility—Evidence for performance consequences of beak variation in the finches comes from two sources. The first is Bowman's (1961) functional analysis of the forces transmitted by bills of different size and shape, and the fracture stresses received at different pressure points along the bill. In theory, in seed-cracking species which apply a biting force at the contact surface between upper and lower mandibles, curvature of the culmen (top edge of the upper mandible) maximizes the pressure component of force and minimizes fracture risk along the bill's entire length. Increased size of the bill and associated musculature also generates a greater overall crushing force. Among the *Geospiza* crushing force is maximized at the greatest angle between gonys (bottom edge of the lower mandible) and culmen, and at the highest ratio of bill depth to length (Bowman, 1961).

Timed field observations of finches attempting to crack seeds of different size and hardness constitute the second line of evidence. These show that the time needed to crack and consume a seed of given size and hardness (handling time) drops steeply with increasing beak size (P. R. Grant, 1981). Each species tends to concentrate on those seed types yielding maximum rates of food intake per unit search and handling time (Schluter, 1982). Handling time and the physical features of the beak and musculature which determine it largely underlie observed correlations between beak dimensions and diet (Fig. 2.2). Handling time influences the relative survival of individuals of different beak size and shape as the abundance of small/soft and large/hard seeds fluctuates (Boag and Grant, 1981; Price *et al.*, 1984*a*; Gibbs and Grant, 1987*a*).

Rapid speciation—The Darwin's finches are not closely related to extant mainland groups. The 15 extant species originate virtually simultaneously at the tip of a long branch connecting them to South American emberizines (Petren *et al.*, 1999). This suggests a recent burst, although extinctions in the past are likely. Speciation rates are heterogeneous within the extant group (Fig. 1.1). The clade of 12 species including all the ground finches and tree finches represents a significant burst when contrasted with the three contemporaneous warbler-like finch lineages ($P = 0.01$; see Box 2.2).

2.3.2 *West Indian* Anolis *lizards*

The *Anolis* on Caribbean islands are diurnal, small-bodied lizards that spend most of their lives perched on vegetation (Williams, 1972). About 140 Caribbean species are recognized (Jackman *et al.*, 1997). They are the only diurnal perching lizards in the

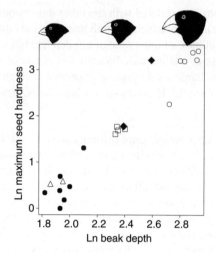

Fig. 2.2 Beak size and maximum hardness of seeds in the diets of granivorous *Geospiza* on Galápagos islands. Symbols refer to different island populations of *Geospiza fuliginosa* (●), *G. difficilis* (△), *G. fortis* (□), *G. conirostris* (◆) and *G. magnirostris* (○). $r = 0.97$, $P < 0.0001$. After Schluter and Grant (1984).

archipelago and are ecologically and morphologically highly diverse, using almost every kind of erect vegetation (Irschick et al., 1997). About 30 of the species inhabit the small islands of the Lesser Antilles where they differ in body size and habitat (Roughgarden et al., 1983). Much recent work however has centred on the roughly 110 species inhabiting the four large islands of the Greater Antilles. Here, most species are recognized as falling into distinct 'ecomorph' categories according to the microenvironments they utilize and their morphological characteristics: trunk-crown, trunk-ground, crown giant, twig, trunk, and grass-bush (Williams, 1972). Within an ecomorph category, lizards on the same island use different habitats. The most closely related species within an island tend to occur in the same 'ecomorph' category but differ in habitat (e.g. Puerto Rico; Fig. 2.3). These lizards constitute an adaptive radiation in every important sense.

Recent common ancestry—The Greater Antilles anoles are all descended from a common mainland ancestor within the past 40 million years or so (Burnell and Hedges, 1990; Jackman et al., 1997). They are not monophyletic, however, as more than one invasion of the islands may have taken place long ago, and at least one island clade of *Anolis* left Central American descendants. Levels of phenotypic differentiation in these mainland taxa are not low, but associations between ecology and morphology are different from those of island *Anolis* (Irschick et al., 1997). Hence it is convenient to consider them apart from the Caribbean anole radiation. Another surprise is that two small Caribbean lizard genera are nested within the *Anolis* clade, so the genus as a whole is not monophyletic (Jackman et al., 1997).

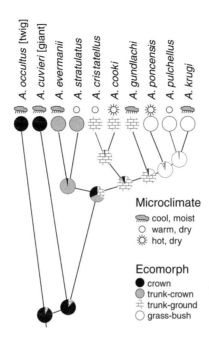

Fig. 2.3 Phylogeny and habitat use of *Anolis* lizards on Puerto Rico (Losos, 1990a, 1992). This fauna is a portion of the full West Indies tree, and is the result of at least three separate invasions of Puerto Rico: the ancestor of the entire clade; the ancestor of *A. cuvieri*; and the ancestor of the largest clade (*A. evermanii* to *A. krugi*) (Hass *et al.*, 1993; Jackman *et al.*, 1997). Shading of pies at internal nodes indicate relative support for alternative ecomorphs as ancestor states (cf. Schluter *et al.*, 1997 and 3.1; all branch lengths were set to 1). Microclimate data are from Williams (1972). A recent molecular phylogeny of the Caribbean anoles included only five of these species and their relationships are not in full agreement with the tree shown (Jackman *et al.*, 1997). However, the differences are only weakly supported in the DNA estimate (J. B. Losos, personal communication) and I have retained the earlier estimates here for convenience.

Phenotype–environment correlation—Within the *Anolis*, members of the same ecomorph that occur on different islands are similar in morphology, ecology and behaviour despite their independent origins (Jackman *et al.*, 1997; Losos *et al.*, 1998). For example, on all three islands the trunk-ground anoles are stocky, have relatively long legs and tail, perch head-down on the lower portions of tree trunks, and leap to the ground to seize prey (Williams, 1972, 1983; Losos, 1990a, 1992). Two key ecological characteristics that vary across all the ecomorphs are perch diameter, which is positively correlated with body mass and size-adjusted lengths of forelimb and hindlimb, and perch height, which is negatively correlated with size-adjusted lengths of hindlimb and tail (Irschick *et al.*, 1997). The relationship involving just two of these variables is shown in Fig. 2.4.

Detecting adaptive radiation

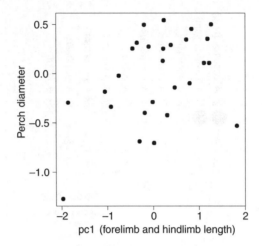

Fig. 2.4 Morphology (pc1) and perch diameter of 27 species of Caribbean *Anolis* ($r = 0.43$, $P = 0.02$). Both axes represent measurements corrected for body length (ln snout-vent length). Perch diameters are residuals from a regression of this trait against body length. pc1 is the first principal component of the correlation matrix based on six morphological traits, all ln-transformed and corrected for body length. It represents mainly the lengths of forelimb and hindlimb. From data in Irschick *et al.* (1997), kindly provided by D. Irschick.

Trait utility—The performance consequences of variation in body size and length of limbs have been explored in most detail using a laboratory gymnasium including racetracks and artificial perches. On broad surfaces (mimicking a wide perch or the ground) large-bodied species sprint faster and jump farther, as do species with relatively long limbs for their size (Losos, 1990*b*). The mechanistic basis for these relationships is the greater acceleration generated by the longer stride of long-limbed species, because of the greater distance over which the muscles apply force. Jumping distance by a lizard of given size and limb length is only weakly affected by perch diameter, but sprinting slows on narrow perches (Losos and Irschick, 1996). Correspondingly, lizards in the field rely principally on jumping when escaping from narrow perches, whereas sprinting is used more often on wide perches and this relationship holds both within and between species (Losos and Irschick, 1996). Long-limbed species also stumble and fall at a higher rate than short-limbed species when sprinting on narrow perches (Losos and Sinervo, 1989). A long tail is also thought to be advantageous as a counterbalance while jumping, but the mechanics of this relationship are less well understood.

Fitness consequences of morphological differences within species have also been inferred from transplants of lizards to new islands having different vegetation characteristics from the source island and from each other (including perch diameter and height) (Losos *et al.*, 1997). A consistent shift in mean phenotype was observed in the direction predicted from the vegetation. A genetic basis to the shift is not confirmed, however, and an influence of phenotypic plasticity is likely (Losos *et al.*, 2000).

Rapid speciation—Speciation rates are highly variable within the genus. For example, the Puerto Rican *A. occultus* (Fig. 2.3) is the outgroup to the clade consisting of all other West Indian anoles and hundreds of mainland species (Jackman *et al.*, 1997; Irschick *et al.*, 1997). The contrast is significant even if we restrict attention to the 109 other species of *Anolis* on islands of the Greater Antilles ($P = 0.003$; see Box 2.2).

2.3.3 Hawaiian silversword alliance

The 28 species that make up the Hawaiian silversword alliance, including the silverswords and greenswords (*Argyroxiphium*) and their relatives *Dubautia* and *Wilkesia*, are perhaps the greatest living example of adaptive radiation in plants (Fig. 1.2). Members exploit a huge array of habitat types including cold arid alpine regions, hot exposed cinder cones, wet bogs where water uptake is nevertheless inhibited, and dimly lit rain forest understories. The species are also greatly varied in their morphology and physiology, such as in growth form (from near-herbaceous mats to small trees) and leaf size, shape, pubescence, conductance, and rates of transpiration (Carlquist, 1980; Carr *et al.*, 1989; Robichaux *et al.*, 1990; Baldwin and Robichaux, 1995). The wet–dry habitat divide has been crossed so many times—the minimum is seven—that the state of the ancestor of all the extant species is impossible to pinpoint.

Recent common ancestry—Molecular and morphological evidence all points to a single colonization of the Hawaiian archipelago by an ancestor closely related to the California tarweeds, *Madia* and *Raillardiopsis* (summarized in Baldwin, 1997). The last common ancestor of the Hawaiian and mainland species is estimated to have lived about 15 mya, but the last common ancestor of the extant members of the silversword alliance is about 5 mya, roughly the age of the oldest modern high island in the archipelago (Baldwin and Sanderson, 1998; Fig. 1.2). The group was undoubtedly present on older high islands that are now eroded or submerged (see Baldwin and Robichaux, 1995). Many evolutionary transitions seem to have occurred prior to, or just after, island colonization: unlike their mainland relatives the Hawaiian species are perennial, polyploid, and self-incompatible. The silversword alliance is monophyletic—there is no trace of descendants in other regions of the Pacific. However, the principal genus, *Dubautia*, is not (Fig. 1.2).

Phenotype–environment correlation—Vegetative characteristics and habitat appear to be correlated, especially along the wet–dry continuum. Within *Dubautia*, for example, leaves of species in wet forest environments are broad and thin-textured with a weakly developed cuticle whereas those from dry environments or bogs (where water uptake is inhibited) tend to be small, succulent, and have a thick cuticle

(Carr et al., 1989). The statistical validity of these correlations between morphology and environment has not been rigorously tested.

Robichaux (1984; Robichaux and Canfield, 1985) compared the tissue elastic properties of *Dubautia* species from wet (excluding bog) and dry environments, using branches cut from wild plants and brought into the laboratory. Tissue elastic modulus is the change in tissue turgor pressure per unit change of leaf water content, measured near full hydration. This measure is lower in plants from dry environments (2.9–3.9 mpa) than from wet environments (9.6–14.5 mpa). Therefore, species from dry environments retain higher turgor pressure as tissue water content decreases. This association is statistically significant after phylogeny is corrected for ($\chi^2 = 10.54$, df = 2, $P < 0.01$, Pagel's 1994 method). The difference between species in tissue elastic modulus is probably genetic, as the measure in *D. ciliolata* is similar between wild individuals and plants raised in the greenhouse (Robichaux, 1984). Species of the silversword alliance are also diverse in floral architecture, but relationships to environment have not been quantified.

Trait utility—Differences in tissue elastic modulus between *Dubautia* species from wet and dry environments reflect differences in performance as much as phenotype: the species from dry environments are more drought-tolerant. Many other physiological processes are known to depend on high turgor pressure in plants in general, including growth, stomatal conductance, and photosynthesis, although this has not yet been established for *Dubautia* or the other members of the alliance (Robichaux, 1984).

Rapid speciation—The modern representatives of the silversword alliance are considerably younger than the tarweeds *Madia* and *Raillardiopsis* that form the likely outgroups (Baldwin and Sanderson, 1998). Compared with these mainland lineages at about 6 mya, the ancestor of the silversword alliance left an excessive number of descendant species ($P = 0.0002$; see Box 2.2).

2.3.4 Columbines

The last example focusses on floral differentiation, which along with physiological diversification represents the dominant theme of research on adaptive radiations of flowering plants (Grant and Grant, 1965; Stebbins, 1970). This example is a substantial departure from the previous cases because the columbines *Aquilegia* are a mainland radiation whose species are geographically widely distributed (Hodges and Arnold, 1994a). Also, the most conspicuous phenotypic differences among the *Aquilegia* are in traits that determine mating success (flower colour and form) (Fig. 2.5). The group is nevertheless considered among the *adaptive* radiations because the floral structures directly exploit features of external environment (i.e. pollinators) and diverge accordingly (Hodges and Arnold, 1994a). It just happens that these traits are also tied to mating and sexual isolation. This contrasts with animals whose divergence in secondary

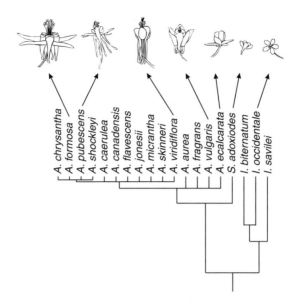

Fig. 2.5 Adaptive radiation of columbines *Aquilegia* and its sister taxa *Semiaquilegia* and *Isopyrum*. The phylogeny is based on nuclear ITS and chloroplast DNA sequences (Hodges and Arnold, 1995). Poorly supported nodes were collapsed. Images were redrawn from Hodges (1997), with permission of Cambridge University Press.

sexual characteristics is often less directly tied to features of external environment. *Aquilegia* has additionally undergone considerable divergence in physiological traits in relation to habitat (Chase and Raven, 1975), though these changes are less conspicuous than those involving the flower.

Common ancestry—*Aquilegia* includes about 70 species (Hodges, 1997) whose gene sequence differences are so slight (less than 1% in nuclear and chloroplast DNA sequences in a sample of 15 species) that relationships within the group are all but impossible to resolve (Fig. 2.5). The 15 species sampled together with species in five related genera indicate that genus *Aquilegia* is monophyletic.

Phenotype–environment correlation—Floral differentiation among the species involves a suite of traits including colour (yellow, blue, red, white, purple, and green), orientation (upright vs pendent) and length of nectar spurs (Fig. 2.5). Variation in these traits is associated with pollinator type; however these associations have not been quantified over the many species. Most effort has focussed on the differences between *A. formosa* and *A. pubescens* (Grant, 1952). Flowers of the former are red and yellow, pendent, short-spurred and pollinated mainly by hummingbirds, whereas flowers of *A. pubescens* are white or pale yellow, upright, long-spurred and pollinated mainly by hawkmoths (Hodges and Arnold, 1994*a,b*).

Trait utility—Experimental manipulations of spur length and flower orientation in *A. formosa* and *A. pubescens* verify the influence of these traits on efficiency of pollination by hummingbirds and hawkmoths. When flowers of *A. pubescens* were manipulated to hang pendent rather than upright, hawkmoths all but abandoned them (Fulton and Hodges, 1999). When nectar spurs on upright flowers of the same species were shortened experimentally, visits by hawkmoths did not decline but pollen removal was significantly reduced. Hummingbird visitation remained low, however.

In an earlier field study Miller (1981) showed that fluctuations in relative densities of pollinators resulted in oscillating relative seed sets of white and blue flowers in the polymorphic *Aquilegia caerulea*. In a year when hawkmoths were rare or absent, diurnal bee pollination was responsible for most seed set and blue flowers had an advantage over white flowers. Hawkmoths were present and visiting flowers around dusk in two other years, and white flowers were favoured instead.

Rapid speciation—The 70 species of the *Aquilegia* clade represents a substantially elevated speciation rate when set against its sister taxon, *Semiaquilegia*, which contains only a single species (Hodges, 1997; $P = 0.005$; see Box 2.2). A lineages-through-time plot by Wollenberg *et al.* (1996) suggested that speciation rates in *Aquilegia* have been much higher in the recent past than earlier in the clades's history. However, their null model assumes that speciation and extinction rates are equal, whereas a null model with unequal speciation and extinction rates might fit the data better (cf. Nee *et al.*, 1994).

2.3.5 Other cases

Many other candidates for adaptive radiation have been proposed, some of which are listed in Table 2.1. The table includes only genera in which at least one of the two criteria of 'fit' (phenotype–environment correlation and trait utility) has been confirmed with quantitative measurements and a test. Common ancestry has been confirmed in most of them. Tests of rapid speciation have not been carried out in most cases. In producing the table, I endeavoured to examine, and if warranted include, at least one genus from all the best-known adaptive radiations from oceanic archipelagos and young lakes. My selection of cases from mainlands is haphazard but biased toward genera in which phylogenies have contributed to tests of association between phenotype, environment, and performance. I have doubtless overlooked some fine candidates for adaptive radiation.

The list (Table 2.1) reveals that substantial progress has been made in testing correlated evolution of traits, environments, and performance in some groups. The absence from the list of many well-known groups is equally telling, however. Many of the best-known adaptive radiations still lack quantitative tests of 'fit' of organisms to environment. For example, I was unable to include most Hawaiian radiations such as the *Drosophila*, the honeycreepers, *Tetragnatha* spiders, *Cyanea*, and other lobelioid genera, and the composites *Bidens* and *Tetramolopium*. Also missing are the *Partula* snails of Moorea.

Perhaps most (or even all) of the notable absentees are true adaptive radiations, as are several mainland examples passed over. After all, many groups not listed show high levels of interspecific variation in phenotypic traits thought to be ecologically relevant, such as leaf form in the Hawaiian lobelioid, *Cyanea* (Givnish *et al.*, 1995). Others exhibit large amounts of ecological variation, such as the larval habitat of the Hawaiian *Drosophila* (Kambysellis *et al.*, 1995; Kambysellis and Craddock, 1997) but the phenotypic correlates of this variation are obscure (I was unable to confirm a statistical association between substrate and egg morphology). Still others exhibited variation in both phenotype and environment but associations between the two were unquantified and remain essentially anecdotal. A final category of cases involve closely related species that show performance and fitness tradeoffs between environments revealed by reciprocal transplants (see Primack and Kang, 1989; Mopper, 1998) but for which phenotypic traits underlying these tradeoffs are unknown. Until these traits can be identified it is not clear whether reduced performance in the novel environment stems from phenotypic adaptations to the native environment or simply a failure to recognize novel resources.

These considerations do not discount the prevalence of adaptive radiation, but they underscore the amount of work remaining to be done of even the most basic kind. I nevertheless make use of many of these uncertain cases and others like them in later chapters, to ensure that our inquiry is not narrowly limited to only a few well-established cases of adaptive radiation. I later address whether a result detected from a wider set of cases really is one that characterizes adaptive radiation.

2.4 Adaptation vs adaptive radiation

Adaptive radiation has a strong connection with *adaptation* but the two concepts are distinguishable on several grounds. The most obvious is that adaptive radiation includes speciation which need not, in principle at least, require natural selection at any stage. However, the concepts also vary in their implications for the study of phenotypic evolution: adaptive radiation is more than diversification of adaptations.

Originally, 'adaptation' applied broadly to trait values that increase fitness compared to alternative trait values (Reeve and Sherman, 1993) but increasingly the term is restricted to derived traits (or trait values) built by natural selection *for their current roles* (Gould and Vrba, 1982; Baum and Larson, 1991). The practical distinction between these concepts may be small when applied to differences between closely related species if new roles accumulate slowly. The definition of adaptive radiation nevertheless has more in common with the first, ahistorical concept of adaptation. Adaptive radiation is a statement about 'fit' of organisms to their diverse environments and not about the historical sequence of steps that produced this fit. Traits that arise in one environment and that later become co-opted for use principally in another ('exaptations'; Gould and Vrba, 1982) still fall within the domain of adaptive radiation. For example, Colbourne et al. (1997) suggested that the elongate setae on the inner margin of the carapace of several species of *Daphnia* originated for clearing debris in association with detritus-based feeding, but possession of the trait subsequently allowed colonization of ponds with turbid waters—a novel environment. The radiation of *Daphnia* into waters of varying turbidity is no less an adaptive radiation because of it.

An ahistorical definition of adaptive radiation in no way diminishes the importance of elucidating trait histories. Rather, our more limited goal here is to arrive at a useful definition that, once fulfilled in a particular case, allows us to pose questions about mechanisms and history. Eventual tests of these questions address the *causes* of adaptive radiation.

Another distinction is that research into adaptation does not typically address the causes of species differences in phenotype, whereas this issue lies at the heart of investigation into adaptive radiation. For example, the goal of comparative study of adaptation is to link phenotype with environment and trait function (e.g. Losos, 1990*a*, *b*; Harvey and Pagel, 1991; Wainwright and Lauder, 1992; Wainwright, 1994; Losos and Miles, 1994). The research program is not typically concerned with whether selection between species is truly *divergent*, which is the most significant question in the study of adaptive radiation.

Adaptation may occur without adaptive radiation. For example, the larvae of the more than 73 species of *Enallagma* damselfly are distinguished mainly by their use of either of only two kinds of aquatic environment (McPeek, 1990*a*). Larvae of most species are confined to lakes where predatory fish are present, whereas larvae of the remaining species occur in fish-free lakes where dragonfly nymphs are the main predators. Species inhabiting the two environments are differently vulnerable to predation by fish and dragonflies, and their phenotypes are strongly correlated with predation regime. However, larvae of *Enallagma* species occurring within the same lake type are ecologically indistinguishable, and the majority of speciation events in the group seem to be associated with no detectable ecological and phenotypic divergence at all (McPeek, personal communication).

2.5 Discussion

I have attempted to present a workable definition of adaptive radiation and applied it to a number of cases. In this final section I briefly review a few difficulties and unresolved issues.

How much divergence is needed?—Detecting adaptive radiation requires evidence of association between phenotype and a range of environments, but attempts to specify a minimum amount of divergence are unlikely to succeed. An example shows why. Bill lengths of the eight species of Old World *Phylloscopus* warblers in the

Himalayas of Kashmir (Table 2.1) span a range of only 2 mm, although body mass ranges from 5 to 10 g. Richman and Price (1992) converted beak and body differences between the warblers plus the goldcrest (*Regulus regulus*, an ecologically similar and phylogenetically not-too-distant species) into three uncorrelated principal components which accounted for 81%, 12%, and 6% of the morphological variance among species. Nevertheless, the strength of each of their associations with environmental variation was roughly the same (Fig. 2.6), implying that in this group even the slightest morphological changes have significant ecological consequences.

Too little replication—The statistical comparative method only works if parallel evolution has occurred, whereas many phenotypic and environmental transitions in

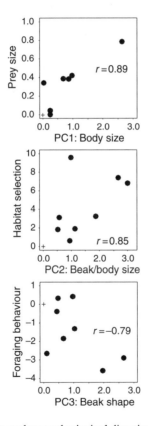

Fig. 2.6 Correlations between ecology and principal directions of variation in external morphology of eight *Phylloscopus* warblers and *Regulus regulus*. PC1 is overall body size; PC2 is the ratio of bill to body size (tarsus length); PC3 is bill shape (ratio of width to length). Points shown are the phylogenetically independent contrasts; all correlations are significant ($P < 0.05$). Redrawn from Richman and Price (1992) with the permission of *Nature* and the authors. (+) indicates origin of regression.

adaptive radiations are unique. Consider the hypertrophied pharyngeal jaw in the two sister species of sunfish, *Lepomis gibbosus* and *L. macrochirus*. The trait likely arose only once, as has its corresponding pattern of niche use: a diet of snails (Wainwright and Lauder, 1992). The association between diet and morphology is nevertheless real, as several tests revealed. First, similarities to other taxa were noted. A hypertrophied jaw occurs also in other snail-eating fishes. Second, the authors summoned the functional sciences to more deeply probe the link between jaw and diet (Wainwright, 1994). Think about what the pharyngeal jaw does. It chews and transports food from the buccal cavity to the esophagous. In most *Lepomis* the major muscles of the pharyngeal jaw accomplish this by repeatedly depressing and retracting the upper pharyngeal jaw bones. But *L. gibbosus* and *L. microlophus* exhibit a novel activity pattern when a snail is passed to the jaw: all the major muscles of the jaw contract together in repeated activity bursts that cease once the shell of the snail is fractured. This pattern stabilizes the position of the lower jaw while simultaneously depressing and retracting the upper jaw against the lower, creating a powerful crushing and shearing force. The bones of the pharyngeal jaw are also more robust in these two species than in other *Lepomis* and together with the hypertrophied muscles are capable of generating a greater force (Lauder, 1983; Wainwright and Lauder, 1992). Clearly, the association between jaw and diet, though not statistically repeatable, is no accident. The unique diet of snails is linked to the force-generating capabilities of the structures that handle the prey items.

Stasis—Phenotypic evolution in adaptive radiation is the antithesis of stasis. Earlier, we considered the *Enallagma* damselflies as having too much of the latter. Unfortunately, the boundaries between adaptive and nonadaptive radiation are not always so clear cut. Even the best-known adaptive radiations include some rather slow sections. The Galápagos finches include two old sibling lineages of warbler finch (*Certhidea*) whose similarities in behaviour, diet, and morphology seem strange and inexplicable next to the extraordinary variety of the rest of the clade (Petren *et al.*, 1999). The radiations of cichlid fishes in the East African lakes also include subclades that show little or no ecological or morphological differentiation (e.g. Greenwood, 1984; Sturmbauer and Meyer, 1992). Apparently, adaptive and nonadaptive radiation are extremes along a continuum, and many lineages may show elements of both.

Monophyly reconsidered—Many adaptive radiations are not monophyletic, i.e. they do not include *all* the descendants of the common ancestor. Geographical and other considerations often justify leaving out parts of a clade when an adaptive radiation is presented. Examples include the *Partula* snails of Moorea (one of the *Partula* of Moorea went on to colonize Tahiti and may even have colonized back again; Johnson *et al.*, 1993; Clarke *et al.*, 1996, 1998), the *Anolis* lizards of the West Indies (at least one lineage recolonized the mainland of Central America; Irschick *et al.*, 1997; Jackman *et al.*, 1997), and the Hawaiian *Drosophila* (which gave rise to the *Scaptomyza*; Baker and DeSalle, 1997). The fifteenth Darwin's finch, *Pinaroloxias* of

Cocos Island, descended from a colonist from Galápagos that never speciated further; Fig. 1.1).

Monophyly is apparently not regarded as essential in adaptive radiation, and this makes sense. There is no reason to think that if a lineage colonizes a new adaptive zone that all its descendants will remain there. Yet, the concept of adaptive radiation surely loses force when applied to a grab-bag of species selected from a larger clade. Deleting subclades makes sense but only if pruning is kept to a minimum.

Rapid speciation—The methods for detecting 'rapid' speciation have difficulties that remain to be worked out. Most of the criteria are relative and hence the outcome of the test depends on which other lineages we decide to include. Then we must decide if rates ought to be adjusted before the test is conducted. For example, should a test correct for geographic area? Speciation rates of Hawaiian *Bidens* are not much greater than that in its putative sister taxon (F. Ganders, personal communication) but this does not take into account the fact that it has achieved this rate in a comparatively tiny geographic area. Another problem is that methods for testing rapid speciation from branching rates of phylogenetic trees (Box 2.2) do not easily distinguish elevated speciation from reduced extinction.

Continued application and development of quantitative approaches for comparing speciation rates will help decide how useful these methods really are to fulfilling the concept of adaptive radiation. At that point, it may be possible to resolve the difference between the current definition of adaptive radiation and others that do not include high speciation rate as a criterion (Givnish, 1997; Barrett and Graham, 1997; Jackman *et al.*, 1997; see Sanderson, 1998 for discussion).

3

The progress of adaptive radiation

> *[Adaptive radiation] has a characteristic course which may be closely similar in different episodes*
>
> —Simpson (1953)

3.1 Introduction

Does a recognizable progression exist in adaptive radiation or is each radiation unique? The idea of a characteristic sequence of events common to many radiations is appealing but has been little tested. One possibility proposed a number of times is that there should be a widespread trend toward greater niche specialization. Another is that the rate of morphological evolution should be highest early in a radiation and then decline steadily through time as species diversity builds. In this chapter I explore these and several other hypothesized trends using comparative and historical sources of data. This inquiry is necessarily descriptive, but its ultimate purpose is to establish whether there are regular features of ecological diversification that a complete theory of adaptive radiation eventually must explain.

I begin by testing Simpson's (1953) view that specialization is a dominant trend in adaptive radiation. He described this trend as the 'differentiation of a more broadly adapted population into ... separate populations each more narrowly adapted to part of the original adaptive range.' I test whether the founders of adaptive radiations are typically niche generalists and whether their descendants become more and more specialized as radiations age and species diversity builds. The alternative hypothesis is that trends in specialization are weak or nonexistent, and that continuous expansion onto new resources and into novel environments is the only truly general feature of niche evolution in adaptive radiation.

Second, I ask whether expansion to new resources and environments follows a characteristic ordering in taxonomic groups that use the same broad classes of resources. For example, do birds diverge initially by habitat, with changes in other resource dimensions coming later? Does floral differentiation precede vegetative divergence in angiosperms pollinated by specialized animal vectors, and is the order reversed in those pollinated by generalist vectors? Third, I ask whether the precise order of expansion to new environments is ever replicated in related lineages undergoing independent adaptive radiation in similar environments. Finally, I address how phenotypic diversity changes during an episode of adaptive radiation. Does it expand most rapidly early in a radiation and does it fall along with species diversity as a radiation nears its end? Or does it continue to increase at a steady pace independently of species diversity?

Answering these questions has wider implications. The notion of a characteristic progression in adaptive radiation is embedded within the broader hypothesis that macroevolution itself is predictable. If we 'rewound life's tape' to a previous time in history and allowed it to run again, would the result be the same as that achieved the first time (Gould, 1989a)? According to Gould the correct answer (no!) is self-evident. Chance historical events inevitably hold sway, and the future is never like the past. 'Alter any early event, ever so slightly and without apparent importance at the time, and evolution cascades into a radically different channel' Gould (1989a, p. 51).

Alternatively, under the same environmental conditions the replay might be recognizable and even strongly reminiscent of the previous go-round (Conway Morris, 1998). If so, to what degree and for how long? Adaptive radiations are fertile testing grounds because lineages undergoing the process have landed many times over in environments at least coarsely similar (e.g. abundant resources with few competitors and predators). If we can someday hope to recognize the features of environment that trigger adaptive radiation, surely we may expect *some* of the features of the ensuing radiation to duplicate those of other radiations set in motion by similar agencies.

3.2 Generalist ancestors, specialized descendants?

Is the founder of an adaptive radiation typically a generalist, exploiting many resources? Does this generalist ancestor give rise to descendants with progressively narrower niches? Many researchers have thought so (e.g. Simpson, 1953; Mayr, 1942; see Thompson, 1994). Simpson identified specialization as one of the two dominant themes of adaptive radiation, the other being spread to new resources and environments.

Some justification for this idea comes from observations that niche breadth is inversely correlated with species diversity. Pollen specialization by bees is most prevalent in environments having greatest bee species richness such as mediterranean ecosystems, and less prevalent in the tropics and other temperate zone habitats where bee diversity is lower (Müller, 1996). Vertebrate species in depauperate communities, such as on remote islands, often have broader niches than their relatives in species-rich mainland communities (MacArthur, 1972; Terborgh and Weske, 1975; Schluter, 1988b; Ricklefs and Bermingham, 1999). This suggests that the average niche breadth within species-rich faunas declined as the faunas were assembled.

However, such patterns may have other explanations. Possibly, adaptive radiations are just as often begun by specialists whose descendants undergo a steady expansion of niche breadth in the novel environment, such as on islands. Such a trend towards decreasing specialization could persist for some time, perhaps even lasting as long as the radiation itself. Additionally, species-rich mainland faunas are typically the product of many radiations of multiple taxa accumulated over long spans of time. Increases in specialization accompanying total faunal buildup need not show up as a strong signal within any of the component radiations at lower taxonomic level. In other words, specialization could occur as a weak trend over the long term without being an integral part of adaptive radiation in the narrow sense.

I approach the problem using phylogenetic methods (Box 3.1). I use the niche breadths of contemporary species and trees of their phylogenetic relationships to estimate whether ancestral niche breadth is generalist or specialist, and possible trends in time. The phylogenetic methods themselves are not without problems. Nevertheless, enough sense can be made of the results to allow a few moderately robust conclusions.

> **Box 3.1** Reconstructing ecological history.
>
> How do we recover the traits of extinct ancestors—what they ate, the habitats they used, how they interacted, and how different they were from one another? Three methods are used: fossil tracing, taxonomic succession, and ancestor reconstruction using phylogenetic trees. All three are relied upon at different points in this book, and I review them briefly here.
>
> *Fossil tracing* —Fossils are potentially the most straightforward window to the past. Character states through time are read directly from the stratigraphic layers preserving them. The niche of a fossil species is invisible but may be inferred if the relationship between morphology and niche use is well worked out for contemporary species. Nehm and Geary (1994) presented a beautiful example of a fossilized niche shift in a marine gastropod. Shell measurements, which map closely onto use of different ocean depths, underwent a clear transition across a series of stratigraphic intervals between about 4 and 6 mya. Other examples of niche shifts documented over reasonably narrow stratigraphic intervals can be found in Carleton and Eshelman (1979), Gingerich (1985) and Geary (1990).
>
> The problem with fossil tracing is that the feat can rarely be carried out for all ancestors. The vast majority of radiations lack a sufficiently detailed fossil record. As well, physiological and behavioural evolution is not easily documented in this manner. For this reason the study of ecological changes during adaptive radiations relies more often on the next two methods.
>
> *Taxonomic succession* —Taxonomic succession pieces together the temporal sequence of steps in adaptive radiation by comparing contemporary lineages assumed to be at different stages. A classic example is Williams' (1972, 1983) study of *Anolis* lizard 'ecomorphs' on the Caribbean island banks of Jamaica, Puerto Rico, and Hispaniola (Table 3.3). Williams noted that the ecomorph lists from the three islands form a nested series. Puerto Rico's five ecomorphs include the four found on Jamaica plus one more. Hispaniola's six ecomorphs include the five seen on Puerto Rico plus a trunk ecomorph. He suggested that the three radiations represented different stages of the same ecological sequence, with Jamaica one step behind Puerto Rico, and Puerto Rico one step behind Hispaniola. Comparison with other islands and the mainland led Williams to propose that the first split produced the crown-giant and twig ecomorphs (both inhabiting the tree crown), and that

Generalist ancestors, specialized descendants? • 39

new ecomorphs were added in the order listed in Table 3.3. Cuba also possesses all six ecomorphs, but the numbers of species in each category are not yet known.

Taxonomic succession uses phylogenetic information to a limited extent (e.g. to establish that separate radiations are indeed independent) but taxonomy is not just a surrogate for phylogeny. The approach has more in common with the study of ecological succession in plant communities, in which temporal sequences are inferred by comparing contemporary plots of different age. Its major assumption is that the attributes of young lineages at the present time mirror early events in lineages that are presently older. This assumption is risky if critical aspects of environments change over time.

Ancestor reconstruction —This is now the most widely used method to infer past changes. The method computes the attributes of ancestors of a clade of species by working backwards from the extant species at the tips of a phylogenetic tree. Once the reconstruction is complete, changes through time are simply read off the diagram depicting ancestral states in much the same way that morphological transitions may be directly read from a complete fossil record. The difference between phylogenetic reconstruction and fossil tracing is that even with perfect information about the traits of extant species and their phylogeny the reconstructed ancestor states are estimates, not observations, and hence are uncertain.

Parsimony (see Maddison and Maddison, 1992) and maximum likelihood (Schluter *et al.*, 1997) are the main approaches used to estimate ancestor states. Parsimony assumes that change is rare and that time (branch length) is irrelevant for estimating change. Likelihood assumes instead that evolution can be mimicked by a random walk between states through time. The two methods usually give nearly equivalent answers (at least when branch lengths are equal), but I prefer likelihood because it provides a measure of the statistical support for each reconstructed state. A weakness of both approaches is the assumption that the phylogeny itself is error-free.

Figure 2.3 in the previous chapter illustrates the likelihood method for discrete traits using the *Anolis* ecomorphs of Puerto Rico. To simplify the analysis the two size morphs within the tree crown are grouped into a single category (tree crown). In this example, ancestor estimates support Williams' (1972, 1983) proposed sequence. For example the grass-bush ecomorph appears to have arisen last. However, the topology of the phylogenetic tree within the *A. evermanii* to *A. krugi* clade is uncertain (Jackman *et al.*, 1997; J. B. Losos, 1996).

3.2.1 *Measuring specialization*

Specialization here refers to a narrowing in the breadth of resources and environments utilized. This is the least ambiguous concept of specialization, conforming to that employed by ecologists (Futuyma and Moreno, 1988). To come up with a meaningful index of each species' niche breadth, however, is not trivial. Is a herbivore that consumes 10 kinds of plants more specialized than an insectivore consuming 20 types of insects? More meaningful comparisons are possible when resources are similar and when those exploited by one species are a subset of resources exploited by another

(Futuyma and Moreno, 1988). This is often true when species are closely related, as in most of the examples discussed below.

Another problem is that only one resource dimension usually figures in the calculation of niche breadth. Rank order of niche breadths along one resource dimension may bear no relation to ordering along another. For this reason many of the examples analysed here may warrant further testing as information on niche breadth over multiple resources becomes available.

Simpson (1953) discussed several additional concepts of specialization that are not covered here. Some, like 'morphological specialization' refer mainly to changes in the mean value of traits rather than to breadth. For example, the statement that 'Miocene ungulates evolved ... dietary specialization in more fibrous vegetation' (Jernvall *et al.*, 1996, p. 1491) refers to a change in mean fibre content but does not imply that the number of diet items or the range of fibre contents eaten also changed. Likewise, the observation that 'a series of related forms, beginning with generalised species and exhibiting successively higher degrees of specialisation to the habit of algal grazing' occurred in Lake Victoria cichlid fishes (Fryer and Iles, 1972, p. 480) does not imply any trend in breadth of resources consumed. A related concept equates generalized with 'primitive' and specialized with 'derived', but this is not very useful.

3.2.2 The generalists-to-specialists hypothesis

At least two reasons lie behind the thinking that founders of adaptive radiations, at least those landing in new or remote environments, should be generalists. First, the generalist is more likely than the specialist to meet its resource requirements in a novel setting. Second, the sample of species likely to immigrate to a new and distant location is biased toward common species having wide dispersal. More often than not these are generalists. Such expectations are based on radiations in remote, depauperate circumstances and do not necessarily apply to radiations initiated by 'key innovations' that enable exploitation of new ecological opportunities within the ancestral geographic range. Even here, however, a rapid widening of the niche might be favoured in the first lineage to tap the new opportunities. In both circumstances the most recent common ancestor to the set of contemporary species that arose by adaptive radiation might still be a generalist.

Once an adaptive radiation is under way, a trend toward greater specialization might be expected for several reasons:

(a) The rise in the number of competing species should lead to an increasingly fine partitioning of available resources.

This idea receives formal treatment in the 'compression hypothesis' of MacArthur and Pianka (1966). They predicted that resource depression should cause competing species to specialize in their use of resource patches. The hypothesis is also central to Sugihara's (1980) 'sequential breakage' model of community development, in which each added species causes one of its predecessors to give up a random fraction of its resource allocation. In both, resource gradients become divided into progressively smaller segments as the number of species rises.

(b) Increasing commitment to exploiting a single resource should limit the ability to evolve a more generalized habit in the future.

Suites of adaptations to specific resources may severely constrain subsequent evolution, inhibiting the evolution of a more generalist lifestyle. Or, genetic variation to exploit other resources might be lost after specialization.

(c) Selection favouring specialization is more powerful than that favouring a generalized lifestyle.

An allele that confers an advantage to its bearer only on one resource will spread more rapidly in a population specializing on that resource than in a population of generalists because only a portion of the latter species experiences the advantage (Whitlock, 1996; Kawecki, 1998). Similarly, fitness gains will occur mainly in the portion of a generalist population that exploits the most frequently-used subset of resources (Holt and Gaines, 1992). Such alleles may spread even if they compromise fitness on rarely-used resources. The result is expected to be increased specialization.

(d) Sympatric speciation creates specialists.

Intermediate genotypes exploiting both of two (or more) resources may experience a consumption disadvantage relative to more specialized genotypes. Selection against the intermediates favours in turn the evolution of reproductive isolation between the specialists. Hence, specialists should accumulate through time if speciation is frequently sympatric (Futuyma and Moreno, 1988).

(e) The spectrum of prey types that can profitably be exploited diminishes if prey species coevolve with predators.

Divergent responses of prey to predation may favour predator divergence and specialization (Brown and Vincent, 1992).

There are counterarguments against each of these points. Generalists are hardly uncommon in nature. Subdivision of resources may not be as important a trend in adaptive radiation as continuous expansion to new resource types. Not all adaptations that assist in exploiting a single resource (e.g. changes in body size) are difficult to reverse (Futuyma and Moreno, 1988). Larger population size in a generalist results in a higher rate of occurrence of advantageous mutations, making up for their weaker selection (Whitlock, 1996). Varying environments may also overcome benefits to specialization. Sympatric speciation is favoured only under special conditions and may not be the dominant mode of speciation in most radiations. Finally, prey defences do not necessarily diverge in response to predation. They may evolve in parallel (Abrams, 2000). For these reasons, one cannot expect a trend toward increased specialization *a priori*. The issue must be decided empirically.

3.2.3 Insights from phylogenies

Closely related species often differ greatly in the breadth of resources used. Many genera of herbivorous mites and insects include some species that use only a single plant species and others that consume many species (Ehrlich and Raven, 1964; Holloway and Hebert, 1979; Fox and Morrow, 1981; Colwell, 1986; Futuyma *et al.*, 1995). In Australian desert skinks of the genus *Ctenophorus* some species consume almost every available type of invertebrate prey, whereas others eat termites almost exclusively (Pianka, 1986). This variation in degree of specialization, when placed on a phylogeny, permits us to estimate which condition is ancestral and hence the subsequent direction of any trend in degree of specialization.

Twenty taxa were found that included both generalists and specialists, and for which phylogenetic trees of relationships were available (Table 3.1). They included congeneric species or, when a genus was not monophyletic or too small on its own, species from closely related genera. Not all cases are confirmed adaptive radiations, only because most have not been tested. Maximum likelihood was used to reconstruct the state of the most recent common ancestor of sets of ecologically differentiated species (Box 3.1). The analysis assumes that the two states 'generalist' and 'specialist' can be modelled as a markov process. To improve stability of estimates I used the 'equal rates' model of (Schluter *et al.*, 1997; Mooers and Schluter, 1999).

In some cases a generalist ancestor was the outcome, such as in the *Partula* snails of Moorea (Fig. 3.1). More often, however, ancestors were reconstructed as specialists. This is particularly true if we confine attention to the statistically significant reconstructions. The Hawaiian *Drosophila* are an example (Fig. 3.2). The larvae of most species are monophagous, exploiting rotting tissues of a single plant family. A few species exploit multiple hosts, and this appears to be the derived condition (Kambysellis *et al.*, 1995; Kambysellis and Craddock, 1997). These basic findings are little changed when we reconsider the same nodes after available outgroups are added to the trees, or when we reconstruct the preceding node instead. The generalists-to-specialists hypothesis therefore finds no support from this first analysis.

The data were also used to test whether the instantaneous transition rate from the generalist state to the specialist state (the 'forward' rate) ever significantly exceeds the transition rate in the reverse direction (the 'backward' rate) (see Pagel, 1994 and Mooers and Schluter, 1999 for technical details), and whether a generalist ancestor was the result. A significantly greater forward rate (from generalist to specialist states) was detected in only 1 of 20 data sets, in the columbines *Aquilegia*. In 18 of the 19 remaining cases the two rates were not significantly different, and in the last the ancestor reconstruction was not a generalist. These results suggest that a trend toward specialization is sometimes seen but is not universal or widespread.

Most of the lineages analysed were continental and do not conform to a situation involving colonists to remote archipelagos. For example, many are of insects on plants. Nevertheless, specialist ancestors turn up even in the lineages on remote archipelagos, such as the *Drosophila* and silverswords (*Dubautia* and relatives) in Hawaii. The last common ancestor of the Darwin's finches is reconstructed as a habitat generalist, but only a single descendant is a habitat specialist (the mangrove finch, *Chamarhynchus heliobates*).

Resource identity may give extra clues about ancestral states under the following circumstances: all specialized descendants use the same resource A, and all generalists include resource A in their repertoire but otherwise overlap little in resource use. In this case, specialization on A is the most likely ancestor state and generalists are derived. For example, 10 of 11 species of Australo-Papuan scrubwrens (*Sericornis*) are largely restricted to wet rainforest, whereas the eleventh resides in wet rainforest plus most other forest and scrub habitats. This suggests that wet forest specialist is the ancestor state, corroborating the result from the earlier analysis that ignored resource identity (Table 3.1).

The opposite pattern is also informative: all extant generalists use the same set of resources, whereas specialists use different subsets of this broad set. Here, the ancestor is most likely a generalist and a narrow niche is the derived state. For example, most species of monotrope (epiparasitic plants that obtain nutrients from other plants via mycorrhizal fungi) associate with many of the same species of fungi, whereas the specialist uses a subset of them (Cullings *et al.*, 1996). This further supports the earlier estimate of a generalist ancestor (Table 3.1).

A potential problem with the methods used here is that if all descendants have the same state (e.g. are specialists) then the ancestor can never be reconstructed as the alternative state (e.g. generalist). Yet, surely it is possible for a generalist ancestor to leave only specialized descendants, and vice versa. The very broad diets and habitat spectra of all the Galápagos ground finches (*Geospiza*) do not rule out the possibility that their ancestor was a specialist instead. It is very likely that the first finch to land on Galápagos was initially somewhat specialized, because continental finches are known to be less catholic (Schluter, 1988*b*).

Losos (1992) used a different approach to address this problem when deciding whether the ancestor to the specialized *Anolis* ecomorphs was also a specialist or a generalist instead. Rather than assume *a priori* that the ancestor must have been one of the present-day ecomorphs (cf. Fig. 2.3), Losos estimated the morphology of the ancestor and from this inferred its ecological state. I used his method to re-estimate the root ancestor of the Puerto Rican species (Fig. 3.3). The point estimate suggests an intermediate phenotype, possibly indicating that the ancestor was a generalist (cf. Losos, 1992). However, a specialist ancestor is not ruled out by this analysis because the confidence bands for the ancestral phenotype overlap phenotypes of all the specialist phenotypes (Fig. 3.3).

Table 3.1 Ancestral niche breadths estimated using maximum likelihood. G or S indicates whether the last common ancestor of the extant species is estimated to be a generalist or specialist, respectively. Letters in bold indicate that a state has significantly better support than the alternative (Mooers and Schluter, 1999; Ree and Donoghue, 1999; Pagel, 1999). '$S = G$' means that the likelihoods of the two states are identical. For stability, the reconstructions assume that forward and backward transition rates are equal (see Schluter et al., 1997 for technical details). All branch lengths were set to 1 because estimates of time between nodes were unavailable in some taxa. Outgroups were not utilized for any comparison. Adding them, when available, made little difference to the overall outcome

Taxon (no. species)	Region	Niche comparison	Ancestor state	Source
Vertebrates				
Sericornis (11)	Australia, New Guinea	no. habitats[a]	**S**	Christidis et al. (1988)
Molothrus, Scaphidura (5)	N., C. and S. America	no. host species[b]	S	Lanyon (1992)
Geospiza, Camarhynchus, and allies[c] (15)	Galápagos	no. habitats	G	Grant (1986); Petren et al. (1999)
Ctenotus (12)	Australia	no. food types[d]	$S = G$	Pianka (1986, 1998)
Invertebrates				
Oreina (23)	Europe	no. host plant families	**S**	Dobler et al. (1996)
Drosophila (42)	Hawaii	no. host families[e]	**S**	Kambysellis and Craddock (1997)
Papilio (9)	N. America	no. host species[f]	**S**	Scriber et al. (1995)
Enchenopa binotata complex (9)	N. America	no. host species[g]	S	Wood (1993)
Anthidium (28)	W. Palearctic	no. pollen sources[h]	S	Müller (1996)
Nucella (6)	N. hemisphere	no. prey and habitat types	**G**	Collins et al. (1996)
Partula (9)	Moorea	no. food types	G	Johnson et al. (1993)
Ophraella (12)	N. America	no. host species[g]	G	Futuyma et al. (1995)
Timema[i] (14)	N. America	no. plant species	G	Crespi and Sandoval (2000)
Heliothis, Heliocoverpa, Australothis (21)	Global	no. host species[j]	G	Cho (1997)
Dendroctonus (19)	N. and C. America, Asia	proport. host species[k]	$S = G$	Kelley and Farrell (1996)

Plants			
Dubautia, Argyroxiphium, Wilkesia (25)	Hawaii	no. habitats[l]	S	Robichaux *et al.* (1990)
Dalechampia (24)	Africa, C. and S. America	no. pollinator spp.	S	Armbruster and Baldwin (1998)
Brocchinia (13)	S. America	no. habitats[m]	S	Givnish *et al.* (1997)
Monotropa, Pterospora, Sarcodes (4)	N. America	no. fungal associates[n]	G	Cullings *et al.* (1996)
Aquilegia (15)	N. hemisphere	no. pollinator species[o]	G	Hodges (1997)

[a] *S. frontalis* is scored as habitat generalist, remaining species are rainforest specialists.
[b] *S. oryzivor* (1 host) and *M. rufoaxil* (7 hosts) scored as specialists, remaining species (71–216 hosts) are generalists.
[c] *C. heliobates* was added to the microsatellite tree of Petren *et al.* (1999) as the sister species of *C. pallida*. *C. heliobates* was regarded as the only habitat specialist. Results are the same when the cactus finch *G. scandens* is also scored as a habitat specialist.
[d] Specialists are species with diet breadth < 3.
[e] Specialists are those species using only one plant family. Scaptomyzoids are excluded.
[f] Specialists include *P. garamas*, *P. palamedes* and *P. troilus*, whereas the rest are generalists. *P. scamander* was excluded because of lack of host information.
[g] Specialists are those species using only 1 host.
[h] Specialists include the oligolectic species.
[i] Includes 17 populations of 14 species.
[j] Specialists include both monophagous and oligophagous species.
[k] Specialists are those species using fewer than half of the potential hosts.
[l] Used the phylogeny in Fig. 1.2.
[m] Specialists include *B. micrantha*, *B. paniculata*, *B. melanacra*, *B. prismatica*, *B. steyermarkii*, and all the *Saxicoles* except *B. maguerei* Givnish (personal communication 1998).
[n] Assumes that the uncoded Monotropoideae are generalists.
[o] Generalists include species with 'open' flowers (*A. ecalcarata*), as they are presumed to be effectively pollinated by many insect species. Polytomies were resolved as a 'comb' placing *A. ecalcarata* at the root.

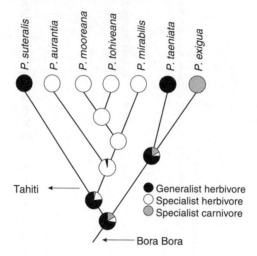

Fig. 3.1 Evolution of specialization in *Partula* land snails from the Island of Moorea. Diet data are from Johnson *et al.* (1993). The four specialist herbivores exploit mainly *Freycinetia* (> 80% of diet) whereas the two generalist herbivores use a variety of host plants. *P. exigua* feeds on small tornatellid snails. Shading of pies at nodes indicate relative support for alternative ancestor states. The phylogeny was extracted from an allozyme-based tree for *Partula* of the whole Society Islands (Johnson *et al.*, 1993). Problematic populations from Mt. Ahutau were excluded. Tree topology is highly uncertain as relationships are complicated by a history of gene flow among Moorea species and between *P. suteralis* and Tahitian species (Clarke *et al.*, 1996). Arrows indicate that the ancestor of the Moorea species colonized from Bora Bora and that a Moorea lineage subsequently colonized Tahiti. All branch lengths for this analysis were set to 1. Results are similar when branch lengths from data in Johnson *et al.* (1993) are used instead.

3.2.4 Remarks

Niche breadth does not fossilize and so we rely almost entirely upon ancestor reconstruction to test the generalists-to-specialists hypothesis. This approach does not always give a clear answer, especially if transitions occur at a high rate. One further consideration needs also to be weighed. The methods for estimating ancestors assume that transitions between specialist and generalist states are independent of speciation and extinction events. The consequences of violations of this assumption have received little attention.

There are at least two scenarios that violate this assumption. First, consider an ancestor specialized on a single host and whose descendants accumulate hosts through time. Speciation may ensue when a lineage acquires a new host, a process that occurs with unequal probability in different parts of the tree. In some parts of the tree acquisition of a new host usually precipitates a speciation event, yielding two new branches both

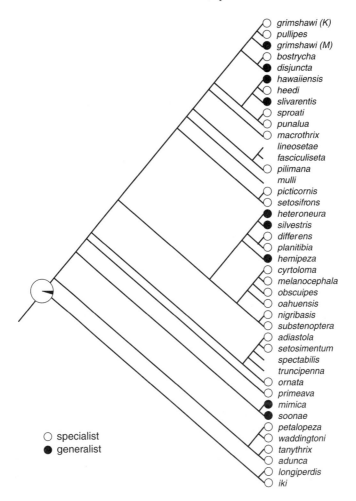

Fig. 3.2 Specialist ancestry of the Hawaiian *Drosophila*. The phylogeny is a composite tree based on nucleotide sequences from 4 mtDNA and 5 nDNA genes (Kambysellis and Craddock, 1997; Baker and DeSalle, 1997). Branch lengths are arbitrary. The node indicated is the common ancestor of the drosophiloids only. Shading of pie at the ancestral node indicates relative support for alternative ancestor states (generalist or specialist). All branch lengths were set to 1 for this analysis. The states of tips lacking a symbol are unknown. The ecological data are originally from Heed (1968) and Montgomery (1975). Modified from Kambysellis and Craddock (1997).

of them specialists. In other parts of the tree, acquisition of a new host sometimes precipitates speciation, but more often it does not. In the resulting phylogeny generalists will tend to occur at the tips of long basal branches, and specialists will sit at the tips of the short branches. In this case the generalist state may tend to appear the

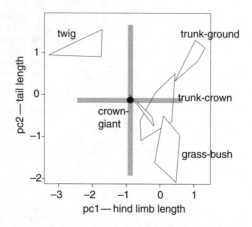

Fig. 3.3 Estimated phenotype of the ancestor to all the *Anolis* ecomorphs on Puerto Rico (cf. Fig. 2.3). Axes are the first two principal components (pc) from an analysis of morphological variation among species (Losos, 1992). Axis labels indicate the traits that load most heavily. Polygons enclose means of species in each ecomorph category (species from Puerto Rico, Jamaica and Hispaniola combined). The ancestor was reconstructed using maximum likelihood and the tree in Fig. 2.3. Shaded lines indicate support limits (analogous to 95% confidence intervals). Modified from Schluter *et al.* (1997) after Losos (1992), with permission of the Society for the Study of Evolution.

more basal even though the ancestor was a specialist. A higher rate of extinction of generalists can induce a similar bias.

A second scenario is the inverse of the first. Consider that descendants of a generalist ancestor gradually lose host species from their repertoire, perhaps because hosts acquire efficient defences with time. Host attrition is (temporarily) reversed only when a new geographic region is colonized, because hosts there are evolutionarily naive and lack defences. Assume that colonization of a new region also leads to speciation. The resulting tree will tend to have specialists at the tips of long basal branches and generalists at the tips of short branches. More often than not this arrangement may erroneously place a specialist at the root of the tree. This bias is the core of the objection of Rothstein *et al.* (manuscript) to Lanyon's (1992) reconstruction of a specialist ancestor to the brood-parasitic cowbird clade of *Molothrus* and *Scaphidura* (maximum likelihood also estimated a specialist ancestor, but support was not strong (Table 3.1)). The geographic distributions of the cowbirds is more in accord with a generalist ancestor (Rothstein *et al.*, manuscript).

The first scenario, in which reconstructions are biased in favour of generalists, is probably more common than the second scenario. If so, then we may have underestimated the frequency of specialist ancestors.

Taken together, the results provide little support for the generalists-to-specialists hypothesis. Specialists appear no less common than generalists as ancestors of extant clades. Indeed, specialist ancestors may predominate. These results are based on a variety of clades not all of which are confirmed adaptive radiations. Nevertheless, the picture is not very different in the smaller set of 'probable' cases. The Galápagos finches alone show evidence of a generalist common ancestor (in habitat use). All but one or two of its descendants have remained generalists, which hardly indicates a trend toward specialization. In contrast, the ancestral Hawaiian silversword is estimated to have been a specialist. These conclusions are necessarily qualified by uncertainties inherent in the methods used. Nevertheless, if a trend toward specialization really does exist then it is clearly very difficult to detect.

If we can be confident of the analyses, then three important implications follow. First, the direction of evolution of niche breadth is not predictable. Secondly, specialization is no impediment to ecological diversification. Lastly, continuous spread to new environments, and not specialization, is the dominant ecological trend of adaptive radiation.

3.3 Repeatable rules of niche spread

Expansion to new resources and environments is a dominant theme in adaptive radiation, indeed it may be the only major ecological theme. This section addresses the possibility that this expansion sometimes follows a recognizable sequence of stages. The question has been dealt with mainly in birds and flowering plants, thus my examples focus on these taxa. Here I am concerned with broad axes of divergence only. Later I consider replication of finer details of niche sequences.

3.3.1 The 'habitat first' rule in birds

Diamond's (1986) comparative study of the New Guinea montane avifauna is the most comprehensive attempt to identify sequences of divergence along major environmental axes. His result: habitat separation comes first, followed by prey size and food type.

Diamond reached this conclusion by measuring pairwise differences between species at different estimated 'stages' following speciation. These stages were calculated indirectly according to degree of overlap of the geographic ranges of closely related, congeneric species. Relatedness was judged by similarity in morphology, behaviour, and song. Diamond assumed that speciation was allopatric, an assumption with a great deal of support in birds (Grant and Grant, 1997; Coyne and Price, manuscript) and that through time the ranges of newly-formed species became gradually more sympatric. Marginal sympatry represented an early stage of divergence, partial sympatry a later stage, and full sympatry an even later stage. Ecological differences recorded included 'habitat' (encompassing all forms of spatial segregation including elevation, foraging height and vegetation type), food type/foraging method, and body mass, an index of prey size.

50 • *The progress of adaptive radiation*

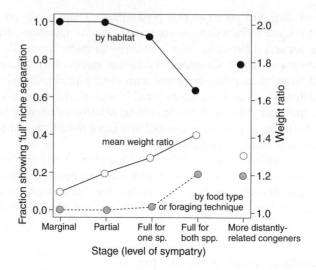

Fig. 3.4 Resource differentiation between closely related bird species of montane New Guinea at different stages after allopatric speciation. The last category groups pairs of less closely related species—pairs just similar enough to be classified as congeners. Data are from Diamond (1986).

All pairs of closely related species sympatric in part of their ranges were substantially different in habitat (Fig. 3.4). Average body size differences accumulated gradually from an initial ratio of about 1.1 (not insignificant but probably representing substantial overlap in prey size) to about 1.4 by the time of full sympatry. Average differences in foraging/food type were initially trivial then slowly increased. Separation along this dimension appeared to show a jump between the third and fourth stages but the data refer to the fraction of species showing full separation. Mean separation between species presumably diverges more gradually. Individual taxa may depart from the average trend shown. For example, habitat segregation was followed by differentiation in body size in some taxa, and by food type/foraging mode in other taxa, whereas significant divergence in both size and foraging/food occurred in a minority of taxa (Diamond, 1986). As differences along these secondary dimensions accumulated, habitat differences decayed somewhat but remained important.

The methods employed were less than ideal. Except for body mass, the ecological measurements are qualitative and simplified. Actual differences between species may be less clear-cut (Christidis *et al.*, 1988). Geographic overlap and taxonomic similarity are poor measures of time even within the confined geographic region of montane New Guinea. The separate pairwise comparisons are not independent since many species are used several times in different combinations. Yet, despite these problems the study established a valuable benchmark for later work. The conclusions were remarkably straightforward considering the heterogeneous mixture of bird taxa included.

How have Diamond's conclusions fared since then? Rather well, according to a few later studies (Christidis *et al.*, 1988; Price, 1991; Richman and Price, 1992; Suhonen *et al.*, 1994; Robinson and Terborgh, 1995). Habitat, broadly defined as in Diamond's survey, is the commonest ecological difference between sympatric sister species of birds. This implies that habitat differentiation also occurred early in the divergence of most bird clades, if rates of trait evolution have not changed dramatically toward the present time. Size differences indeed tend to characterize more distantly related congeners.

Why should habitat so often be the first axis of differentiation? Some insights may be gained by a study of exceptions to the rule. For example, closely related Galápagos finch species tend to occupy the same habitats and the same elevations (Lack, 1947; Grant, 1986) but differ greatly in beak and body size (e.g. Fig. 2.2). Major shifts in food type occurred both early and late in diversification (Schluter *et al.*, 1997). This reversal of the typical trend may reflect the very low species diversities of Galápagos bird communities, the different types of habitats and resources available, or intrinsic attributes of the original colonists.

3.3.2 A divergence rule for flowering plants

Verne Grant (1949) pointed out a difference between groups of flowering plants in the sorts of diagnostic characters used to classify them to species. Roughly 40% of taxonomic characters refer to floral structures (except the calyx) in lineages pollinated by specialized animal vectors, whereas the number is only about 10% in more 'promiscuous' angiosperms pollinated by unspecialized insects and abiotic agents (wind and water). Vegetative characters assumed greater importance in the second group. Grant's claim may reflect the behaviour of taxonomists as much as real differences between kinds of flowering plants, but it might easily be reformulated as a two-part hypothesis for niche differentiation: (1) differentiation in floral traits precedes vegetative divergence in lineages pollinated by specialized animal vectors; (2) the reverse sequence holds in lineages pollinated by unspecialized animal vectors or by abiotic agents. I group physiological traits with vegetative traits. The question regarding the order of niche differentiation is then whether habitat/physiology or pollination goes first.

Interest in divergence rules is motivated by two broader issues. One concerns the immediate basis of reproductive isolation. Does reduced gene flow between young sympatric species result principally from different pollen vectors or from divergent morphological and physiological adaptations to contrasting habitats? The other issue concerns the ultimate causes of speciation, whether it results from contrasting natural selection on vegetative structure and physiology between environments (e.g. Bradshaw, 1972) or instead by divergent sexual selection on floral structures mediated by pollinators (Crepet, 1984; Kiester *et al.*, 1984).

The second part of the rule may be more realistic than the first. Many angiosperm lineages whose species attract a broad array of unspecialized animal vectors indeed exhibit vegetative differentiation, often extraordinary levels (Heslop-Harrison, 1964). Most Hawaiian plant radiations fall into this category. Divergence in morphology, life history, and physiology is pronounced in the silversword alliance (e.g. Fig. 1.2) whereas their small, simple flowers with nonshowy colours are comparatively undifferentiated (Carlquist, 1974). Striking morphological and physiological differentiation in the absence of dramatic variation in flower structure is also seen in Hawaiian *Bidens* (Ganders, 1989) and *Tetramolopium* (Lowrey, 1995). Other radiations in this class include the Macaronesian *Argyranthemum* (Francisco-Ortega *et al.*, 1996, 1997) and the *Brocchinia* of the sandstone table mountains of Guyana (Givnish *et al.*, 1997).

Nevertheless, even simple flowers may undergo dramatic changes that outstrip vegetative transitions, especially in traits affecting mating system and mode of pollination. Some of these changes have immediate consequences for reproductive isolation. The Hawaiian *Schiedea* have experienced multiple transitions between unspecialized-insect pollination and wind pollination (Fig. 3.5). These transitions are associated with conspicuous changes in floral phenotype and mating system (Sakai *et al.*, 1997). Another drastic floral shift that occurs frequently in plant populations, even in lineages with relatively unspecialized flowers, is a reduction in flower size associated with increased reliance on self-fertilization (Stebbins, 1970; Ritland and Ritland, 1989; Wyatt, 1988; Husband and Barrett, 1993; Macnair and Gardner, 1998). This is frequently manifested in plants inhabiting novel or marginal habitats.

The two closely related columbines *Aquilegia formosa* and *A. pubescens* served as an initial test case for the first part of the rule, that floral divergence is faster than vegetative divergence in plants with specialized animal vectors (Grant, 1952). The elevational ranges of these two species abut and divergent pollinators appear crucial to their persistence in the zone of contact. Without them gene flow might cause their collapse (see Chapter 2). A number of other specialized animal pollination systems also exhibit levels of floral variation between closely related species that greatly outdo superficial vegetative changes. Well-studied examples include the tropical vines *Dalechampia* (Armbruster, 1988, 1993; Armbruster and Baldwin, 1998) and many orchids (e.g. Benzing, 1987; Gill, 1989; Hapeman and Inouye, 1997; but see Chase and Palmer, 1997).

However, absence of superficial morphological differences does not rule out vegetative divergence of a physiological nature (Heslop-Harrison, 1964), as further analysis of *Aquilegia formosa* and *A. pubescens*, the initial test case, reveals. Although nearly identical in leaves and stems, the species occur on different soils (Chase and Raven, 1975) and probably possess distinct physiological tolerances not easily measured from external morphological measurements. Tests of the rule therefore require reciprocal transplants of closely related species between the places they inhabit to determine whether loss of performance stems mainly from changes in pollinators (implicating floral evolution) or additionally involves lowered photosynthesis and growth (implicating vegetative differentiation).

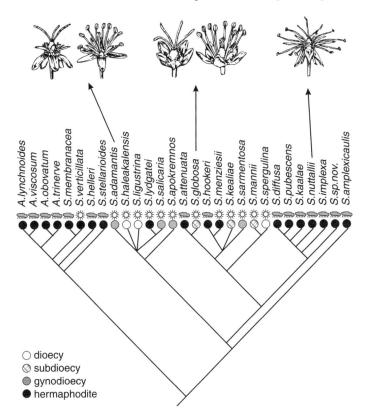

Fig. 3.5 Evolution of mating systems in the Hawaiian *Schiedea*. Hermaphrodites (sexually monomorphic) are pollinated by unspecialized insects or are autogamous. Dioecious (with males and females), gynodioecious (with females and hermaphrodites), and subdioecious species (with males, females, and hermaphrodites) are wind-pollinated. The tree is based on morphological characters (Sakai *et al.*, 1997). Branch lengths are arbitrary. Symbols above tips indicate habitat, whether wet or dry. Habitat and mating system (monomorphic vs dimorphic flowers) are significantly correlated ($\chi_2^2 \approx 12.0$, $P < 0.01$; tested using Pagel's 1994 likelihood ratio test, using equal branch lengths and equal forward and backward rates between pairs of states). Flower illustrations are from Sakai *et al.* (1997), with permission of Cambridge University Press.

3.3.3 Speciational habitat trends and taxon cycles

Rules of niche divergence go a step further in a habitat model of adaptive radiation inspired by the taxon cycle hypothesis. First I summarize the taxon cycle and then its derivative.

The taxon cycle hypothesis, originally formulated by Wilson (1959, 1961) and modified by Ricklefs and Cox (1972), was invented to explain observed patterns of distribution and differentiation of taxa on archipelagos. These patterns were grouped into four sequential stages. Species at stage 1 of the cycle are recent colonists from the mainland. They are widespread and undifferentiated. Stage 2 species are descended from stage 1 and are still widespread, but they are older residents of the archipelago and show some phenotypic differentiation between islands. Species in stages 3 and 4 are older still and have gone extinct on progressively more islands. They also exhibit higher levels of between-island differentiation than earlier stages. The usual endpoint of the cycle is extinction of the whole lineage but a late-stage species sometimes expands its range again, recreating stage 1 and beginning the cycle anew. Pregill and Olson (1981) suggested other explanations for the distributional patterns of the West Indian birds that partly inspired the hypothesis. While many further tests of the hypothesis are still required, recent molecular studies of birds on the islands of the Lesser Antilles confirm that species in the later stages of the proposed taxon cycle are indeed older residents of the archipelago than taxa in stages 1 and 2 (Ricklefs and Bermingham, 1999).

Of particular interest here is the habitat sequence superimposed upon the cycle (Fig. 3.6(a)). Stage 1 species tend to occupy 'marginal' habitats that fringe the islands, such as savanna and scrub forest, whereas species at later stages occupy more 'interior,' wetter forests (Wilson, 1961; Ricklefs and Bermingham, 1999). This suggests that each taxon began in the marginal habitats at stage 1 and with time shifted to interior environments.

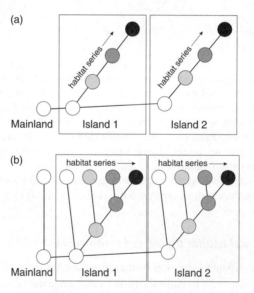

Fig. 3.6 Habitat trends on islands in species undergoing a taxon cycle (a) and in a lineage undergoing a speciational habitat trend (b). Open symbol represents a lineage in a 'marginal' habitat. Darker shading represents progressively more 'interior' habitats.

The strict taxon cycle does not leave much room for adaptive radiation unless allopatric populations form separate species and undergo divergence in phenotype and resource use. This has not been described. Sympatry between differentiated species that formed within the archipelago is also uncommon. In the best-known examples of taxon cycles, the ants of Melanesia (Wilson, 1961) and birds of the Lesser Antilles (Ricklefs and Cox, 1972; Ricklefs and Bermingham, 1999), close relatives occur infrequently on the same islands. The few examples that do exist in these taxa appear to represent separate invasions from the mainland (Wilson, 1961).

But the same archipelagos are also populated by taxa that have undergone speciation and habitat differentiation *in situ*, often within single islands. *Anolis* of the Greater Antilles fall into this category as do some beetles, geckos, and frogs (Liebherr and Hajek, 1990; Hass, 1991; Hass and Hedges, 1991). Liebherr and Hajek (1990) wondered whether such taxa might follow a series of habitat shifts analogous to those seen in the taxon cycle (Fig. 3.6(b)). They called their habitat sequence a 'taxon cycle' but this is confusing because habitat trends are not a necessary part of a true taxon cycle. Rather, their hypothesis is that a 'speciational trend' (Grant, 1989) takes place along the marginal-interior habitat gradient. At any point in time, speciation is most likely in that taxon whose habitat position is farthest inland (Fig. 3.6(b)), the point most different from the ancestral state. A trend is recognizable by a positive correlation between a species' position along the habitat gradient and the number of nodes connecting that species to the root of the phylogenetic tree.

Do radiations on archipelagos typically follow a speciational trend along the marginal-interior habitat gradient? The evidence so far is scant but not promising. First, while island radiations sometimes begin in the marginal habitats fringing the island (e.g. the *Partula* of Moorea), this does not appear to be the rule. The most basal species of *Anolis* on islands of the Greater Antilles occupies wet interior environments (Fig. 2.3). The ancestral Hawaiian *Tetramolopium* probably came from tropical montane New Guinea (Lowrey, 1995). The radiation of carabid beetles on Hispaniola (*Platynus fractilinea* group) seems to have begun in upland forest (Liebherr and Hajek, 1990). Second, spread to new habitats during the insular radiation seems to occur without regard for position on the marginal-interior gradient. For example, *Tetramolopium* on Hawaii (Lowrey, 1995), *Anolis* on islands of the Greater Antilles (Fig. 2.3), and carabid beetles on Hispaniola (Liebherr and Hajek, 1990) have spread to, and radiated in, dry marginal habitats.

3.4 Replicate radiations

Any trend in specialization across many taxa would represent a weak version of repeatable evolution involving only a general niche metric that ignores details of niche use. Rules of resource-axis differentiation in more narrowly circumscribed taxa deal with a potentially stronger but still coarse form of predictability. Here an even stronger form of repeatability is considered in which the same niche sequence is undergone by lineages inhabiting similar environments.

3.4.1 Postglacial fishes

The simplest case concerns young pairs of fishes inhabiting lakes of previously glaciated areas. These represent lineages that colonized lakes in the northern parts of North America and Eurasia after the recession of the immense sheets of ice that covered the region until about 15 000 years ago. Colonization of these lakes was made difficult by the limited number and duration of passage routes. Marine access to coastal lakes was limited to salt-tolerant species and changes in sea level meant that many lakes were accessible for only a short period (McPhail and Lindsey, 1986). Species intolerant of salt water dispersed inland via infrequent changes in the drainage patterns of lakes and rivers during deglaciation (McPhail and Lindsey, 1986). Fish faunas of these lakes remain depauperate.

A number of the fish lineages that colonized postglacial lakes quickly gave rise to sympatric species pairs that are among the youngest found in any taxon (Chapter 7 summarizes evidence that speciation rates are accelerated in these environments). Table 3.2 lists cases for which a genetic distance greater than zero was detected between sympatric forms, independent evidence of assortative mating exists, and morphological differences are likely to be inherited rather than just environmentally induced (Schluter, 1996*b*). Many other similar cases are not included here (Behnke, 1972; Svärdson, 1979; Schluter and McPhail, 1993; Robinson and Wilson, 1994) because genetic confirmation of species status is lacking or because reproductive isolation between forms is too incomplete (e.g. brook charr in E. Canada; Dynes *et al.*, 1999).

These cases provide evidence of parallel diversification on a wide scale. Sympatric species are typically highly differentiated ecologically and independent pairs tend to divide resources in the same way. Typically, one species is a pelagic zooplanktivore whereas the other consumes benthic invertebrates or larger prey from the littoral zone or deeper sediments. A consistent set of morphological differences is associated with this habitat split. Planktivores are usually smaller and more slender than benthivores with narrower mouths and longer, more numerous gill rakers. While genetic constraints have probably assisted the parallel evolution of suites of traits (Schluter, 1996*c*), strong selection is also implicated. Finally, the repetitive pattern of displacement indirectly implicates competition as a factor driving divergence (see Chapter 6).

Does predictability extend beyond the two-niche stage? Fewer examples are available to evaluate this question, but these suggest that the three-niche stage may reliably incorporate the two-niche stage as a subset. For example, three-species communities of round whitefish (*Prosopium*) in Alaska and Idaho include a planktivore and two benthivores, one occupying deeper waters than the other (Lindsey, 1981; Smith and Todd, 1984). The three species of brown trout (*Salmo trutta* complex) in some Irish

Table 3.2 Examples of young sympatric species pairs in lakes of recently glaciated areas. Genetic differences are given as $x/y/z$ where x is Nei's allozyme distance, y is percentage mtDNA sequence divergence based on restriction enzymes, and z is percentage mtDNA nucleotide divergence. The last measure combines differences in both nucleotide sequence and haplotype frequency. Modified from Schluter (1998), with additional diet data on *C. clupeaformis* from Bernatchez et al. (1999)

Nominal species	Region	Trophic characteristics	Genetic difference
Three-spine stickleback *Gasterosteus aculeatus*	British Columbia	Limnetic (planktivore) Benthic (benthivore)	0.02/—/ 0.02–0.10
Lake whitefish *Coregonus clupeaformis*	E. Canada, Maine	Dwarf (planktivore) Normal (benthivore)	0.01/0.5/—
Lake whitefish *C. clupeaformis*	Yukon, Alaska	High gill rakers (planktivore) Low gill rakers (benthivore)	0.01–0.02/—/ 0.02–0.30
Brown trout *Salmo trutta*	Ireland	Sonaghen (planktivore) Gillaroo (benthivore)	0.04/—/0.08
Brown trout *S. trutta*	Sweden	Planktivore Benthivore	0.03/—/—
Arctic charr *Salvelinus alpinus*	Scotland	Planktivore Benthivore	0.02/—/—
Arctic charr *S. alpinus*	Iceland	Planktivore and piscivore Small and large benthivore	0.001/—/0.01
Rainbow smelt *Osmerus mordax*	E. Canada, Maine	Dwarf (planktivore) Normal (benthivore, piscivore)	—/—/0.01–0.10

lakes include a planktivore, a benthivore, and a piscivore (Fergeson and Taggart, 1991). The four ecomorphs (possibly species) of Arctic charr (*Salvelinus alpinus* complex) in an Icelandic lake join the divergent elements of the trios in possessing a planktivore, two benthivores (one living at greater depths than the other), and a piscivore (Skúlason et al., 1989).

The extent of parallel evolution is even more dramatic than indicated in Table 3.2 because there is evidence of multiple origins of benthic and limnetic pairs within several of the groups, including three-spine sticklebacks and lake whitefish (Schluter and McPhail, 1993; Bernatchez et al., 1999; Taylor and McPhail, manuscript). In the sticklebacks parallel evolution has even occurred in the mate recognition traits, leading to the parallel origin of species (parallel speciation; Schluter and Nagel, 1995; Rundle et al., 2000) (see Chapter 8).

Table 3.3 The number of *Anolis* lizard species in each of the ecomorph classes. Species counts exclude two recent colonists of Jamaica and Hispaniola, both trunk-ground: the extinct *A. roosevelti* on Puerto Rico, a crown-giant, and several other species that do not fit into the standard ecomorph categories

Ecomorph	No. species		
	Jamaica	Puerto Rico	Hispaniola
Crown-giant	1	1	3
Twig	1	1	3
Trunk-crown	2	2	4
Trunk-ground	1	3	5
Grass-bush	—	3	6
Trunk	—	—	5

3.4.2 Anolis *lizards*

A spectacular case of replicate radiations involves the *Anolis* lizard ecomorphs on large islands of the Greater Antilles (see Chapter 2). Williams (1972) described six major ecomorphs differing in their use of perching substrates: trunk-crown, trunk-ground, crown giant, twig (also in the tree crown), trunk, and grass-bush. Lizards belonging to the same ecomorph category are morphologically similar (Fig. 3.3). Williams proposed that morphological and behavioural similarity between species falling into the same ecomorph category but inhabiting different islands is mainly the result of parallel and convergent evolution, not shared ancestry. This idea was recently confirmed by molecular sequences. Every ecomorph has arisen multiple times (Losos *et al.*, 1998).

Williams also noticed that ecomorph sets on different islands were nested (Table 3.3). Puerto Rico has the same four as Jamaica plus the grass-bush ecomorph. Hispaniola has all five of the ecomorphs found on Puerto Rico plus the trunk ecomorph. Williams (1972) proposed that the faunas on the three islands are at successive stages of the same evolutionary sequence. Thus, the grass-bush ecomorph was the last to appear on Puerto Rico, whereas the trunk ecomorph originated on Hispaniola after the other five. The molecular phylogeny of Losos *et al.* (1998) is not incompatible with this idea but it does not allow a strong test. However, the data suggest that later ecomorphs did not always arise from the same ancestral type, a possibility needing confirmation.

3.4.3 *Other cases*

The possibility of replicate adaptive radiation exists wherever faunal convergence is seen or where related taxa diversify independently in very similar environments.

The multiple instances of size displacement in sympatric *Cnemidophorus* lizards are another example (Radtkey *et al.*, 1997), as are the replicate evolution of arboreal, semiarboreal and terrestrial *Mandarina* snails on the Bonin islands of the western Pacific near Japan (Chiba, 1999*a*). More complex cases include the independent origin and radiation of foliage-gleaning warblers in Asia and North America (Price *et al.*, 1998, 2000) and the striking cases of convergent evolution in separate radiations of cichlid fishes in different East African great lakes (Fig. 1.3) (Fryer and Iles, 1972; Greenwood, 1974; Meyer, 1993; Kocher *et al.*, 1993). The repeated origin of similar phenotypes in the East African Lakes has been confirmed with molecular studies, but as in the *Anolis*, the cichlid ecomorphs may not have arisen from the same ancestral phenotype each time (Kocher *et al.*, 1993).

The smaller radiations of 9–11 cichlid fish species in different crater lakes of West Africa did not yield equivalent ecological assemblages (Trewevas *et al.*, 1972; Schliewen *et al.*, 1994). Interesting differences also emerge between the radiations of warblers of Asia (*Phylloscopus*) and the New World (*Dendroica*), groups which on the whole show striking convergence in many aspects of their morphology and ecology (Price *et al.*, 2000). For example, the *Dendroica* are heavier and more frugivorous than *Phylloscopus*.

3.5 Phenotypic evolution near the end of adaptive radiation

Many lineages undergoing adaptive radiation exhibit bursts of speciation that later subside, perhaps as the supply of new niches is exhausted. A reasonable expectation is that ecological expansion should also slow for the same reasons, reducing phenotypic differentiation. Such events might mark the end of an adaptive radiation. In agreement with this expectation, net rates of morphological evolution calculated from distantly-related species are lower than those based on species that diverged in the more recent past (Gingerich, 1983; Lynch, 1990).

In this section I explore the connection between patterns of phenotypic evolution and declining rates of species accumulation. Most information on trends comes from the fossil record and deals mainly with long time scales and high taxonomic levels. These studies provide a picture of how speciation and morphological change might be connected in the latter stages of diversification. Do similar trends prevail in more recent adaptive radiations at lower taxonomic levels?

3.5.1 Macroevolutionary patterns

The pattern of rise and fall of species numbers and total morphological diversity through time differs widely among higher taxa in the fossil record (Foote, 1992, 1993; Roy and Foote, 1997). Nevertheless, three elements are seen again and again in different groups, as illustrated by the trilobites (Fig. 3.7). Species richness and morphological variety rose rapidly at about the same time early in the history of

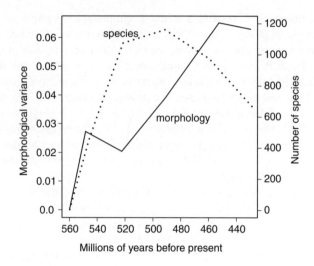

Fig. 3.7 Morphological diversity in the trilobites peaked after species richness. Both declined after this period. Morphological diversity is the total phenotypic variance ('disparity') among species in two-dimensional outline. Modified from Foote (1993).

the clade. Rates of increase of both morphological diversity and species richness eventually slowed, but the decline was not simultaneous. Species diversity peaked well before morphological diversity.

The last feature is perhaps the most puzzling. Purely random extinction would maintain phenotypic diversity at a high level in the face of declining species diversity, at least for a while (Foote, 1993; Roy and Foote, 1997). But it is not easy to see how phenotypic diversity could continue to *rise*. Presumably, at the beginning of the end of the radiation, species having the most extreme phenotypes persist longer than 'interior' species. Morphological expansion may even continue at the outside edge of the range of phenotypes even as extinction began to generate holes in the middle of the phenotype distribution (see also Jernvall *et al.*, 1996). Even if new species are no longer being generated, species not facing extinction may continue to adapt to their current environments, yielding a net divergence.

The study of phylogenetic trees has not yet seen any large-scale tests of the association between phenotypic and morphological diversity through the course of diversification. Of course, such study would be limited to those species which have survived to the present, as extinctions go unrecorded (with the possible exception of mass extinctions; Harvey *et al.*, 1994). Phylogenetic trees are nevertheless able to detect a slowing-down in rates of species accumulation toward the present time. For example, the number of avian lineages accumulated at a declining rate between the time birds originated and the origin of most contemporary avian families (Nee *et al.*, 1992). This trend was not seen in all parts of the avian lineage, however. Rates of speciation accelerated in the Passeri (songbirds) and Ciconiiformes (waders, hawks, and most seabirds) toward the time marking the origin of modern bird families.

3.5.2 Trends in recent groups

How faithfully do macroevolutionary patterns predict those of adaptive radiations at lower taxonomic levels? One pattern that seems to hold is a decline in the net rate of phenotypic divergence toward the present time (Fig. 3.8(a)). Below I use measurements of continental mammals and birds to show that at low taxonomic levels this decline is mainly an artifact of scaling and that phenotypic diversification in recent lineages continues apace even as species diversity plateaus or declines.

Available measurements of mammals and birds suggest that morphological variance between species has risen steadily over the past 5 million generations or so, perhaps even at rates faster than linear (Fig. 3.8(b)). This conclusion appears to contradict that reached by Lynch (1990) using the same mammalian data set. His

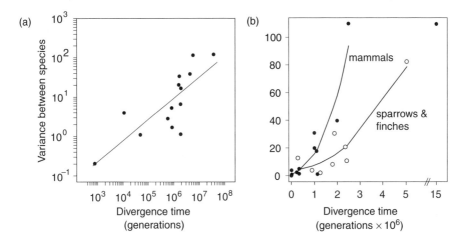

Fig. 3.8 Morphological variance between species as a function of time since they diverged from a common ancestor. (a) The log-log plot. Each point is an average of several comparisons within a given mammal clade. Morphological variance is scaled relative to the variance within populations (Lynch, 1990). The slope of the regression line is 0.5, which suggests that morphological variance lags behind as one considers progressively older taxa. Modified from Lynch (1990). (b) Plot of the same variables as in (a) but on the untransformed scale. Filled points are the mammals. The fitted curve is a cubic spline fit to all points but the very oldest. Open circles are based on five external beak and body dimensions of several continental sparrow and finch clades (D. Schluter, unpublished data). Time was estimated using molecular clocks and assume a generation time of 1 year. The Y-measurements are not on the same scale as the mammalian data; values were multiplied by 300 to allow plotting on the same graph. The fitted line is a cubic spline.

log-log plot of cranial variation between mammal species against divergence time shows a declining net rate in older groups (Fig. 3.8(a)). The regression slope in his plot is about 0.5, which is half the slope expected if rates of divergence are constant in time. The resolution of this apparent contradiction lies in the fact that relationships between morphological variance and time effectively have a Y-intercept greater than zero (Fig. 3.8(b)). Initial differences between species, although very small, appear to take no time at all. This leads to very high estimates of evolutionary rate in the youngest comparisons, a rate which is not sustained as time increases above zero. The log-log plot is misleading if variance is necessarily bounded above zero, and it obscures the longer-term increase which in the mammals shows signs of abating only after many millions of generations have elapsed (Fig. 3.8(b)).

Three sets of 'replicate' adaptive radiations allow further examination of trends in morphological divergence. The first set includes the *Anolis* lizard ecomorphs on three large islands of the Greater Antilles: Jamaica, Hispaniola, and Puerto Rico (Table 3.3). Puerto Rico's radiation is the oldest, Jamaica's is the youngest, and Hispaniola is of intermediate age (Jackman *et al.*, 1997). The second set comprises the three cichlid 'flocks' of Lake Victoria (youngest), Malawi, and Tanganyika (oldest) in East Africa (Fryer and Iles, 1972). The third set includes the foliage-gleaning warblers of the Old and New Worlds (Price *et al.*, 1998, 2000). Radiations within each set are highly comparable but of different age, and as far as morphological variance is concerned they may represent different stages of the same sequence (I use the term radiation loosely here, as several of the individual groups are polyphyletic and not entirely independent).

Morphological variance is highest in the oldest groups of all three sets (Fig. 3.9), suggesting that phenotypic variance continues to increase with age. It is reasonable to assume that morphological divergence has not yet reached a peak even in the oldest groups. If these later radiations truly represent later stages of a common sequence, then the New World warblers, the cichlids of Lakes Malawi and Victoria, and the *Anolis* ecomorphs on Jamaica and Hispaniola can look forward to a future in which the magnitude of species differences will more than double.

The continuous rise of morphological variance suggested by these data contrasts with trends in species diversity. Despite its great age, Lake Tanganyika has fewer species (ca. 200) than Lakes Victoria (ca. 500 currently described; O. Seehausen, personal communication) and Lake Malawi (ca. 600), suggesting a downturn. The number of warbler species in Asia is not greater than that in the New World despite the greater age of the former clade. Long branches leading to the tips of the phylogenetic tree of the Old World warblers also suggest a recent decline in the rate of speciation in this group (Price *et al.*, 1998). Species diversity on Greater Antilles islands correlates with island area (Losos, 1998), being greatest on the largest island, Hispaniola, despite its relative youth. However the oldest island, Puerto Rico, has twice the number of species as Jamaica despite its slightly smaller size. This suggests that diversity may still be rising on the Greater Antilles.

Phenotypic evolution near the end of adaptive radiation • 63

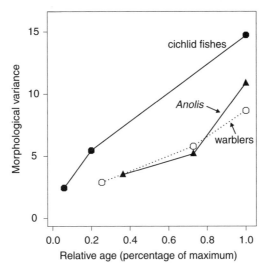

Fig. 3.9 Morphological variance among species in sets of 'replicate' radiations varying in age. Cichlid data were obtained from Table 32 of Fryer and Iles (1972) which gives the range of phenotypes in Lakes Victoria, Malawi, and Tanganyika for each of five traits: vertebrae, longitudinal scales, dorsal fin spines, anal fin spines, and body length. Variance in each lake was calculated as the square of the difference between the log-transformed upper and lower limits of the range, summed over the five traits. The ages of the radiations in each lake are based on mtDNA molecular clocks (Meyer, 1993; for discussion see McCune, 1997). *Anolis* lizard measurements were obtained from Losos (1992). Only the four ecomorphs present on all three islands are included (Table 3.3). For each ecomorph on each island mean values were calculated for the first two principal components based on several external body dimensions including mass, snout-vent length, number of digital lamellae, and the lengths of the forelimb, hindlimb, and tail (higher pc's were discarded). Total variance on a given island is the sum of variances among ecomorph means for pc1 and pc2. Relative age was determined for a given island from the oldest node of the tree in Jackman *et al.* (1997), one of whose branches includes only descendants occurring on the given island. Time was counted as number of nodes, counting backwards from one of the tips on Jamaica to the root of the tree. The three warbler points refer, in order, to the *Dendroica* alone, all New World wood warblers, and the Old World *Phylloscopus* warblers. Variance is the sum of variances among species means in five external dimensions, log-transformed, including tarsus length, wing length, and the length, depth and width of the beak (Price *et al.*, 2000). Time is measured as the time of the first split within each of the three groups estimated using revised mtDNA sequences (Price *et al.*, 2000) (sequences in Price *et al.*, 1998 for New World warblers were erroneous; Price, personal communication).

Overall, the results above suggest that rates of species accumulation may decline even while morphological diversity is still rising, in agreement with the pattern from the fossil record (Fig. 3.7). This raises the possibility that declines in species numbers and morphological diversity at the end of adaptive radiation may have different causes.

3.6 Discussion

As Simpson surmised, adaptive radiations often repeat one another in broad outline and occasionally in their fine details. This is most noticeable when lineages undergoing adaptive radiation in similar environments are closely related. At a coarser scale, species may diverge along resource dimensions (pollinators, habitat, food size) in similar sequence even in more distantly related taxa. Replays of life's tape thus yield outcomes similar to that achieved the previous time, provided we refrain from rewinding too far back. Losos (1992) suggested that predictability of faunal evolution is facilitated by strong interspecific interactions, whereby earlier species limit the options available to later ones.

Repetition may also extend to patterns of species accumulation and morphological variance through time. In particular there is a hint that these two measures may increase together early in a radiation but that expansion to new resources and environments continues even as diversity begins to fall near the end of adaptive radiation. Morphological variance among species continues to grow with time, if not indefinitely then at least over the lifespan of most radiations. Admittedly, this is not the same as showing that niche expansion also continues unabated. Perhaps species merely continue to acquire morphological adaptations to their niches long after diversity has peaked and niche shifts are a thing of the past. Alternatively, the pattern may indicate that changes in species number and morphological variation may have different causes toward the end of adaptive radiation, in contrast to the beginning of the radiation where the two increase together. This feature of adaptive radiation has received very little study.

In contrast a trend toward greater niche specialization, thought by Simpson (1953) to be one of the most predictable features of adaptive radiation, may not hold in general. Very little indication exists from phylogenies that the founders of adaptive radiation are typically generalists whose descendants gradually specialize. Frequently the founders are estimated to have been specialists instead that gave rise to both specialist and generalist descendants. These results are based on phylogenetic methods whose accuracy remains to be proven. For this reason they need to be corroborated using other approaches and data sets.

These findings on specialization do not rule out the possibility of a trend toward specialization over longer time spans involving multiple radiations of many taxa. However, they reveal that transitions between specialist and generalist states occur often, significantly weakening the case for a strong trend in one direction. If the results herein hold up, they will necessitate a dramatic revision of Simpson's view of the sequence of ecological changes that take place during adaptive radiation, at least at a low taxonomic level. In particular, expansion to new resources and environments rather than niche subdivision would emerge as the predominant ecological trend of adaptive radiation.

4

The ecological theory of adaptive radiation

Most differentiations of either subspecies or species ... represent occupation of subzones within an adaptive zone, or niches of a biotype ... or of adjacent minor peaks on one part of the selection landscape
—Simpson (1953)

4.1 Introduction

What causes adaptive radiation, elevated rates of speciation together with correlated divergence of phenotype and use of environment? In this chapter I review the body of ideas that make up the 'ecological theory' of adaptive radiation. This theory holds that adaptive radiation is ultimately the outcome of divergent natural selection stemming from environments and resource competition. The theory was built mainly by Simpson (1944, 1953), Lack (1947), and Dobzhansky (1951), but it brought together views of many other naturalists back to Darwin. It survives to the present as the last major synthesis of ideas to explain the processes driving ecological diversification of lineages.

I begin with a summary of this theory and a few of the observations that inspired it. I then review the most significant additions and alternatives that have been developed since then. Some of these alternatives predate the ecological theory but have been elaborated in the meantime. They include especially genetic drift, the spread of alternative advantageous mutations unrelated to differences in environment, sexual selection, and selection arising from ecological interactions other than competition. Evidence bearing on these ideas is addressed in subsequent chapters.

4.2 The ecological theory

The ecological theory incorporated three main processes that I discuss in turn. The first is phenotypic divergence of populations driven by divergent natural selection between environments. The second process is phenotypic divergence mediated by competition for resources, which has two elements. Competition between closely related species drives them to exploit different environments where contrasting selection pressures prevail. Conversely, absence of competing species from distantly related taxa creates the ecological opportunities that permit phenotypic diversification. The third process of the ecological theory is 'ecological speciation', whereby new species arise by the same processes that drive differentiation of phenotypes, namely divergent natural selection stemming from environments and resource competition.

4.2.1 Divergent natural selection between environments

Divergent natural selection between environments was viewed as the principal cause of differentiation of populations and species in traits used to exploit those environments (Huxley, 1942; Mayr, 1942; Simpson, 1944, 1953). The idea was that each environment subjects its consumers to unique selection pressures arising from the advantages of distinct combinations of traits for efficient resource exploitation.

This hypothesis seems to be a straightforward inference from various early observations that the phenotypes of different species, especially traits involved in extracting, subduing, and masticating foods, appear well-built to perform those tasks. The species of Galápagos ground finches *Geospiza*, most of whom consume mainly seeds, have stout conical bills whereas the beaks of the insectivorous warbler finches (*Certhidea*) are small and slender and are suited to gleaning small insects from leaves and twigs (Fig. 1.1; Lack, 1947; Bowman, 1961). The flowers of closely related species of flowering plants (e.g. Fig. 2.5) seem well-designed to distribute and receive pollen via different kinds of effective animal pollinators (Grant, 1949).

But divergent natural selection is a much stronger claim than just that populations experience different selection pressures. The crux of the hypothesis is that intermediate phenotypes have lower fitness. Simpson (1944, 1953) illustrated the idea in his famous depictions of divergence on 'selection landscapes' (Fig. 4.1), which were inspired by Wright's (1931, 1932) fitness landscapes for gene frequencies. The axes of Simpson's surface represented phenotypic traits and fitness was the height of the surface. The contours of the surface were shaped by features of environment. Populations diverged because they were pulled toward different adaptive peaks and away from the valleys of low fitness between them (Fig. 4.1).

Under the ecological theory, then, divergent natural selection explains the correlation between phenotype and environment in adaptive radiation (Fig. 4.2). If environments are discrete, or points along environmental gradients provide unequal rewards, then selection landscapes may have multiple peaks and valleys. For example, gaps in the frequency distribution of perch diameters available to *Anolis* lizards would give rise to multiple optima of perch use and limb length. This assumes that no one phenotype can best exploit the entire spectrum of perches (i.e. there must be trade-offs). The correlation between limb length and perch diameter arises as separate populations and species evolve toward the different optima. For this scenario to work some mechanism is needed to bring populations across fitness valleys, from the domain of one adaptive peak to another. Simpson proposed that changing environments was one way this was accomplished in nature, an idea I explore further in Chapter 5.

The ecological theory • 67

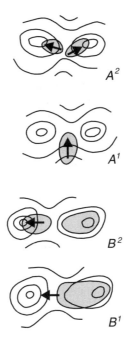

Fig. 4.1 Simpson's depiction of divergence on a selection landscape. Arrows indicate the direction of evolution. In A, a population advancing uphill splits to occupy two adaptive peaks. In B, an ancestral population occupying one peak gives rise to a daughter population that evolves toward the second. Redrawn from Simpson (1953, p. 156), with permission of Columbia University Press.

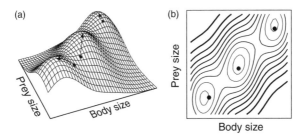

Fig. 4.2 Divergent natural selection as the cause of a correlation between mean phenotypes and environment. (a) The adaptive landscape. Mean fitness of a phenotype and resource combination is indicated by the height of the surface. Arrows indicate paths of steepest fitness ascent for three populations in the vicinity of different peaks. (b) Mean fitness contours for the surface in (a). Thin contour lines represent highest mean fitness. Means of the three populations have approached the peaks.

Most evolutionists working after Wright (1931, 1932) nevertheless recognized that nonadaptive processes, including genetic drift, accidental allele loss following bottlenecks in population size, and correlated response to selection, also played a role in phenotypic differentiation (Huxley, 1942; Mayr, 1942; Dobzhansky, 1951). Simpson (1944, 1953) suggested that genetic drift might contribute to shifts between adaptive zones. Even Fisher (1936), who insisted that 'evolution is progressive adaptation and consists of nothing else', allowed incidental differences between populations and species to accumulate as unselected by-products of natural selection on correlated traits. In his study of the adaptive radiation of Darwin's finches, Lack (1947, p. 79) argued that 'conditions are extremely similar on many of the Galapagos islands, and the differences in beak and wing-length shown by island forms so often seem haphazard and pointless, that it is probable that many of these differences are genuinely unrelated to possible environmental differences.'

Yet, despite this caveat Lack's final conclusions relegated only a small portion of interpopulation variation to chance, and indeed his overall analysis of the Galápagos finches helped sway majority opinion to the view that selection's role in divergence was paramount (Gould, 1983). It is fair to say that by about 1950 few would have argued that nonadaptive processes played a significant role in generating population differences in phenotype when they matched ecological differences—i.e. when phenotypic divergence fit the criteria of an *adaptive* radiation.

4.2.2 Competition and ecological opportunity

Phenotypic divergence resulting from resource competition is the second process of the ecological theory. By competition I mean the antagonistic interaction between phenotypes that arises from depletion of shared resources. Mayr (1942), Lack (1947), Simpson (1953) and others viewed resource competition as the most important interspecific interaction in adaptive radiation. Other interactions such as direct predation received only casual mention. Competition was seen as having two roles depending on its source. Competition among forms within the radiating lineage promoted divergence, whereas that from species in other taxa inhibited it.

Lack (1947) championed competition's role in divergence between close relatives. He regarded divergence between competitors as the last stage of the evolutionary cycle that produced two coexisting species from a single ancestor. 'The meeting of two forms in the same region to form new species must ... result in subdivision of the food or habitat, and so to increased specialization. The repetition of this process has produced the adaptive radiation of Darwin's finches' (Lack, 1947). As evidence Lack presented multiple examples from the Galápagos finches in which phenotypic differences between similar species were greater where they occurred together (sympatry) than where they occurred separately (allopatry). The most impressive case concerned the two ground finches, *Geospiza fuliginosa* and *G. fortis*. Both have a similar beak size on islands where each occurs alone, but *G. fuliginosa* has a smaller beak and *G. fortis* a larger beak on islands where the two occur together. Brown and Wilson

(1956) later compiled a list of similar examples from other avian taxa and labelled the phenomenon 'character displacement'. The impact of these papers was profound. Although examples were then known only from birds, resource competition became regarded as fundamental to adaptive radiation, equal in standing to extrinsic features of environment.

Under this hypothesis competition is an agent of divergent natural selection, as are the differences between environments discussed previously. Competing species are simply part of the environment. Nevertheless, separate treatment is warranted by the special status that competition has attained in the ecological theory, and by the magnitude of the eventual controversy surrounding its role in divergence (see Section 4.3.2). It is also useful to distinguish the impact of interactions between species of the radiating lineage itself from that of external factors. For this reason I evaluate the influence of interactions between species (Chapter 6) separately from the effects of differences between external environment (Chapter 5). I use the term 'character displacement' to describe the general process of phenotypic change induced or maintained by resource competition. I apply the term even to those cases in which allopatric populations are secondarily derived from those in sympatry. This is because phenotypic shifts that result following release from competition imply that displacement is an ongoing process in sympatry.

The flip side of competition between relatives is 'ecological opportunity', which was viewed as the major regulator of the rate and extent of phenotypic differentiation and also speciation. Ecological opportunity may be loosely defined as a wealth of evolutionarily accessible resources little used by competing taxa. The concept was the cornerstone of Simpson's (1944, 1953) theory of 'adaptive zones' (a zone is a set of related resources or niches constituting a particular 'way of life'). For adaptive radiation to occur 'the zone must be occupied by organisms for some reason competitively inferior to the entering group or it must be empty' (Simpson, 1953, p. 207). The novel selection pressures thus encountered would stimulate adaptive radiation. The absence of competitors '... apparently facilitates the crossing of valleys between one adaptive peak and another' (Mayr, 1942, p. 271). Nature abhors an empty niche.

There are many putative examples. Absence of competition from other birds was viewed as the major ecological factor behind the radiations of finches in the Galápagos and Hawaiian archipelagoes in 'directions which would otherwise have been closed to them' (Lack, 1947, p. 118). No other avian taxa of equivalent age were known to exhibit such a wide array of ecological types. High rates of phenotypic evolution in Hawaiian and Galápagos plants may be explained similarly by the low diversity of native floras (Carlquist, 1974).

At a larger scale the rise of placental mammals in the early Cenezoic was seen as permitted by the mass extinction of reptiles at the end of the Cretaceous, and the diversification of marsupial mammals of Australia by the scarcity of placental mammals there (Huxley, 1942; Simpson, 1953). Compilations of more recent evidence from the fossil record indicate that accelerated rates of speciation and morphological diversification follow mass extinctions (Jablonski, 1986, 1989; Sepkoski, 1996).

These bursts tend not to involve surviving members of lineages previously dominant, but rather they frequently involve taxa that comprised a small fraction of the pre-extinction fauna. A reasonable interpretation is that the dominant incumbent taxa had suppressed diversification in other groups, which were able to radiate only after the dominants were removed by extrinsic forces (Simpson, 1953; Stanley, 1979; Benton, 1983, 1987; Jablonski, 1986, 1989; Sepkoski, 1996).

Simpson (1944, 1953) also stressed a third route to ecological opportunity, one that did not require a change in distribution or the creation of a competitive vacuum through extinction. Sometimes a change in traits possessed by a lineage, a single 'key character' or a whole block of characters, would confer access to an array of new niches or would bestow competitive superiority over taxa already using them. The extent of diversification initiated by the trait was viewed as 'the extent of adaptive opportunity provided by the change' (Simpson, 1953, p. 223). As an example, Simpson suggested that the huge diversification of rodents, to the point of constituting a higher taxon, may be explained by one new feature: their 'persistently growing, chisel-like incisors' (Simpson, 1953, p. 346). However, explosive radiation may be correlated with appearance of a new trait for a great variety of reasons including 'chance' (Heard and Hauser, 1995). To confirm the mechanism that Simpson envisioned two steps are necessary. First, one must confirm the correlation between appearance of a key character and diversification. Second, one must distinguish ecological opportunity from other mechanisms.

4.2.3 Ecological speciation

Speciation is essential to adaptive radiation in cross-fertilizing organisms, for without reproductive isolation the accumulated morphological differences between populations would vanish in sympatry by hybridization. To Simpson (1944, 1953) a vacant adaptive zone was as stimulating to speciation as to morphological differentiation. High rates of speciation in adaptive radiation were seen as driven by divergent natural selection, which prevailed when ecological opportunities arose, such as via relaxed competition from other taxa (Huxley, 1942; Mayr, 1942; Lack, 1947; Simpson, 1953). Simpson himself was not explicit about how genetic isolation arose, but he felt that it was somehow linked to phenotypic evolution. Both occurred as separate populations within a species accumulated adaptations to different ecological niches.

Mayr (1942) and Dobzhansky (1951) were more precise about mechanism. They viewed pre- and postmating isolation as arising incidentally between populations that diverged in phenotype in response to environment: 'many of these differences, particularly those affecting physiological and ecological characters, are potential isolating mechanisms' (Mayr, 1942, p. 59). Dobzhansky (1951) was concerned mainly with postzygotic isolation arising from genomic incompatibilities, and he argued that these 'physiological isolating mechanisms may be a product of natural selection.... The genotype of a species is an integrated system adapted to the ecological niche in which the species lives. Gene recombination in the offspring of species hybrids may lead to formation of discordant gene patterns'. This is the 'by-product' mechanism of speciation, so called because natural selection never directly favours reproductive isolation. The other mechanism of speciation was reinforcement, whereby partial postmating isolation directly favoured the evolution of premating isolation (Fisher, 1930; Dobzhansky, 1937). Reinforcement is a part of ecological speciation if environmental selection pressures are responsible for the evolution of partial hybrid breakdown and initial levels of premating isolation.

Under this view species are ultimately the result of divergent natural selection, whether reproductive isolation evolved solely as a by-product of selection on other traits or additionally involved reinforcement. These selection pressures arise from the same processes that drive phenotypic and ecological differentiation, namely environments and resource competition.

This thoroughly selectionist perspective on speciation did not enjoy as large a following as did the similar perspective on phenotypic differentiation discussed earlier. This is because researchers then and now realized that many nonecological mechanisms may also cause speciation. Nonecological modes of speciation include: genetic drift in stable populations, genetic drift through founder events and population bottlenecks, some modes of speciation by sexual selection, and the fixation of alternative advantageous genes in allopatric populations experiencing similar selection pressures. The last of these mechanisms is 'adaptive' in the sense that selection drives the genes to fixation, but differences between environment are not required and selection is not divergent. Polyploidy is a nonecological mechanism of speciation common in plants (Niklas, 1997; Ramsey and Schemske, 1998). Polyploid speciation does not require contrasting environments but is nevertheless often associated with niche shift (Ramsey and Schemske, 1998).

These considerations make it clear that given enough time, allopatric populations should eventually acquire genetic differences via nonecological processes. Many of these changes would result in new species if reproductive isolation were not already present. Therefore the main question of interest is whether, in adaptive radiation, reproductive isolation typically evolves much sooner as a consequence of divergent natural selection.

4.3 Extensions and alternatives

The ecological theory has inspired most of the last half-century of research into adaptive radiation. Nevertheless, the past decades have also witnessed great advances in the development of new hypotheses that extend or challenge all three components of the theory. Taken together these ideas raise the possibility that a wider diversity of processes underlie adaptive radiation. At the very least they provide useful alternatives against which the mechanisms of the ecological theory should now be tested. In the rest of this chapter I provide an overview of the most significant of these alternative ideas.

4.3.1 Alternatives to divergent natural selection

Meeting the phenotypic criteria for adaptive radiation strongly implicates natural selection in differentiation because stochastic fluctuations in trait means generated purely by mutation and genetic drift are unlikely to yield a significant correlation between mean phenotype, use of environment, and performance in those environments. The ecological theory provides an explanation for these observations, but in general such observations by themselves do not establish the theory's central claim, that natural selection is divergent. Other tests are first needed to demonstrate that intermediate phenotypes have reduced fitness. Without these additional tests, correlations between mean phenotype and use of environment, the hallmark of adaptive radiation, may have other explanations instead.

One such alternative is random drift along adaptive ridges (Emerson and Arnold, 1989), a model that incorporates strong selection but only off the ridge top (Fig. 4.3). Chance perturbations of a population mean away from the ridge top are returned to it by selection because phenotype and resource use are otherwise mismatched. However, the mean is not generally expected to return to the same point on the ridge as before, since its height is the same at many locations. The result is gradual divergence of populations even though selection does not directly favour divergence. Similarly, a set of populations all beginning at the same point on one side of the ridge may diverge from one another as they ascend it, owing to genetic drift and chance difference in their genetic makeup. A correlation between phenotype and use of environment is a predicted outcome, with the expected slope of the relationship depending on the orientation of the ridge.

This alternative explanation for patterns of species differences is reasonable if the environment variable is a quantity describing utilization rather than an externally imposed feature of the habitat. Short-legged lizards seek out narrow perches (Fig. 2.4), but this does not mean that a combination of slightly longer legs and wider perches is in any way inferior. Drift along adaptive ridges is easiest to imagine when resources are

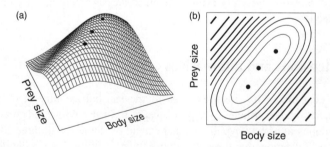

Fig. 4.3 Phenotypic and ecological differentiation via genetic drift along an adaptive ridge. (a) An adaptive landscape having a single broad ridge connecting phenotypes and environment combinations having equivalent fitness gains. Filled symbols are the means of three hypothetical species that have differentiated along the ridge. (b) The surface in (a) plotted using mean fitness contours. The thickness of contour lines declines with increasing mean fitness.

continuous but it can also occur when they are discrete (e.g. habitats) if utilization by a phenotype is sufficiently broad to span the gaps. Spatial separation of environments (e.g. drier lowlands vs wetter highlands) is also no guarantee of multiple adaptive peaks, because the limits of a species' geographical range are themselves evolvable and may drift (Kirkpatrick and Barton, 1997).

A modification of the previous model is the rising ridge, whereby fitness is low at one end of the ridge and rises toward the other end. At equilibrium, all populations are expected to have the same mean, but differences between populations in the rate of input of advantageous mutations (e.g. because of differences in population size) will strand them for some time at different points along the crest of the ridge. This too will yield a (temporary) correlation between phenotype and environment without divergent natural selection.

A related hypothesis is that divergence has occurred in response to uniform selection (Cohan and Hoffman, 1986; Cohan et al., 1989). For example, Lenski and Travisano (1994) subjected 12 replicate populations of *E. coli* to 10 000 generations of selection in identical environments. All lines were started from the same clone, so genetic variance within each line was initially zero. Populations diverged substantially in body size despite the uniformity of external conditions (Fig. 4.4). If the separate lines are nevertheless using the same niche then no correlation between phenotype and environment results, and the pattern would not be mistaken for an adaptive radiation by our criteria. However, if correlated niche differentiation has indeed occurred, then the results would represent either different points along an adaptive ridge or (more likely) different adaptive peaks. The latter possibility is not necessarily inconsistent with the hypothesis of divergent natural selection, but we would need to incorporate different mutation histories to explain why initial divergence was toward different peaks.

While the plausibility of adaptive ridges may be debated in each circumstance, the crucial point is that tests of divergent natural selection are needed and comparative

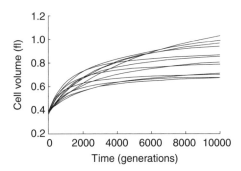

Fig. 4.4 Change in mean cell size in 12 replicate *E. coli* populations propogated for 10 000 generations in identical laboratory environments. From Lenski and Travisano (1994), with permission of the National Academy of Sciences, U.S.A., and the authors.

evidence alone cannot provide them. Designing and implementing such tests present an interesting challenge to students of adaptive radiation. A second task is to establish whether the contours of the selection surface are indeed shaped by features of external environment, for this too is the theory's claim. The alternative is that fitness peaks and valleys are the outcome of some sort of internal constraints. Fitness peaks might represent the most feasible configurations, architecturally or developmentally, of phenotype and resource use rather than optimal combinations specific to environment. A final challenge is to test alternative mechanisms responsible for shifts between adaptive peaks, particularly the relative importance of genetic drift and varying environments. Evidence bearing on these issues is reviewed in Chapter 5.

4.3.2 Alternatives to competition

A strong consensus was in place by about 1950 that character displacement was crucial to adaptive radiation and that external environment alone was not enough to explain phenotypic differences between sympatric descendants within a radiating lineage. The Galápagos finches provided the best examples, and Lack's (1947) results fit well with the general perception that differences between sympatric species were influenced by competition.

This consensus fell apart among later biologists and was replaced by an enduring skepticism in many minds over the prevalence of character displacement in nature. Two main arguments produced this turn of events. One came from field studies, which suggested that natural environments were temporally highly variable and that food was limiting only infrequently. This led to the proposition that the sort of selection pressures needed to cause character displacement would be too weak and intermittent to be effective (Wiens, 1977). Whether or not intermittent selection is an impediment to divergence is debatable (see Gotelli and Bossert, 1991).

The second argument was ultimately the more damaging because it centred on the weakness of the evidence for character displacement itself. Competition is not the only explanation for exaggerated divergence in sympatry, and in some putative cases of character displacement one or more alternatives was found to be more consistent with the facts. For example, Grant (1975) showed how enhanced beak length difference between sympatric populations of nuthatches could be explained as an incidental outcome of parallel trends of geographic variation in response to a common environmental variable. In almost every other case too little information was available to distinguish among alternative explanations. Finally, doubts were raised over whether phenotypic differences in sympatry were exaggerated at all. New statistical analyses showed that morphological differences between species within communities were not typically greater than those seen in randomly generated 'null' assemblages of species (Strong et al., 1979; Simberloff and Boecklen, 1981; see also Grant, 1975). These studies did not address character displacement in adaptive radiation specifically, but many of the analyses focused on closely related species and are therefore relevant to it.

These critical analyses prompted many counterarguments and counter-analyses (see Gotelli and Graves, 1996 for a review of the technical issues and contrasting perspectives). Much of the ensuing debate concerned the validity of statistical approaches used to generate 'null' frequency distributions of species differences and especially whether these null distributions were already contaminated by competition (Grant and Abbott, 1980; Colwell and Winkler, 1984). Whatever their limitations, however, the new methods exposed serious weaknesses in the case for character displacement.

Among the added perspectives of more recent decades was the realization that interactions besides competition may also promote divergence between sympatric closely related species. For example, intraguild predation, the mutual consumption of vulnerable life stages between species sharing similar food requirements, may be common (Holt and Polis, 1997) and should readily favour divergence. Unfortunately, the theory for divergence under these alternative interactions remains undeveloped.

'Apparent competition' is one of the more interesting of the other interactions, one that is receiving increased attention from ecologists for its potential role in structuring communities (Holt and Lawton, 1994). The mechanism is a mutually antagonistic one that may arise between species sharing a common predator or parasite. Under apparent competition, an increase in the density of one prey species leads to an increase in the numbers of its predators (or parasites), indirectly increasing predation pressure on other species also consumed. This is similar to competition for food and may be easily mistaken for it in ecological experiments, but the interaction is mediated through a higher trophic level rather than a lower one. One might view it as a kind of competition, since 'enemy free space' behaves like any other resource (Ricklefs and O'Rourke, 1975; Jeffries and Lawton, 1984).

Character displacement via apparent competition is theoretically possible (Fig. 4.5). Prey species may diverge in phenotype and habitat use such that the degree to which they share predators, and hence predation's total impact on fitness, is reduced (Ricklefs and O'Rourke, 1975; Brown and Vincent, 1992; Abrams, 2000). The idea significantly expands earlier views on how interactions may cause divergence because apparent competition may apply to species that do not compete for food, such as those whose numbers are kept down by predation pressure. It may even apply to species that never encounter one another, such as insects on different host plants.

Apparent competition is not inevitable between species having common enemies. Indirect mutualism stemming from the dilution of predation pressure is perhaps just as likely (Abrams and Matsuda, 1996). Moreover, divergence is not assured even when apparent competition occurs (Abrams, 2000). Prey that interact antagonistically via a shared predator may exhibit parallel shifts in antipredator defence, such as increased body armour, the net result of which may be convergence not divergence. The role of shared enemies in adaptive radiation is currently highly uncertain.

76 • The ecological theory of adaptive radiation

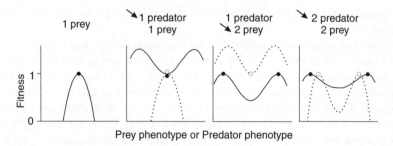

Fig. 4.5 Faunal buildup of two interacting trophic levels, predators and prey, according to Brown and Vincent (1992). Fitness functions for prey are indicated by the solid lines, those for predators by the dotted lines. The first panel depicts a single consumer species at demographic and evolutionary equilibrium. The symbol '●' indicates its mean phenotype and mean fitness. In the second panel prey fitnesses change following invasion by a specialized predator. The predator mean phenotype (indicated by '○') evolves to exploit the most frequent prey phenotype, and in doing so generates 'enemy free space' to either side of the original mean. This allows invasion or evolution of two new prey species that displace the first (third panel), a change that disrupts the predator fitness function in turn. The last panel indicates demographic and evolutionary equilibrium in a community of two predators and two prey.

4.3.3 Alternative mechanisms of ecological opportunity

New ideas have also emerged on the importance of other interactions mediating ecological opportunity. Depauperate archipelagoes lack predators as well as competitors, and perhaps this has as great an impact on adaptive radiation as the absence of competing taxa. Little attention was originally paid to this topic. Predation hardly receives mention in the early classic texts on adaptive radiation. One can imagine circumstances under which predation would be inimical to adaptive radiation of prey lineages and others in which it is likely to facilitate adaptive radiation.

Predation may slow morphological divergence and speciation in prey lineages in several ways, such that a release from predation would spur adaptive radiation. Most simply, predation may weaken interspecific competition (e.g. Gurevitch *et al.*, 2000) and thereby reduce selection that would otherwise promote divergence between prey species. Second, lower prey density may reduce the long-term viability of prey populations. This would translate into fewer prey populations persisting long enough to undergo divergence and speciation. The effect may be especially severe in novel, peripheral environments where populations are barely succeeding without the added burden of predation.

Third, predation might constrain the types of morphological changes that are feasible and thereby elevate the costs of niche shift (Benkman, 1991). The likelihood or speed of a successful shift from an ancestral resource type to a novel one is potentially reduced when improvements in both resource consumption efficiency and predator evasion are required. Presence of predators may thus add to the prerequisite 'key' adaptations needed to enter an adaptive zone, slowing if not halting altogether diversification of prey lineages.

The repressive effects of predation are central to Ehrlich and Raven's (1964) hypothesis of 'escape and radiation' coevolution between interacting lineages. They proposed that host and parasite lineages would exhibit cycles of diversification and stagnation through time. Their idea was inspired by the extraordinary diversity and relatively high degree of host specialization of insects that eat vascular plants and by the apparently contemporaneous rise of diversities of phytophagous insects and angiosperm plants in the Cenozoic (Strong *et al.*, 1984; Farrell, 1998). Their idea was that insects and the plants they feed upon are locked in a perpetual arms race of defence and counterdefence and that diversification in each lineage speeds up whenever it temporarily gains the upper hand. When a plant species acquires a novel chemical defence that largely rids it of its insect burden, the fortunate lineage undergoes accelerated adaptive radiation. In time the defence is overcome by the evolution of some form of immunity in a species of insect. Faced with a diversity of untapped plant resources this insect lineage undergoes an adaptive radiation in turn. The gradually increasing pressure by herbivores slows further diversification of the host lineage. The cycle is renewed when one of the descendant plants evolves another novel defence.

In support of a role for predation, the paleontological record indicates that long periods of escalation have taken place between the ability of predators to defeat prey (e.g. better shell-crushing) and in the capacity of prey to avoid predation (e.g. more breakage-resistant shells; Vermeij, 1987, 1994). The trend seems to result from the differential proliferation of prey clades that possess novel defence capabilities, not the wholesale conversion of lineages lacking them (Jablonski and Sepkoski, 1996). The rise of these more predator-resistant clades, and the comparative decline (although apparently not extinction) of the less-protected clades may have been facilitated by the 'release' from predation experienced by the former group. If the better-defended clades assisted in the numerical rise of more effective predators, then the mechanism driving the decline in less well-defended clades is 'apparent competition' rather than resource competition.

In the above scenarios predation quells adaptive radiation of prey, but it may be a stimulus to adaptive radiation in other circumstances. Predation risk may intensify competition between prey species for resources in refuges (Holt, 1987; Mittelbach and Chesson, 1987) and thereby promote divergence in other aspects of resource use. Predation may also increase the diversity of niches available in an area, because different strategies for prey avoidance are favoured in different environments (e.g. McPeek, 1990*b*, 1995). In support, most genera of aquatic invertebrates include species that occupy different points along gradients of increasing predation risk (Wellborn *et al.*, 1996). Finally, predation may favour divergence of prey species via 'apparent competition' (see above).

The net outcome of predation on adaptive radiation is likely to depend strongly on the specific characteristics of predator and prey lineages, such as the extent to which predators regulate prey dynamics; whether predators are limited by prey abundance; prey vulnerability at different life stages; breadth of predator diets; behavioural flexibility of prey in different environments; the diversity of strategies that reduce predation risk; and the magnitude of trade-offs in predation risk between environments. We should not be surprised to find a variety of outcomes in nature.

Simpson's concept of ecological opportunity is fairly static and therefore simplistic. His 'adaptive zone' consists of a finite number of related niches which, when underutilized, comprise the ecological opportunities that spurs adaptive radiation in the lineage exploiting it. As the radiation proceeds these niches become filled, causing the process to eventually grind to a halt. This niche-filling model of adaptive radiation contrasts sharply with a later, more dynamic concept emphasized by Whittaker (1977) but echoed by other writers as well.

Whittaker (1977) promulgated the alternative view that species diversity evolves as a self-augmenting phenomenon and that the number of niches is probably indeterminate. The basis of his conjecture is the idea that adding a new species to a community (by colonization or evolution) constitutes further resources facilitating entry by species of another trophic group (e.g. a predator). The impacts wrought by this later addition permits the further addition of species to the first group, and so on. 'Diversity of two interacting groups, together representing resources and controls, can thus increase *para passu* through evolutionary time' (Whittaker, 1977, p. 24). The two-trophic-level model of Brown and Vincent (1992) perhaps comes closest to Whittaker's intent (Fig. 4.5). In their model, an array of competing prey species and a suite of predators coevolve in a community that without the interactions between trophic levels permits a stable community of only one prey species.

Whittaker's idea predicts that rates of diversification should rise with increasing species diversity, and for this reason the idea receives skepticism from many paleontologists. Rates of diversification recorded in the fossil record are generally highest in depauperate environments, such as after mass extinctions, not when diversity is highest (Jablonski and Sepkoski, 1996). In Brown and Vincent's (1992) model each new addition to the community is increasingly precarious, requiring ever more special conditions. The idea of continuous diversity increase is therefore probably wrong. Even dynamic niches eventually run out.

However, the static niche-filling model of the original theory overlooks many interesting ecological dynamics that may take place during adaptive radiation and during diversity buildup in general. Species do provide resources for other species, either directly (as prey, direct mutualists, or commensal hosts) or indirectly (e.g. via nutrient recycling), and such mechanisms might substantially augment diversification of many kinds of organisms, at least for a time. Whittaker's hypothesis of mutual facilitation therefore stands as a significant extension to the ecological theory, one requiring serious attention by students of adaptive radiation.

4.3.4 *Alternative mechanisms and roles of speciation*

Nonecological speciation—Genetic changes causing reproductive isolation may evolve by a variety of nonecological mechanisms, and these are the alternatives against which the ecological theory must be tested. I summarize these alternatives briefly.

Genetic drift may cause the slow buildup of genetic changes causing reproductive isolation between allopatric populations of stable (not necessarily small) size (Lande, 1985; Barton and Rouhani, 1987). Or, genetic changes causing reproductive isolation may be precipitated by founder events, stochastic gene sampling events caused by the founding of a new population by very few individuals (Mayr, 1954; Carson and Templeton, 1984; but see Barton, 1989, 1998; Rice and Hostert, 1993; Coyne *et al.*, 1997; Rundle *et al.*, 1999). Populations may accumulate different advantageous alleles regardless of their environment because the same beneficial mutations would never arrive in the same order in different populations (Fisher, 1930; Barton, 1989; Orr, 1995). Natural selection is the cause of fixation of alleles in the third scenario but chance is the initial cause of divergence. I consider it distinct from ecological speciation because different environments are not necessary for the process to take place.

Polyploidy is a relatively common mode of speciation in plants (Stebbins, 1950; V. Grant, 1981; Rieseberg, 1997; Ramsey and Schemske, 1998), but its importance in adaptive radiation at low taxonomic levels has not been quantified. While its occurrence cannot be denied, probably most speciation events during adaptive radiation are not polyploid speciation events. For example, the Hawaiian silversword alliance is tetraploid, the result perhaps of an ancient cross between different species of tarweeds *Madia* and/or *Raillardiopsis*, but there is no indication of hybrid or polyploid speciation events within the alliance (Robichaux *et al.*, 1990; Baldwin, 1997).

Population persistence and the rate of speciation— If speciation in adaptive radiation most often involves nonecological mechanisms, then we need to think of alternative explanations for the elevated rates of speciation that occur. One hypothesis is that of population persistence: ecological processes affect speciation mainly through their influence on the viability of populations undergoing speciation (Mayr, 1963; Farrell *et al.*, 1991; Allmon, 1992; Heard and Hauser, 1995) rather than on the rate at which reproductive isolation evolves (Schluter, 1998). Under this view species accumulate rapidly in novel environments because more populations there avoid extinction long enough to evolve reproductive isolation. For example, the absence of competitors and predators in novel environments may lead to large population densities that reduce the chances of extinction, with the result that more populations undergo the speciation process. Reproductive isolation itself may evolve via any number of mechanisms not necessarily connected with environment. This is in contrast to the hypothesis based on ecological speciation, that rates are high in adaptive radiation because reproductive isolation evolves most quickly when divergent natural selection is strongest (Schluter, 1996). Distinguishing these processes may require estimates of the rate of evolution of reproductive isolation, not just of the rate of speciation as measured from branching rates of phylogenetic trees.

Sexual selection— In many adaptive radiations the species that are its products differ not only in the phenotypic traits useful for exploiting their external environments, but also in the secondary sexual characteristics that determine mating success within each species. The magnitude of these interspecific differences in 'secondary' sexual traits may be so dramatic that they eclipse perceptible variation in all other visible characters, yet appear arbitrary from the standpoint of exploiting environments. What is their relation to adaptive radiation?

Elaborate ornaments and displays, particularly in males, are often especially characteristic of explosive adaptive radiations. For example, in the Hawaiian *Drosophila*, ostensibly an adaptive radiation of more than 800 species (Kambysellis and Craddock, 1997), males are often adorned with head, mouthparts or tarsi having bizarre shapes along with striking body colouration and/or patterning, all of which are used in conjunction with elaborate behavioural displays during courtship (Spieth, 1974). The haplochromine cichlid fishes of Lake Victoria, representing perhaps the fastest adaptive radiation on earth (McCune, 1997), are a second case. Sympatric closely-related species of these cichlids differ in colour, and colour is an essential component of premating isolation (Seehausen *et al.*, 1998). The situation for cichlids in other East African lakes is similar. Such patterns have led to the proposition that speciation and sexual selection, defined as differential mating success of phenotypes, are causally related (West-Eberhard, 1983; Dominey, 1984). West-Eberhard (1983) lists several other arguments for considering sexual selection (and more generally, 'social selection') in speciation, including the strength of the process and the sheer variety of outcomes that yield reproductive isolation as a by-product.

The builders of the ecological theory of adaptive radiation attached little especial significance to divergence of arbitrary secondary sexual characters. Theories for sexual selection and the evolution of reproductive isolation were developed only later (Lande, 1981, 1982; West-Eberhard, 1983; Schluter and Price, 1993; Ryan and Rand, 1993; Liou and Price, 1994; Pomiankowski and Iwasa, 1998; Payne and Krakauer, 1997; Rice and Holland, 1997; Higashi *et al.*, 1999). Prior to this time, differences in secondary sexual characters were regarded mainly as 'isolating mechanisms', traits that evolved to minimize the probability of hybridization (West-Eberhard, 1983). They were seen as products of reinforcement or reproductive character displacement that became important for assortative mating relatively late in the speciation process. However, it is now clear that sexual selection may play a far more active role in the speciation process. For example, sexual selection facilitates reinforcement (Liou and Price, 1994; Kirkpatrick and Servedio, 1999).

Sexual selection may also play a role at a much earlier stage of speciation. Consider first sexual selection in an ecological context. A simple mechanism for the evolution of secondary sexual characters involves the evolution of 'receiver bias' in the choosing sex, usually females, and its exploitation by the opposite sex (males). For example, water mites detect copepod prey by sensing their vibrations in the water column, and males of the species attract (or lure) females by mimicking these vibrations (Proctor, 1992). Here, latent preference (or susceptibility) of the female evolves as an incidental by-product of natural selection on sensory modalities. The role of this mechanism in speciation is uncertain, however, because in the best examples of the process females of closely related species share the same biases (Ryan, 1998). In a second model, female preferences diverge rapidly if the best signals of male fitness, or the transmission properties of these signals, change with environment (Schluter and Price, 1993).

Premating isolation may also evolve by sexual selection independently of ecological environment. The classic model is the Fisher process in which female preference evolves as a correlated response to natural and sexual selection on the male trait (Lande, 1981; Kirkpatrick, 1982a). Because the model assumes that natural selection on the female preference is absent, an infinite variety of equilibria are possible, between which populations are free to drift (Fig. 4.6). This theory was developed for a single trait, but divergence is even easier when additional traits are included. Variations of this theory yield small-scale spatial differentiation of female preferences (Turner and Burrows, 1995; Payne and Krakauer, 1997) or continuous cycling of male traits and female preferences (Iwasa and Pomiankowski, 1995), underscoring the ease with which premating isolation may evolve between different populations. In other variations of the theory females may evolve resistance to male exploitation of receiver bias or to other male strategies for increasing fertilization success ('chase-away'; Rice and Holland, 1997). The multitude of possible mechanisms of resistance in females and counter-exploitation by males also inevitably lead to reproductive isolation between populations without differences between their environments (Palumbi, 1998; Rice, 1998).

These theories apply principally to animals. The distinction between natural selection and sexual selection is less precise in flowering plants, which require wind

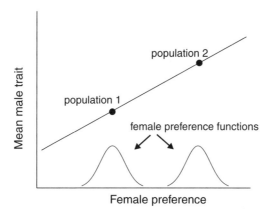

Fig. 4.6 The Fisherian model of sexual selection yields a line of equilibria along which population means may drift. At every point on the line sexual and natural selection on the male trait are in balance. Direct natural selection on the female trait is assumed absent, but it evolves as a correlated response to sexual selection on the male trait (Lande, 1981). Populations sufficiently different in their means would be reproductively isolated. Modified from Price (1998), with permission of The Royal Society.

or animals to move pollen between individuals. This is because enhancements to disbursement or receipt of pollen that increase mating success are also adaptations to external environment. Nevertheless, differential pollen tube growth or selective fruit abortion are ways in which females may exercise preferences for potentially arbitrary male characters or evolve resistance to mechanisms of male exploitation of female biases (Charlesworth et al., 1987).

Under the ecological mechanisms, speciation by sexual selection is an important extension of the ecological theory. Under the alternatives, however, sexual selection enhances the rate of nonecological speciation. The validity of the ecological theory, then, rests on testing the roles of sexual selection in speciation during adaptive radiation.

Adaptive radiation driven by speciation?—The issue of sexual selection's role in speciation goes beyond the question of whether environment is involved in the divergence of preferences. It points to a potentially larger role for speciation in other aspects of adaptive radiation.

Phenotypic diversification in adaptive radiation would be extremely limited without a mechanism promoting speciation. In the ecological theory the by-product mechanism (possibly in combination with reinforcement) serves this function. Speciation, under this theory, is therefore inevitable whenever an adaptive zone is occupied. However, what if speciation is actually more difficult than that, and is not inevitable, or at least is not so rapid or easily accomplished even in the face of strong divergent natural selection? In this case speciation itself becomes a rate limiting step, and any mechanism that facilitates the evolution of reproductive isolation, including sexual selection, would constitute an independent promoter of adaptive radiation.

If speciation is a rate limiting process in adaptive radiation, then any mechanism promoting speciation in one group but which is not available to other groups should drive adaptive radiation more readily in the former lineage, all else being equal. This possibility provides an explanation for why sexual selection and adaptive radiation so often appear to be associated. In a similar way, speciation may be limited by a shortage of geographical circumstances promoting allopatry (Ross, 1972; Cracraft, 1985). The fifteenth Darwin's finch, *Pinaroloxias inornata* of the lone Cocos Island, never underwent an adaptive radiation. Presumably this is because there are no other islands nearby and Cocos is too small for physical barriers to arise that impede gene flow between populations in different parts of the island (Lack, 1947).

This view of speciation as a rate limiting step should also cause us to look carefully at the role of 'key' mutations or key evolutionary innovations in adaptive radiation. Perhaps many such innovations are associated with adaptive radiation not because they promote ecological opportunity but because they elevate the rate of speciation. For example, the diversification of columbines is associated with the evolution of nectar spurs (Hodges, 1997; Chapter 7). One interpretation of this association is that nectar spurs conferred access to an adaptive zone consisting of a large array of empty pollination and habitat niches. A contrasting perspective is that the innovation increased speciation rate by virtue of the high degree of specialization of long-tongued pollinators. This allowed the buildup of a greater number of new species in different habitats in close proximity than would otherwise have been possible in such a short time. For these reasons, tests of the ecological theory of adaptive radiation require that alternative roles for speciation in adaptive radiation be distinguished.

4.4 Discussion

The ecological theory of adaptive radiation together with its later additions and alternatives present a host of empirical challenges. Information on the shapes of selection surfaces on phenotypes is vital to testing the hypothesis of phenotypic differentiation by divergent natural selection. A full test of the ecological theory requires that divergent selection be demonstrated and that alternative hypotheses such as drift along adaptive ridges are rejected. The comparative evidence linking phenotype and environment simply does not address this point. Also required is evidence linking the shapes of selection surfaces to particular features of the resource environment.

The role of competition between close relatives in phenotypic differentiation requires a thorough re-evaluation in light of the controversies. Is competition for resources a significant factor in divergence or has its importance been substantially overblown? If we removed interspecific competition from nature and left environments otherwise intact, would adaptive radiation be qualitatively different? As well, the contributions of other interactions between coexisting species of a radiation need to be addressed.

Is ecological opportunity the main rate limiting step in adaptive radiation? Or has speciation an additional role other than that anticipated by the early theorists, as a generator of new species subsequently available for phenotypic differentiation across a suite of ecological niches? Of particular interest is the role that novelties play in adaptive radiation: as tickets to adaptive zones (i.e. key innovations) or as promoters of speciation?

Finally, the mechanisms producing new species in adaptive radiation must be addressed. Is speciation purely an ecological process, or is it a fortuitous process that occurs independent of environment? Is divergent natural selection the main force behind the evolution of reproductive isolation or is its role diminished by the large number of other ways in which reproductive isolation evolves. If the latter, how do we explain the elevated rates of speciation that mark adaptive radiations?

In the next four chapters I summarize the extent to which these challenges have been met, and the most significant questions that remain.

5

Divergent natural selection between environments

> *To complete the representation of nature, all these elements [of the selection landscape] must be pictured as in almost constant motion—rising, falling, merging, separating, and moving laterally, at times more like a choppy sea than like a static landscape.*
> —Simpson (1953)

5.1 Introduction

Divergent natural selection is the ultimate cause of phenotypic differentiation in adaptive radiation, according to the ecological theory. The two proposed sources of divergent selection are differences between populations and species in external environments and interspecific resource competition. In the present chapter, I investigate the first of these claims. I begin below by clarifying the concept of an adaptive landscape for phenotypic traits. I then address the evidence for divergent natural selection between environments.

An ideal test of this hypothesis of divergent natural selection would satisfy two aims. The first is to establish whether selection on phenotypes is truly divergent, pulling apart the means of populations exploiting different environments. If so, then the mean values of traits distinguishing populations in nature should be associated with distinct peaks in fitness landscapes separated by valleys of lower fitness (Fig. 4.2). Under alternative hypotheses, mean phenotypes diverge even though divergence itself is not directly favoured by selection (e.g. Fig. 4.3). The second aim is to identify the mechanisms of divergent selection: how features of environment generate it. I review tests of divergent selection and summarize our understanding of its mechanisms. The data are organized by the approach used to generate them, allowing also an evaluation of the strengths and weaknesses of these approaches.

Finally, I address a potential problem raised by the finding that divergent natural selection is common in nature: selection alone cannot move a population from one adaptive peak to another across a valley of low fitness. Therefore, how is phenotypic differentiation usually initiated?

5.2 Natural selection and the adaptive landscape

Natural selection affects a population if some phenotypes have significantly higher survival or reproductive success than others. This process can be visualized by a surface whose axes are the phenotypic traits and whose height at each possible phenotype is fitness. To make this analogy successful we need to distinguish between two kinds of surface, the *fitness function* and the *adaptive landscape*. They are related but their differences are crucial. I explain them below so as to recast Simpson's 'selection surface' in light of this difference. As will be made clear, the adaptive landscape for phenotypes is more appropriate for understanding divergent natural selection than is the underlying surface for gene frequencies as conceived by Wright (1932).

To develop the hypothesis of divergent natural selection, I assume for now that external environments are unchanged by the species inhabiting them. That is, I temporarily ignore the complication that a species may change its environment and thus the pattern of selection on itself and on other species. Although simplistic, there are advantages of initially treating environment in this way. Many key aspects of environment *are* unchanged by species, at least for a time, and treating them as such is a reasonable first approximation. Evidence that species modify their environments and therefore interact is covered in Chapter 6.

5.2.1 Natural selection is a surface

The fitness function (or fitness surface) f describes expected survival or reproductive success of *individuals* as a function of their phenotypes (Fig. 5.1(a)) (Pearson, 1903; Lande, 1976, 1979; Phillips and Arnold, 1989; Schluter, 1988a; Schluter and Nychka, 1994). For a single trait z, the survival or reproductive success W of individuals can be modelled as

$$W = f(z) + \text{error}.$$

$f(z)$ is the height of the surface above the value z: it is the expected fitness of all individuals having exactly that phenotype. W is realized fitness of an individual having phenotype z, which is rarely the expected value because of chance (represented by the error term). If W is survival (0 or 1) then $f(z)$ is the *probability* of survival of individuals of phenotype z. An example of such a function is shown in Fig. 5.1(a). The equivalent formula for selection on a suite of traits, $z = \{z_1, z_2, z_3, \ldots, z_m\}$, is

$$W = f(z) + \text{error}.$$

The function f completely describes selection on individuals over the full range of possible phenotypes, including phenotypes well outside the ranges spanned by individuals of any single population of interest. Hence inquiry into the nature of f is the starting point of any investigation of natural selection on populations and its evolutionary consequences. Fitness functions are potentially very complex with multiple maxima (peaks), minima (valleys), and inflection points.

86 • Divergent natural selection between environments

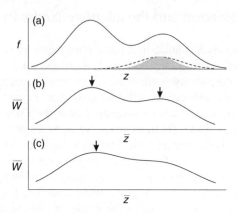

Fig. 5.1 Natural selection on a continuously varying trait z. (a) A fitness function describing f, the expected survival or reproductive success of individuals, as a function of their phenotype z. The shaded area indicates the distribution of phenotypes in a hypothetical population. The dashed line indicates a second, broader phenotype distribution. (b) The adaptive landscape corresponding to the shaded phenotype distribution in (a). Arrows indicate the location of the two adaptive peaks. $\overline{W}$ at a given value of $\bar{z}$ was calculated by centring the phenotype distribution at that $\bar{z}$ and computing the average $f(z)$ of all individuals present. (c) The adaptive landscape corresponding to the second (broader) phenotype distribution in (a). Only a single adaptive peak is present, indicated by the arrow. Modified from Kirkpatrick (1982b).

It is useful to think of the fitness function f as describing the external environment in terms of the traits involved in exploiting it (this view is inadequate when selection is density- or frequency-dependent; see Chapter 6). For example, the probability of dry-season survival of an individual Darwin's ground finch depends on the size and shape of its beak because these traits determine its ability to harvest seeds of given size and hardness (e.g. large beaks can crack larger and harder seeds), and such seeds are not equally abundant (Boag and Grant, 1981; Price *et al.*, 1984a; Gibbs and Grant, 1987). The fitness function for beaks thus mirrors the abundance of seeds along a seed size/hardness gradient. Fitness functions can be partially estimated from data on the survival and fecundity of measured individuals (Lande and Arnold, 1983; Endler, 1986; Manly, 1985; Phillips and Arnold, 1989; Schluter, 1988a; Schluter and Nychka, 1994; reviewed in Brodie *et al.*, 1995) (see Box 5.1).

5.2.2 The adaptive landscape for phenotypic traits

The usual net effect of f on a distribution of phenotypes is to pull the population mean in some direction. This direction is not necessarily toward the highest peak in the fitness function or toward those phenotypes within the population that have highest fitness. The population mean is instead pulled in the direction of the most rapid rise in population *mean* fitness (Lande, 1976, 1979). This direction, and its rise, represent the selection gradient, β. In the absence of genetic bias the direction of the gradient is also the direction of evolution of the mean phenotype. (Chapter 9 discusses the contribution of genetic bias to phenotypic divergence.)

This direction favoured by natural selection, and others not favoured, can be visualized by the surface depicting population mean fitness. The dimensions of this surface are the phenotypic traits as before, but its height at the mean phenotype $\bar{z}$ is

---

Box 5.1 Estimating fitness functions.

The fitness function, f, describes expected fitness W of individuals according to their phenotype z:

$$W = f(z) + \text{random error}.$$

$z = \{z_1, z_2, \ldots, z_m\}$ is the set of m traits under selection. The goal is to estimate f, or some of its features, from data on survival and reproductive success of individuals in populations. There are two approaches to achieving this goal. One directly estimates the function f as accurately as possible. The other focusses on coefficients that describe the most critical features of f.

The best statistical techniques for estimating f directly are those which make the fewest assumptions about its shape because *a priori* knowledge of its shape is usually lacking. Recent years have seen the development and application of several nonparametric regression techniques, of which the cubic spline (summarized in Schluter, 1988a; Hastie and Tibshirani, 1990) and locally weighted regression (LOWESS; Chambers *et al.*, 1983) are most widely used. 'Nonparametric' here does not refer to the distribution of residuals around the estimate $\hat{f}$, but rather to the fact that we are not interested in estimating parameters of an equation to describe f. A visualization of the estimated surface, $\hat{f}$, is the final product.

Estimating fitness functions for suites of traits, z, is more difficult. One solution is to reduce the size of the problem by visualizing only the few most interesting (univariate) cross sections of the full surface obtained using an exhaustive search (Schluter and Nychka, 1994). This approach is a special case of fitting 'generalized additive models' to complex regression surfaces (Hastie and Tibshirani, 1990).

An alternative approach is to ignore the details of the fitness surface and focus instead on coefficients that describe its consequences for changes in the population distribution. The most useful coefficients are those describing directional selection, the selection differential, $s = \{s_1, s_2, \ldots, s_m\}$, and the selection gradient, $\beta = \{\beta_1, \beta_2, \ldots, \beta_m\}$ (Lande and Arnold, 1983). The vector β points to the steepest uphill direction on the adaptive landscape. Its length gives the steepness of the landscape in that direction, and hence the strength of the pull on the population mean (Lande, 1979). Each β_i indicates the force of directional selection on the trait i, whereas each s_i is the net selection on trait i resulting from direct selection on it and from selection on all other traits with which it is correlated. β indicates the multivariate direction favoured by selection, whereas s indicates the direction toward which the population is actually headed. The two directions are not the same when phenotypic traits are intercorrelated (Lande and Arnold, 1983).

> s and β can be computed from data on survival and reproductive success of individuals W and their phenotypic measurements z using the multiple regression module of any basic statistics package, following the recipes in Lande and Arnold (1983). The accuracy of estimates of β so obtained may not be high when traits are strongly correlated. Moreover, the β_i are biased if correlated traits under selection are left out of the analysis. These problems and possible solutions are discussed more thoroughly in Endler (1986) and Mitchell-Olds and Shaw (1987). When assessing whether selection on two populations is divergent it is wise to estimate s and β, as both give useful information about the direction of selection and divergence.
>
> Related coefficients are available for describing selection on population variance (Lande and Arnold, 1983). Confusingly, these coefficients were originally referred to as differentials and gradients of disruptive, stabilizing and correlational selection. They are not, because the coefficients do not indicate the presence of modes or dips in the fitness surface. It is more accurate to call them coefficients of curvilinear selection (Schluter, 1988a; Phillips and Arnold, 1989; Brodie et al., 1995). These coefficients are not as useful for our purposes; direct estimates of f using the regression methods described yield more information about the form of fitness functions.

$\overline{W}(\bar{z}) = \overline{W}(\bar{z}_1, \bar{z}_2, \ldots, \bar{z}_m)$, where $\bar{z}_i$ is the mean of trait i in the population. This surface $\overline{W}$ is the *adaptive landscape* for phenotypes (Lande, 1976, 1979; also called the adaptive surface, the adaptive topography, and the mean fitness function). The selection gradient, β, is the multidimensional tangent (slope) of the adaptive landscape at the population mean.

Adaptive landscapes resemble their corresponding fitness functions but are less rugged with fewer, less sharply defined peaks and valleys. For example, the adaptive surfaces in Fig. 5.1(b) and 5.1(c) were calculated from the fitness function in Fig. 5.1(a) by assuming that the dispersion of phenotypes in the population (greater in Fig. 5.1(c) than in 5.1(b)) remains constant whatever the mean phenotype. The landscape in Fig. 5.1(c) has only a single peak, in contrast to the fitness function (Fig. 5.1(a)) which has two. In this case selection would favour a smaller mean phenotype in the population, that corresponding to the single optimum on the left side of the surface in Fig. 5.1(c). Mean fitness would rise throughout, even though the population traverses a valley in the fitness function f.

Like the fitness function, the adaptive landscape for phenotypic traits reflects features of the external environment. The connection between $\overline{W}$ and the external environment is looser than that between f and environment because mean fitness depends also on the frequency distribution of phenotypes in the population. Simply increasing the phenotypic variance of a trait can completely eliminate adaptive peaks and valleys (Kirkpatrick, 1982b; Whitlock, 1995; compare Fig. 5.1(c) with Fig. 5.1(b)). Despite its sensitivity to changes in the shape of the population distribution, the adaptive landscape for phenotypes remains a useful tool for visualizing the process of selection. Its features will usually be interpretable because they depend so greatly on f and hence on features of environment.

5.2.3 Adaptive landscapes for gene frequencies

The theory of adaptive landscapes was first formulated for gene frequencies and not phenotypic traits (Wright, 1932). In its genetic version each dimension of the adaptive landscape corresponds to allele frequency at a different di-allelic locus. Population mean fitness at a particular set of allele frequencies is represented by the height of the surface above that point. Simpson's (1944, 1953) landscape for phenotypes is an adaptation of Wright's genetic concept. Here I point out the crucial difference between the phenotypic and genetic versions, and explain why the former is more useful in the present context.

The phenotypic and genetic adaptive landscapes are connected in a causal sequence because differential fitness of phenotypes may favour some alleles over others, determining a direction of increasing mean fitness in the underlying genetic landscape. However, the correspondence between genotype and phenotype is not one-to-one. A peak in the adaptive landscape for phenotypes may correspond to multiple peaks in the genetic landscape, since multiple genotypes may produce the same phenotype. A simple example in Fig. 5.2 involves a trait controlled by two loci whose effects are additive. In this case intermediate gene combinations may have lower mean fitness

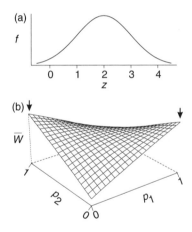

Fig. 5.2 An adaptive landscape for gene frequencies at two loci underlying a phenotypic trait under stabilizing selection. (a) Selection for a single optimum phenotype. (b) Mean fitness as a function of allele frequencies at two unlinked loci that make up the additive genetic component x for the phenotypic trait. Each locus has two alleles with states 0 and 1. p_1 and p_2 are the frequencies of the '1' allele at the first and second loci. x is the sum of states of all four alleles, and ranges from 0 to 4. Loci are assumed to be in Hardy–Weinberg equilibrium with no linkage disequilibrium between them. z is determined as the sum of x and an environmental deviation e, assumed to have a normal distribution with mean 0 and variance 1. Alternative optima are indicated by arrows. The fitness of intermediate frequencies is reduced because genetic (and phenotypic) variance is elevated.

but this reduction does not correspond to any minimum in the phenotypic adaptive landscape or to a resource minimum in an environmental gradient. Nonadditive gene interactions complicate the match between phenotypic and genetic adaptive peaks even further but do not change this basic conclusion. This lack of correspondence between the genetic adaptive landscape and environment is one reason why it is so important to establish the agents of environment responsible for natural selection on populations.

Another vital difference is that regions of a phenotypic landscape may not map to any dimensions of the genetic landscape. The phenotypically variable but nonheritable trait is a trivial example. A more interesting case is the novel phenotype created artificially to more accurately measure selection on extreme phenotypes. For example, eggs may be added to (or removed from) nests to measure selection on clutch size in birds, accomplishing what the birds themselves can do only after alteration of morphology or reproductive metabolism. Many variants so constructed might not be attainable by natural selection on genetic variation, at least in the time frame imagined. As a third example, consider the bimodal fitness function in Fig. 5.1(a). A population with a bimodal phenotype distribution would have even higher mean fitness than the unimodal populations illustrated. Nevertheless, bimodality may be difficult to accomplish genetically (except by the evolution of a polymorphism).

These examples remind us that whereas the phenotypic adaptive landscape gives a picture of selection pressures on traits, it does not by itself predict how populations will evolve. Genetic constraints on evolution by natural selection are discussed in Chapter 9.

5.2.4 Divergent natural selection

Divergent natural selection can now be restated as natural selection that pulls the means of two or more populations toward different adaptive peaks. Under the ecological theory of adaptive radiation, populations diverge because they are pulled toward distinct peaks in the adaptive landscapes for phenotypes (e.g. Fig. 4.2). These peaks are themselves generated by uneven fitness gains at different positions along environmental gradients.

In the next four sections, I review the four main approaches to testing this hypothesis: comparison with the neutral expectation; reciprocal transplant experiments; direct measurements of divergent natural selection; and estimates of adaptive landscapes from resource distributions. As in other chapters, I cover the literature as broadly as possible, which means that cases are not restricted to well-documented cases of adaptive radiation in the narrow sense.

5.3 Comparison with the neutral expectation

Several methods test divergent natural selection by comparing the level of population differentiation observed with the amount expected solely by mutation and genetic drift (the neutral expectation). Here I summarize results of three methods: the Q_{ST} method, the neutral rates test, and the quantitative trait locus (QTL) sign test.

The Q_{ST} method (Spitze, 1993) was designed for sets of conspecific populations that became differentiated while still exchanging migrants. It is based on the idea that divergence of quantitative traits should be similar to that of allele frequencies at nuclear marker loci if neither are subject to selection. Under the influence of migration (which tends to make populations more alike) and mutation and genetic drift (which lead to differentiation of populations), the among-population proportion of total genetic variance (variance among populations plus that within populations) in phenotypic traits is expected to equal that of nuclear marker loci (Lande, 1992).

The proportion of among-population genetic variance in the quantitative trait is measured as

$$Q_{ST} = \frac{V_{AB}}{V_{AB} + 2V_{AW}}$$

where V_{AB} and V_{AW} are the additive genetic variances between and within populations, respectively (Spitze, 1993). This quantity is compared with F_{ST} (or its multiple-allele version, G_{ST}), a measure of among-population variance in allele frequency at nuclear loci (Nei, 1972). Divergent selection on the quantitative trait is implied if Q_{ST} exceeds G_{ST}. This method assumes that genetic variance in the quantitative trait is purely additive and that migration rates are low.

Q_{ST} frequently exceeds G_{ST} in natural populations (Fig. 5.3), implying that divergent natural selection commonly drives phenotypic differentiation. Divergent selection is implicated for at least one trait of most species tested. Nevertheless Q_{ST} is highly variable and several traits have clearly diverged more slowly than the neutral expectation. There is little room for Q_{ST} to exceed G_{ST} when the latter is high, and so the method is useful mainly for $G_{ST} < 0.5$ or so (as well, gene flow is virtually nil beyond this range). Traits correlated with body size especially tend to fall above G_{ST} (e.g. this is the case in both species of *Daphnia* tested; Spitze, 1993; Lynch *et al.*, 1998).

The neutral rates test compares observed rates of divergence in quantitative traits with the rate expected under the null hypothesis of mutation and drift (Lande, 1977; Turelli *et al.*, 1988; Lynch, 1990; Martins, 1994). The neutral rate is calculated from estimates of the mutational variance in the quantitative trait. Under the influences of mutation and genetic drift alone (i.e. in the absence of migration or selection), divergence of means of populations or species should mimic a random walk in time (Brownian motion). In this case the variance V_B between mean phenotypes μ_1 and μ_2 of two species 1 and 2 (i.e. $V_B = \frac{1}{2}(\mu_1 - \mu_2)^2$) is expected to increase linearly with the number of generations t since the species last shared a common ancestor:

$$V_B = \delta t$$

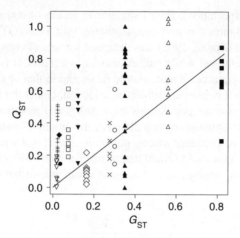

Fig. 5.3 Population subdivision in quantitative traits (Q_{ST}) compared with that in ostensibly neutral nuclear loci (G_{ST}, or F_{ST} averaged over loci). Points having the same symbol represent different traits for the same species. The line indicates $Q_{ST} = G_{ST}$. This plot is modified from Fig. 6 in Lynch *et al.* (1998; see references therein) by adding values for individual traits, and adding additional data on *Carduelis chloris* (Merilä, 1997) and two species of *Scabiosa* (Waldmann and Andersson, 1998). From left to right the species are: *Pinus contorta* (▽), *Carduelis chloris* (+), *Quercus petraea* (♦), *Drosophila buzzatii* (●), *Clarkia dudleyana* (□), *Scabiosa columbaria* (▼), *S. canescens* (◇), *Daphnia obtusa* (×), *D. pulex* (○), *Medicago truncatula* (▲), *Drosophila melanogaster* (△), *Arabadopsis thaliana* (■).

(Lynch, 1990; Martins, 1994). The expected rate of divergence δ is proportional to V_M, the increase in genetic variance added per generation from mutation:

$$\delta/2 = V_M$$

(Lande, 1977; Lynch and Hill, 1986; Turelli *et al.*, 1988). For log-transformed traits, V_M/σ^2 is typically in the range 10^{-4}–10^{-2}, where σ^2 is the phenotypic variance in the trait within species (Lande, 1976; Lynch, 1988). Hence under the neutral expectation we expect the rate

$$\Delta = \delta/2\sigma^2$$

to also fall within this range. Assumptions of this method include heritable differences between species means, additive genetic variance within populations, no genotype by environment interactions, and no gene flow between populations.

The rate of divergence δ may be estimated for a clade of species if times of divergence are known (Box 5.2), and Δ may then be calculated as above. Divergent selection is implicated if Δ exceeds the neutral expectation.

Seemingly in contradiction to findings of the Q_{ST} method, application of the neutral rates test indicates that divergence is usually slower than the neutral expectation, often

Box 5.2 Estimating the rate of phenotypic evolution from a phylogeny.

The overall rate of phenotypic evolution in a clade may be estimated from species means and a tree of their phylogenetic relationships. The simplest methods assume that phenotypic evolution approximates a random walk in time (Brownian motion), such that the variance between any two species increases linearly with time since they shared a common ancestor

$$V_B = \delta t.$$

Our goal is to estimate the rate, δ. One approach first calculates the maximum likelihood ('least squares') ancestor state values μ_i at every interior node i in the tree (Martins and Garland, 1991; Schluter *et al.*, 1997). An unbiased estimate of δ is then obtained as a weighted sum of squared changes along each branch of the tree:

$$\hat{\delta} = \frac{1}{N} \sum_{i=1}^{N} \sum_{j=1}^{2} \frac{(\mu_i - \mu_j)^2}{t_{i,j}}$$

(Schluter *et al.*, 1997). N is the number of nodes (ancestors) in the phylogenetic tree. j refers to one or the other of the two descendants of each ancestor i; these may be other ancestors or species at the tips of the tree. The length of branches is time, $t_{i,j}$. If t is measured as generations rather than years then the observed value of $\hat{\delta}$ may be compared with the neutral expectation (see text). This calculation assumes that sampling error in the species means is negligible, i.e. that sample size for each species is large. Martins (1994) presents an alternative way to compute $\hat{\delta}$ that includes a correction for sampling error.

$\hat{\delta}/N\delta$ has a χ^2 distribution with N degrees of freedom (Schluter *et al.*, 1997), which leads to the 95% confidence interval for δ:

$$\frac{\hat{\delta}}{2N\chi^2_{0.975,N}} \leq \delta \leq \frac{\hat{\delta}}{2N\chi^2_{0.025,N}}.$$

by orders of magnitude (Fig. 5.4). The cases tested include a number of mammal clades, yet mammals on the whole are thought to have diversified rapidly under divergent natural selection (Simpson, 1953). The lowest rate estimate comes from the cardueline finches, which are nonetheless ecologically diverse (Newton, 1972). They include *Carduelis chloris* which, as we have seen (Fig. 5.3), is made up of populations that have diverged comparatively rapidly.

A third method tallies the direction of the effects of genes underlying a quantitative trait that differs between two populations or species (Orr, 1998*a*). The data are obtained from an analysis of QTLs, which reveals the number of plus and minus factors contributing to a given difference between two species. An excessive number of plus factors relative to that expected by drift, and taking into account the number needed to produce the observed difference between the species in the trait, is regarded as evidence for divergent selection. Orr (1998*a*) applied the method to one natural case, the difference between *Drosophila simulans* and *D. mauritiana* in a male

94 • Divergent natural selection between environments

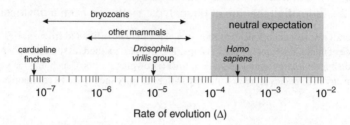

Fig. 5.4 The rate of divergence between species compared with the neutral expectation. The mammal results, including *Homo*, are based on cranial and tooth dimensions of species in fossil and extant clades, mainly of species in the same genus or in related genera (Lynch, 1990). Bryozoan results are from Cheetham and Jackson (1995) and represent rates of divergence in zooidal morphology. The *Drosophila* data are from Spicer (1993) and represent rate of divergence in head and wing dimensions of adult flies. I used his modal P-value based on all pairwise comparisons between species to backcalculate the modal value of Δ, assuming an average divergence time of 3×10^7 generations. The point from cardueline finches was calculated from data in Björklund (1991). The rate is based on the variance between genera in average beak and body dimensions, on the assumption that σ within species is 0.05, and assuming that genera diverged simultaneously at 20 mya. All rates were estimated from ln-transformed variables.

genital structure. The number of plus factors (8 of 8 loci) was greater than expected by chance, suggesting divergent natural selection (in this case, probably divergent sexual selection). This method assumes complete additivity of effects across loci (no epistasis).

5.3.1 Remarks

Comparisons with the neutral expectation suggest that divergent natural selection is frequently a cause of phenotypic differentiation in a subset of traits. The evidence in favour comes mainly from the Q_{ST} method applied to conspecific populations. In contrast the neutral rates test provides little support for divergent selection, judging from analyses of morphological differences between species. Instead, selection appears to constrain rather than promote divergence. How can this apparent contradiction be resolved? Perhaps after an initial burst of divergent selection species attain different adaptive peaks and then differentiation slows, with the result that the neutral expectation is not exceeded in the long term.

However, the more likely explanation is that the neutral rates test is weaker at detecting divergent natural selection than the Q_{ST} method. The mutational variance used as a baseline for rejecting the null hypothesis in the neutral rate test ($\approx 10^{-4}$–10^{-2}) lumps all sorts of mutations including those that are harmful because of their effects on the values of other traits affecting fitness (Turelli et al., 1988). Excluding deleterious mutations would lower the expected rate of divergence under neutrality, probably drastically.

Both the Q_{ST} and neutral rates tests are limited by the fact that neutrality is not really the most appropriate null hypothesis for testing divergent natural selection. Drift along an adaptive ridge (Fig. 4.3) would predict a lower rate of divergence than the neutral expectation because the mutations contributing exclusively to drift along the ridge top are more rare. A null model based on the ridge concept would thus be rejected more easily by existing data. The neutral models overestimate the level of phenotypic differentiation expected in the absence of divergent selection, and consequently they may not be exceedingly powerful. The QTL sign test is still untested and its power cannot yet be compared with the other two.

None of these three methods by itself provides information on the targets or the mechanisms of divergent natural selection. Consequently they can provide only a partial test of the ecological hypothesis and are best used in combination with the more 'phenotypic' approaches described below.

5.4 Reciprocal transplant experiments

Measures of mean performance of phenotypes transplanted between environments provide an abundant supply of information on the different selection pressures experienced there. The logic of the approach is straightforward. Under the hypothesis of divergent natural selection, phenotypic divergence takes place because the optimal value of morphological and physiological traits differs between environments (Fig. 5.5). Selection in one environment brings about a rise in mean fitness there as trait means approach new optima, bringing about a decline in mean fitness with respect to the old environment. This predicted decline is detectable by transplanting.

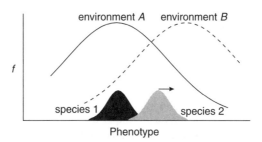

Fig. 5.5 Divergent natural selection leading to trade-offs in performance of populations transplanted between environments. Fitness functions are plotted for a single trait z in two environments, A and B. Shaded areas indicate trait distributions for two species. Species 1 occurs in environment A and species 2 occurs in B. Species 1 is assumed to be at the optimum in environment A. Species 2 is recently descended from species 1 and has colonized the second environment. The arrow indicates its direction of divergence toward the new optimum. A trade-off results when, as divergence proceeds, the rise in average fitness of species 2 in environment B is accompanied by a drop in its mean fitness when transplanted back to A.

This section reviews results from reciprocal transplant experiments in which performances of phenotypically differentiated morphs, populations, or closely related species are measured in both their environments (Fig. 5.5). I remark briefly on a related method in which genetic correlations between fitness in two or more environments are measured using transplanted individuals of a single population. This second approach is inferior to the reciprocal transplant of already-differentiated forms for reasons explained below, but I discuss it here because it is still often used.

5.4.1 Four examples of reciprocal transplants

Encelia—Eleven species of *Encelia* plants (Asteraceae) are distributed along a thermal and moisture gradient in southwestern North America, from coastal habitats to inland desert. Ehleringer and Clark's (1988) study of photosynthetic capacity in coastal and desert species illustrates how morphological divergence in a small number of traits can influence performance under contrasting environmental conditions.

The photosynthetic capacities of the species react similarly to heat. The photosynthetic rate in both cool coastal plants and hot desert species is maximized at about 30 °C and tissue damage occurs at about 45 °C. However, species differ conspicuously in pubescence of their leaves. Those from cool, wet habitats have few or no hairs on the leaf surface whereas the density of leaf hairs is higher in species from drier habitats. In the desert species, *E. farinosa*, a dense covering of leaf hairs reduces absorptance of incident solar radiation to about 17%, compared with 50% absorptance in the hairless coastal *E. californica* and in shaved leaves of *E. farinosa* (Ehleringer and Clark, 1988). The result in *E. farinosa* is a substantially lower leaf temperature and water loss rate and a higher rate of photosynthesis under hot and dry conditions.

Reciprocal transplants show that growth of *E. farinosa* is much higher than *E. californica* in the desert whereas the reverse is true on the coast (Fig. 5.6(a)). Pubescence may be the main cause of the growth advantage in the desert site, as a hairless mutant of *E. farinosa* performs as poorly as *E. californica* there. However, this mutant strain performs no better than the wild-type on the cooler, wetter coast, suggesting that differences in unknown traits give *E. californica* the advantage there.

Intertidal snails—Intertidal gastropods from exposed shores with heavy surf are usually small, thin-shelled, and have a large aperture and foot compared with relatives on protected shores. Boulding and Van Alstyne (1993) investigated the mechanisms responsible for this difference in two species of *Littorina* of coastal Washington. *L. sitkana* inhabits protected areas of the coast whereas *L. sp.* (an undescribed species) is restricted to the wave-exposed shore.

A reciprocal transplant experiment involving mark–release–recapture showed that rank-order of survivorships of the species is reversed between environments (Fig. 5.6(b)). This pattern persisted over sites, size classes, and years despite considerable variation in absolute survivorship (Boulding and Van Alstyne, 1993). Differences in survival on protected shores are mainly the result of heavy predation by shore

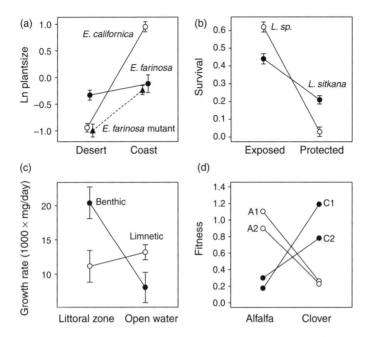

Fig. 5.6 Mean performance and fitness in four reciprocal transplant experiments. Error bars represent 1 SE. (a) Mean sizes of two species of *Encelia* shrubs (and a hairless mutant of one of them) in common gardens under desert (Phoenix, Arizona) and coastal (Irvine, California) conditions. Size is surface area of ground covered by individuals, in m^2, eight months after planting outdoors under natural precipitation levels. Plant size and seed production are strongly correlated. From data in Table 9.2 of Ehleringer and Clark (1988). (b) Summer survival in two species of intertidal *Littorina* snails introduced to a protected shore and a shore exposed to heavy surf. Data are from Table III in Boulding and Van Alstyne (1993); numbers are from sites FW and ND in 1987. (c) Mean growth rates of subadult benthic and limnetic stickleback species (*Gasterosteus spp.*) transplanted between their preferred lake habitats using enclosures. Data are from Schluter (1995), and combine results from two years. (d) Mean fitness of aphid clones (*Acyrthosiphon pisum*) obtained from two alfalfa fields (A1 and A2) and two clover fields (C1 and C2) and transplanted between these same two crops. Fitness is population growth rate calculated from measurements of clonal longevity and fecundity. Modified from Via (1991).

crabs, especially on *L. sp.*, which has a thin shell and large aperture. Many recovered *L. sp.* shells from protected sites had been 'peeled' in the manner diagnostic of crab predation.

In the laboratory, *L. sp.* is preferred by shore crabs and can withstand less force. In contrast, *L. sp.* survived better than *L. sitkana* on exposed shores (Fig. 5.6(b)). Laboratory experiments show that *L. sitkana* is more easily dislodged by wave action than *L. sp.* (Boulding and Van Alstyne, 1993). This alone does not explain why *L. sitkana* is absent from the exposed shore, as its survival there is not less than in protected areas. The authors suggest that a greatly reduced growth rate in winter is partly to blame.

Freshwater sticklebacks—Many lakes in previously glaciated areas of the northern hemisphere contain very young pairs of fish species (Table 3.2) morphologically highly differentiated. Sympatric species tend to divide resources in a similer way. Usually one species (the limnetic) is a pelagic zooplanktivore whereas the other (the benthic) consumes benthic prey from the littoral zone or deeper sediments. A consistent set of morphological differences is associated with this habitat split. Planktivores are smaller and more slender than benthivores and tend to have narrower mouths and longer, more numerous gill rakers.

Habitat transplants in pairs of three-spine sticklebacks reveal a steep trade-off in growth rates (Fig. 5.6(c)). Each species grows at about twice the rate of the other in its preferred habitat (Schluter, 1995). This pattern matched differences in foraging success observed in reciprocal transplants between sections of habitats transported to large laboratory aquaria. This suggests that growth is limited mainly by rates of prey consumption. The volume of prey ingested per strike in littoral sediments was over five times higher in the benthic species than the limnetic species, mainly because they ingested larger prey. Conversely, the rate of prey capture in open water was three times higher in the limnetics than the benthics, mainly because of superior rates of seizure and retention of small evasive plankton (Schluter, 1993).

Pea aphids—Via (1991) collected pea aphid clones (*Acyrthosiphon pisum*) from established fields of red clover and alfalfa and measured the fitness of each type when transplanted to the alternative host (Fig. 5.6(d)) in the field. Her experiment differs from the above three because there was no prior indication of morphological or physiological differentiation between the forms. I included it anyway as my fourth case because it illustrates how substantial physiological divergence may accumulate between species in different environments, and yet are not associated with conspicuous morphological differences and therefore go unnoticed until a reciprocal transplant is carried out. Fitness of the native on each host plant was five times that of the immigrant. The phenotypic basis of fitness differences is still unknown.

5.4.2 General patterns

I searched the published literature for examples of reciprocal transplants involving congeneric species, closely related populations within a species, or discrete morphs within a population (hereafter 'type') (Table 5.1). I restricted attention to cases in which phenotypic differentiation had been recognized beforehand (since this is what we are trying to explain). Reciprocal transplants carried out between environments without prior knowledge of phenotypic differences between the populations inhabiting them (e.g. Via, 1991) were excluded because a failure to detect trade-offs in such cases would be uninformative. Trade-offs are not predicted until morphological or physiological differentiation is underway (Fig. 5.5). Even so, some of the largest trade-offs yet recorded between close relatives fall into the category of cases not included herein.

I used measures of performance or fitness available. If more than one measure was provided I multiplied them (e.g. survival × fecundity). All measures were then ln-transformed ($\ln(x + 1)$ in the case of proportions). Trade-offs are quantified as the proportion of nonerror variance attributable to the interaction between 'phenotype' and 'environment' in two-way ANOVAs. This quantity measures the extent to which a difference in mean fitness between types depends on environment. An interaction was given a negative sign if each phenotype performed best in the environment of the other, the opposite of the expected pattern.

Trade-offs were common and often relatively large (Table 5.1; Fig. 5.7). In contrast, performance differences between types (averaged over environments) were usually relatively small. Consequently, the lines connecting means of each type between environments typically crossed (observed in 36 of 42 studies), as seen in the examples in Fig. 5.6. In such cases the native type was almost invariably superior to the foreign transplant. Desert sparrows *Amphispiza* provide the most glaring exception. Each fed best in the habitat of the other, but the absolute magnitude of the difference was small (Repasky and Schluter, 1996).

Finally, one of the environments was typically more profitable than the other (averaged over phenotypes), as indicated by a high variance component for 'environment' in Fig. 5.7. This effect is noticeable in three of the four illustrated examples (Fig. 5.6). Consequently, performance of one type varies more between environments than the other type. Therefore as a rule, phenotypic differentiation leads to significant divergence in degree of specialization as measured by performance.

Trade-offs, quantified using the phenotype × environment interaction, were stronger in invertebrates (median = 69.5%) than vertebrates or plants (23.5% and 40%, respectively; KW-test, $\chi^2 = 6.27$, $P = 0.04$). As well, trade-offs were stronger when survival was among the fitness components measured (median = 69%) than when survival was not measured (24%; $\chi^2 = 12.97$, $P = 0.0003$). These last two effects are confounded, as survival was measured in all the invertebrate studies but in only a third of studies of remaining taxa. Trade-offs were not different among the different categories of phenotypes compared (morphs, populations, or species).

5.4.3 Genetic correlations between environments

A related method to detect trade-offs calculates the genetic correlation between performance measured in two (or more) environments. Performance in environment A and that in environment B are treated as separate 'traits' of an individual, and the genetic correlation between them is estimated from the resemblance between relatives placed in the different environments (Falconer, 1981). A negative genetic correlation indicates a trade-off. This method is flawed, but I mention it here because it is still frequently used.

Table 5.1 Summary of reciprocal transplants of closely related morphs, populations, or species differing in phenotype. Variance components for phenotype (P), environment (E), and the interaction between them (P × E) are expressed as a percentage of the sum of all three (i.e. excluding the error term, which contributes equally to these three variance components). For simplicity, results were reduced to two phenotypes in two environments. Transplants involving multiple phenotypes and environments were analysed by averaging results of separate analyses contrasting each phenotype with all the others combined. 'Lines cross' indicates whether lines connecting means of each type across the two environments cross. P × E interactions in the opposite direction to that expected (fitness of the native form lowest in its own environment) are given a negative sign

Taxon	Comparison[a]	N[b]	Measure	E	P	P × E	Lines cross	Source
Vertebrates								
Amphispiza spp.	A	2	Feeding rate	22	22	−56	Yes	Repasky and Schluter (1996)
Anolis spp.	A	2[c]	Sprint speed × fault rate	77	5	17	Yes	Losos and Sinervo (1989)
Artedius spp.	A	2	Attack success	98	2	1	No	Norton (1991)
Gasterosteus spp.	A	2	Growth rate	32	1	67	Yes	Schluter (1995)
Lepomis gibbosus	C	2	Body condition	26	3	71	Yes	Robinson et al. (1996)
Lepomis spp.	A	2	Feeding rate	1	72	27	No	Werner (1977)
Loxia spp.	A	4[d]	Feeding rate	82	6	12	Yes	Benkman (1993)
Pseudacris spp.	A	2	Larval survival × size	14	1	85	Yes	Skelly (1995)
Pseudacris spp.	A	2	Larval survival × size	73	7	20	Yes	Smith and van Buskirk (1995)
Pyrenestes ostrinus	C	2	Feeding rate	89	0	11	Yes	Smith (1987)
Scaphiopus multiplicatus	C	2	Survival	29	2	69	Yes	Pfennig (1992)
Sceloporus undulatus	B	2	Growth rate	21	23	56	Yes	Niewiarowski and Roosenburg (1993)
Invertebrates								
Daphnia pulex	B	2	Survival	24	64	13	No	Hebert and Emery (1990)
Dolerus spp.	A	2	Larval survival × growth	2	7	91	Yes	Barker and Maczka (1996)
Enallagma spp.	A	2	Survival	7	34	59	Yes	McPeek (1990b)
Heliconius erato	B	2	Longevity	0	8	92	Yes	Mallet and Barton (1989)
Heliconius cydno	C	2	Longevity	14	19	67	Yes	Kapan (1998)
Jadera haematoloma	B	2	Survival × fecundity	9	19	72	Yes	Carroll et al. (1998)
Littorina saxatilis	C	2	Survival	19	9	72	Yes	Janson (1983)
Littorina saxatilis	C	2	Survival	3	5	92	Yes	Rolán-Alvarez et al. (1997)

Species	Class[a]	No.[b]	Trait				Ref.
Littorina spp.	A	2	Survival	83	0	16	Yes	Boulding and Van Alstyne (1993)
Neochlamisus bebbianae	B	2	Survival	0	0	100	Yes	Funk (1998)
Papilo spp.	A	2	Survival	28	23	49	Yes	Thompson et al. (1990)
Timema cristinae	C	2	Survival	32	2	66	Yes	Sandoval (1994)
Plants								
Anthoxanthum odoratum	B	6	Survival × size	4	21	75	Yes	Davies and Snaydon (1976)
Artemisia tridentatum	B	2	Fitness measure	65	9	26	Yes	Wang et al. (1997)
Diodia teres	B	2	Survival × fecundity	56	0	44	Yes	Jordan (1992)
Encelia spp.	A	2	Growth rate	59	3	38	Yes	Ehleringer and Clark (1988)
Gilia capitata	B	2	Fitness	0	45	55	Yes	Nagy and Rice (1997)
Impatiens pallida	B	2	Fitness	89	1	9	Yes	Bennington and Mcgraw (1995)
Impatiens pallida	B	2	Pollen removal	35	42	23	No	Wilson (1995)
Impatiens capensis	B	2	Pollen removal	21	69	10	No	Wilson (1995)
Iris spp.	A	2	Survival × size	7	37	56	Yes	Emms and Arnold (1997)
Iris spp.	A	2	Survival	1	1	99	Yes	Young (1996)
Mimulus spp.	A	2	Growth × survival	0	18	82	Yes	Hiesey et al. (1971)
Phlox drummondii	B	7	Fitness	64	13	24	No	Schmidt and Levin (1985)
Polemonium viscosum	B	2	Survival	11	3	86	Yes	Galen et al. (1991)
Plantago lanceolata	B	3	Survival × fecundity	44	14	42	Yes	Van Tienderen and van der Toorn (1991a)
Plantago lanceolata	B	5	Survival to 13 mo.	92	2	−5	Yes	Antonovics and Primack (1982)
Ranunculus lingua	B	2	No. flowers	52	0	48	Yes	Johansson (1994)
Ranunculus repens	B	2	Size (number of leaves)	63	5	32	Yes	Lovett Doust (1981)
Solidago virgaurea	B	2	Photosynthetic capacity	98	0	2	Yes	Björkman and Holmgren (1963)[e]

[a] A = among species; B = among populations within a species; C = among morphs within a species.
[b] Number of phenotypic classes tested.
[c] Compares A. gundlachi with the others combined, between the two most extreme perch diameters.
[d] Average of three pairwise comparisons.
[e] Cited in Heslop-Harrison (1964).

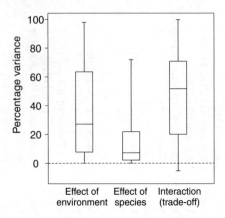

Fig. 5.7 Boxplots showing the percentage variance in performance (or fitness) explained by phenotype, environment, and their interaction in reciprocal transplant experiments. Interactions opposite in direction to that expected (fitness of the native form is lowest in its own environment) are assigned a negative value. Vertical lines span the range of values observed. Boxes span the first and third quartiles, and the horizontal bar within each box gives the median. One interaction value of −56 is excluded. Data are from Table 5.1.

The logic of the method is similar to that underpinning the reciprocal transplant. If phenotypic traits have different optima in different environments (i.e. trait z in Fig. 5.5) then within any population individuals with high z values should have high fitness in environment B and low fitness in environment A, whereas those with low z values will have high fitness in A and low fitness in B. If the phenotypic trait z is heritable then its contribution to the total genetic correlation between performance in A and B will be negative. If the majority of traits behave in a similar way, then the overall genetic correlation should also be negative.

However, the test is flawed because two of its assumptions are unlikely to be generally met. First, each trait z contributes to a negative genetic correlation in performance only if the mean $\bar{z}$ of the population being tested lies between the two environmental optima (Price and Schluter, 1991). This is the only region in which variation in z affects fitness in A and B in opposite ways. If the population is close to equilibrium in its own environment (e.g. in A) then the expected correlation between performance in A and B is zero, because only half the population lies in the range over which fitness in A decreases as fitness in B increases. The other half, to the other side of the optimum, occurs over a range in which fitness in A increases with increasing fitness in B (Fig. 5.5), cancelling the former effect. If the population lies outside the region spanned by the two optima, as might happen if the optima fluctuate or if both environments are novel (Service and Rose, 1985; Price and Schluter, 1991; Joshi and Thompson, 1995), then variation in z affects fitness in A and B in the same way and a positive genetic correlation is the likely outcome.

Second, the test assumes that the contribution of z to the genetic correlation between fitness in A and B is not masked by other sources of genetic variation that may cause performance in the two environments to covary positively (Fry, 1993). The assumption will not be met if, for example, individuals also vary in their load of slightly deleterious mutations, as is generally expected (Charlesworth, 1990; Houle, 1991). Any tendency for individuals with low mutational load to perform well in both environments could cancel the contributions of traits like z, and lead to a nonnegative genetic correlation in fitness between environments (Joshi and Thompson, 1995).

Results from a number of studies show that the genetic correlation between environments is usually near zero or positive (Rausher, 1984; Jaenike, 1990; Joshi and Thompson, 1995; Fry, 1996). Via's (1991) study (Fig. 5.6(d)) is sometimes regarded as an exception because the genetic correlation between fitness of aphids on alfalfa and clover is strongly negative. However, her study is really a reciprocal transplant like those summarized in Section 5.4.2 because the genetic correlation was calculated among clones sampled from different host plants (i.e. between populations). Genetic correlations across environments, among clones sampled from the same species of host plant, were much weaker (Via, 1991).

5.4.4 *Remarks*

Unless the published literature is seriously biased, it can be concluded that trade-offs are evident in most situations involving phenotypically different morphs, populations or species inhabiting different environments. Such trade-offs are the predicted outcome of contrasting selection pressures between environments. Their near-ubiquity (Fig. 5.7) is therefore consistent with the hypothesis of divergent natural selection. In contrast, within-population genetic correlations between performance in two or more environments are usually close to zero or even weakly positive. However, negative genetic correlations are expected only under circumstances unlikely to hold in most populations.

A further strength of the transplant approach is that it often provides additional information on the ecological mechanisms of selection. Many transplant studies have identified traits contributing to performance differences between environments, and have at least partly elucidated environmental mechanisms responsible for trade-offs. Indeed, most of the vertebrate and noninsect invertebrate cases were carried out with specific traits and environmental features in mind (gape size and prey size; leg length and perch diameter; shell thickness and predation pressure) and most such trials were designed to test predictions of biomechanical hypotheses for the observed trait differences between environments. In a few cases, surgical alteration of phenotypes of transplanted individuals, by shaving leaf hairs (Ehleringer and Clark, 1988) or snipping beak tips (Benkman and Lindholm, 1991), further tested the contribution of specific traits. The situation for insects and plants is much poorer because the phenotypic basis of performance differences between phenotypes is usually unknown.

The main disadvantage of reciprocal transplants is that they do not test the fitness of intermediate phenotypes in intermediate environments. This is a problem because drift along adaptive ridges also predicts trade-offs in certain circumstances. Small-beaked finches may eat small seeds more efficiently than large-beaked finches, and large-beaked finches may eat large seeds best, but selection is not divergent unless intermediate-sized beaks consuming medium-sized seeds have lower mean fitness. Ruling out the adaptive ridge hypothesis therefore requires additional evidence that intermediate environments are infrequent, absent or inaccessible to intermediate phenotypes (if these exist). Such evidence may not be available.

Closely related species of host-specialized insects using different hosts are probably least affected by this problem, as intermediate environments may not exist. Species of limited mobility (e.g. plants) inhabiting different, spatially disjunct environments may also be little affected. Finally, tests involving genetically distinct morphs that produce no intermediates when crossed (e.g. because alleles distinguishing them show strong dominance) are free of the problem. In contrast, an adaptive ridge cannot be ruled out for most studies of sympatric vertebrate phenotypes 'transplanted' between different resource classes. The magnitude of trade-offs detected was not different between these two groups of studies (Kruskall–Wallis test, $\chi^2 = 0.845$, $P = 0.36$), but additional tests of divergent natural selection are nevertheless required for the latter group of studies.

5.5 Direct measurements of natural selection

Direct measurement of fitness, fitness components (survival and reproductive success), and performance of phenotypes in nature allow a third test of divergent natural selection. If selection indeed drove population differentiation in quantitative traits then one of two results is predicted. If equilibria have already been attained then each population should experience stabilizing selection around its mean. Or, if differentiation is still underway then divergent selection should be detectable as contrasting directional selection. A combination of these patterns of selection is expected if only one population is at equilibrium. Methods for estimating selection on quantitative traits are summarized in Box 5.1. Below, I review the directions of selection on differentiated populations.

5.5.1 Adaptive peaks revealed by disruptive selection

Disruptive selection refers to the presence of a fitness minimum within the range of phenotypes in a single population. The process is not the same as divergent selection, which is selection pulling population *means* in different directions (this distinction is blurred when disruptive selection splits a population in two). Nevertheless, disruptive selection on a highly variable population can be informative about the shape of adaptive landscapes in less variable populations.

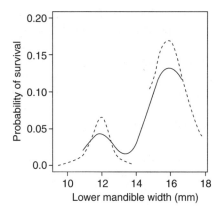

Fig. 5.8 Survival probability of juvenile finches (black-bellied seedcracker *Pyrenestes ostrinus*) in relation to beak width. The population is dimorphic for beak width. The two dashed curves estimate fitness functions f on each morph separately. These curves are cubic spline estimates from Fig. 1(a) in Smith (1993). The solid curve estimates the adaptive landscape for an equivalent monomorphic species in the same environment. The landscape was calculated from f as in Fig. 5.1 by assuming that beak width in a monomorphic population would have a normal distribution with a 5% coefficient of variation.

Survival patterns in the African finch, *Pyrenestes ostrinus*, illustrate this idea (Fig. 5.8). The species is dimorphic in beak dimensions, determined by variation at a single locus with two alleles (Smith, 1993). Each morph eats seeds of a different sedge, with the larger-beaked form eating the harder seeds. Variation in survival of thousands of individually marked juveniles over a seven-year period indicated bimodal selection on beaks (Fig. 5.8). A deep fitness valley was evident between two peaks centred roughly over the means of each beak morph. This fitness function can be used to construct an adaptive landscape for a typical monomorphic species (coefficient of variation ≈ 5%). The result is two peaks (Fig. 5.8).

Other examples of disruptive selection on highly variable populations include morphological traits in pumpkinseed sunfish of species-poor lakes in northwestern USA (Robinson *et al.*, 1996; Robinson and Wilson, 1996) and antipredator traits of garter snakes (Brodie, 1992; Brodie *et al.*, 1995). Adaptive landscapes for population means have not been constructed in these cases.

5.5.2 *Stabilizing selection about different means*

If populations lie at the apex of distinct adaptive peaks, then stabilizing selection should be apparent within each population. Phenotypes closest to their corresponding mean should have higher fitness than extreme phenotypes.

For example, four 'types' (possibly species) of red crossbills (*Loxia curvirostra* complex) were studied by Benkman (1993) in northwestern North America. Each

106 • *Divergent natural selection between environments*

is dependent upon a different conifer spatially separated from others in distribution: western hemlock, Douglas fir, ponderosa pine, and lodgepole pine. These finches use their crossed mandibles to separate overlapping scales of conifer cones, permitting extraction of the seed. Rates of seed extraction from closed cones depend strongly on beak depth, and rate on a given cone is maximized by a beak size close to that of the 'type' mainly exploiting it. A similar pattern holds for width of the palate groove, where seeds are held as they are being husked. Feeding efficiency is not fitness, but rate of energy gain on closed cones is probably the strongest agent of selection on beak size and shape (Benkman, 1993).

Other examples are surprisingly scarce. Beak dimensions of the cactus finch *Geospiza scandens* on Isla Daphne Major, Galápagos, are usually under stabilizing selection (Boag and Grant, 1984; Price *et al.*, 1984*b*). In contrast, the beak of its congener *G. fortis* has experienced directional selection that fluctuates between life stages, years, and generations (Boag and Grant, 1981; Price and Grant, 1984; Price *et al.*, 1984*a*; Gibbs and Grant, 1987*a*). The long-term outcome is oscillation within bounds, effectively broadly stabilizing selection.

Stabilizing selection appears to be less common than directional selection generally, but the compilation may be misleading (Endler, 1986). Most evidence is based on change in trait variance only, which could be caused by directional selection instead. Presence of a fitness maximum within the range of phenotypes in a population has been directly tested only rarely (Schluter, 1988*a*). Additionally, most data on natural selection come from snapshots at one point in time and at a single life stage. When directional selection has been measured at a second life stage, it frequently opposes that detected in the first (Schluter *et al.*, 1991). The consequences for selection over a lifespan, whether stabilizing or not, are not usually known.

5.5.3 Measures of selection in two environments

Directional selection on quantitative traits is occasionally measured simultaneously in spatially separated natural populations inhabiting different environments, or on sympatric morphs or closely related species exploiting different resources (transplants are dealt with separately below). Such measures permit a test of the prevalence of divergent natural selection (Table 5.2). Results are based on selection differentials for single traits rather than for suites of traits. This is because even random selection coefficients for suites of traits must usually differ in direction, complicating detection of patterns. The data are few and represent 'snapshots' of selection not necessarily indicative of the long-term picture. Nevertheless, trends are suggestive. In some cases the mechanisms of selection are known, but usually they represent an educated guess based on environmental correlates and known trait functions. In at least two cases the hypothesized agents of selection were tested using experimental manipulation (*Pseudacris* and *Enallagma*).

Divergent natural selection is relatively common, being detected in over half the cases (Table 5.2). Convergent selection is relatively rare, whereas it should be as common as divergent selection if coefficients are random in direction. Convergent

Table 5.2 The direction of natural selection on quantitative traits for which means are significantly different between closely related species (A), populations (B), or morphs (C) ('types'). Results are based on directional selection differentials for traits analysed one at a time. Divergent[1] ('convergent[1]') indicates directional selection in one type but not the other, with divergence (convergence) the net result. Parallel denotes selection differentials that were in the same direction in the two types, not necessarily the same magnitude

Taxon	Comparison	Trait	Fitness measure	Direction of selection	Agent of selection	Source
Differences between types heritable						
Pseudacris triseriata	C	Body shape	Growth and survival	Divergent	Predation and resource acquisition	Van Buskirk *et al.* (1997)
Vermivora spp.	A	Nest habitat (locust stems)	Nest survival	Divergent	Predation	Martin (1998)
Enallagma spp.	A	Caudal lamella size	Survival	Divergent[1]	Predation	McPeek *et al.* (1996)
Gammarus minus	B	Body size and antenna size	Mating success and fecundity	Parallel	Male–male competition and resource acquisition	Jones *et al.* (1992)
		Eye size		Divergent		
Hyalella azteca	B	Body size	Survival	Divergent	Predation	Wellborn (1994)
Impatiens spp.	A	Sepal aperture	Pollen removal	Parallel	Pollinators	Wilson (1995)
Impatiens pallida	B	Height and leaf area	Fitness	Parallel	Resource acquisition	Bennington and Mcgraw (1995)
		Flowering time (two flower types)		Divergent[1]	Drought avoidance	
Plantago lanceolata	B	Size, growth form	Seed yield	Parallel	—	Van Tienderen and van der Toorn (1991*b*)
Differences between types not heritable						
Pseudacris triseriata	C	Tail shape	Growth and survival	Divergent	Predation and resource acquisition	Van Buskirk *et al.* (1997)
Lacerta vivipera	B	Body size	Survival	Parallel	—	Sorci and Clobert (1999)
Cakile edentula	B	Water-use efficiency, leaf size and stomatal conductance	Fruit mass	Convergent[1]	Water availability	Dudley (1996*a*)
		Photosynthetic rate	Fruit mass	Parallel	Water availability	

selection dominated only one study (in which population differences were environmentally induced, not genetically based; Dudley, 1996a). Parallel selection (selection in the same direction in both populations) is no less common than divergent selection.

The prevalence of divergent natural selection indicated by direct measurements is consistent with the hypothesis of multiple adaptive peaks. However, the frequency of cases showing parallel selection is unexpected and requires explanation. It could indicate that differences between species in the affected traits are neutral even when the traits themselves are not. On the other hand, perhaps we should not expect present-day selection pressures to always mirror those which led to divergence. Closely related species should react similarly to many kinds of environmental fluctuations during and after diverging. Other possible explanations include: incomplete measurement of fitness (e.g. mortality selection on *Gammarus* probably opposes the advantages of large body size and antenna length in one of the two environments; Jones *et al.*, 1992; Culver *et al.*, 1995); and failure to remove correlated effects of general vigour on both traits and fitness (see Price and Liou, 1989; Rausher, 1992).

5.5.4 Selection and evolution in transplanted populations

A potential problem with tests based on measures of selection on wild populations is that the direction of current selection (postdifferentiation) may not be the same as that which led to divergence. One way to recreate the historical selection pressures is to transplant one or both populations to the environment of the other and determine whether the direction of selection in the transplant is toward the mean of the native. Results for discrete polymorphic traits were presented earlier (reciprocal transplant comparisons of type 'C'; Table 5.1) and they invariably support the hypothesis of divergent natural selection (nine of nine studies). Here I summarize the smaller body of results for traits varying continuously.

The guppy, *Poecilia reticulata*, provides a classic demonstration of selection favouring the native phenotype in introduced populations (Reznick *et al.*, 1997). Two separate introductions of guppies from high-predation localities to low-predation sites led within a few years to an evolutionary increase in size and age at reproduction, the native phenotype in low-predation sites. Escape performance likewise declined (O'Steen *et al.*, manuscript). Survival and reproductive success of phenotypes after transplant was not followed. Similarly, Losos *et al.* (1997) introduced *Anolis sagrei* from a single source population to islands with a lower average diameter of available perches. Parallel shifts toward shorter hind-limbs were detected, the direction expected from lizard species using such small perches (cf. Fig. 2.4). However, these shifts may represent phenotypic plasticity not evolution (cf. Losos *et al.*, 2000).

In contrast, the results of Bennington and Mcgraw (1995) are not in full agreement with expectation. They transplanted *Impatiens pallida* ecotypes between mesic floodplain and dry hillside environments and measured selection on traits in the transplants.

Selection strongly favoured earlier flowering in floodplain plants transplanted to the hillside, as predicted from the early mean flowering date of the native ecotype. However, no selection on flowering date was observed in hillside plants transplanted to the floodplain. Selection also favoured taller plants with greater leaf area in both sites in both natives and transplants, contra expectation. This parallel selection may be partly the outcome of correlations between these two traits and general vigour, which was not accounted for.

5.5.5 Selection and evolution in hybrid transplants

Another solution to the problem of recreating ancestral selection pressures is to make hybrids, which are often intermediate in phenotype between parent forms, and to plant them in the environments of both parents. This has the further advantage of a common 'probe' for measuring selection in both parental environments and, if F2 hybrids are used, often a high degree of genetic and phenotypic variation for traits that distinguish the parents.

The first study of this kind (Hiesey *et al.*, 1971) examined evolutionary changes in populations established when F2 hybrids between *Mimulus cardinalis* and *M. lewisii* were transplanted to experimental sites within the elevational ranges of natural distributions of the parent forms (Stanford and Timberline sites, respectively). In both sites trait values of 'spontaneous seedlings' descended from uncontrolled pollination between F2's showed an evolutionary shift toward the native type. At a third site at intermediate elevation (Mather) the descendants of F2's showed a bimodal distribution of trait values, suggesting disruptive selection.

Jordan (1991) transplanted F2 hybrids between coastal and inland populations of *Diodia teres*, an annual herb, to native sites. Six divergent morphological traits were tested for selection. The number of seeds produced was the measure of (female) fitness. Results were mixed: four of six traits were selected in the same direction in both environments, whether judged by selection gradients or selection differentials. Selection tended to favour the native traits on the coast but not inland. Evolutionary response was not measured.

Nagy (1997) extended the design by measuring also evolutionary response in F3's produced under more carefully controlled pollination. F2's of two subspecies of the annual, *Gilia capitata*, were planted in native coastal dune and inland chaparral habitats in California. Fitness (female only) was recorded as the number of inflorescences. F2's were then cross-pollinated by hand so that a plant's contribution to the pollen pool was roughly proportional to its flower number. A random sample of resulting seeds was grown in the laboratory so that means for morphological traits in this F3 generation could be compared with those in the original F2's in a common garden. The difference between means of F2 and F3 plants measured the evolutionary response to selection (Fig. 5.9). The result: selection favoured the native phenotype in seven of eight tests (four weakly correlated traits in two sites). Response in all eight cases was in the direction of the native (Fig. 5.9).

110 • *Divergent natural selection between environments*

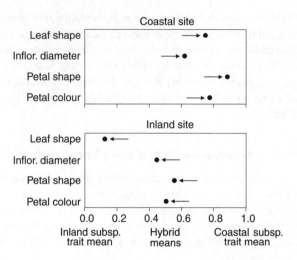

Fig. 5.9 Relative trait means of F3 hybrids between morphologically divergent subspecies of *Gilia capitata*. Trait values were transformed to a common 0–1 scale, with the pure subspecies located at the extremes. F3's were descended from F2's planted in the native sites of both parent subspecies—coastal dunes and inland chaparral. Arrows indicate the direction of evolutionary response in the F3 generation to selection on the F2's in each site. In all cases, the experimental hybrid population evolved toward the native subspecies of the site. All measurements were made on plants raised in the greenhouse. Modified from Nagy (1997), with permission of the Society for the Study of Evolution.

5.5.6 *Remarks*

Measurements of directional selection and short-term evolutionary response are more often than not consistent with the hypothesis of divergent natural selection. The strongest evidence comes from experimental studies in which parental forms or their hybrids were transplanted between environments. Experimental transplants of discrete morphs between environments invariably showed divergent selection. Transplant studies measuring selection on quantitative traits usually, although not always, provided support. Observational studies of selection on already-differentiated forms show that divergent natural selection is more common than convergent selection, but parallel selection is also common. The latter finding is consistent with the idea that species differences are often neutral, but it may instead reflect common response of closely related species to environmental fluctuations. Limited evidence of stabilizing selection on separate populations, and disruptive selection on single, variable populations, also supports the idea that adaptive landscapes have multiple peaks and valleys.

A strength of the experimental approach over observational studies of selection lies in its ability to recover historical (rather than contemporary) selection pressures, those selection pressures hypothesized to have led to divergence. Transplant studies measuring evolutionary response to selection also provided stronger support than those simply observing selection on wild populations, but the total number of studies is small. Short-term evolution may be a more reliable measure than selection alone because it integrates selection over a longer period, all fitness components are accounted for, and the problem of trait associations with overall vigour is eliminated.

5.6 Estimating adaptive landscapes from environments

With sufficient knowledge of the characteristics of environments, and of the dependence of performance on phenotype, it should be possible to estimate fitnesses of alternative mean phenotypes from first principles. Such a landscape would be constructed from the ecological mechanism upwards, rather than by pasting together and interpolating fitness surfaces estimated for the species that presently inhabit local sections of the landscape. The landscape may be constructed for a broad range of phenotypes, many of which may not occur in nature. Predicted phenotypes are then those corresponding to peaks in this resource landscape. Comparison of predicted mean phenotypes to those observed simultaneously tests divergent natural selection and the proposed environmental mechanism.

5.6.1 Adaptive landscapes for beak size

This approach was used to predict mean beak sizes in populations and species of Galápagos ground finches of the generalist granivore guild (Schluter and Grant, 1984). The main principle guiding this effort is that natural selection acting on individual beaks in a hypothetical solitary species on an island should lead to a mean beak size of the population that approximately maximizes population density. This assumption is justified by theory only under special conditions, that appear to be more or less satisfied in the Galápagos finches, but not generally (Brown and Vincent, 1987; Taper and Case, 1992*a,b*; Fear and Price, 1998).

Briefly, the steps to calculating expected population density for a hypothetical solitary finch were as follows. First, abundance of seeds was estimated on 15 different Galápagos islands in one or more dry seasons. The density of each seed species was recorded along with its mean size and hardness. The next step determined the upper and lower limits to the feasible range of seeds that a population of finches with a given mean beak size can consume. The upper limit is the most crucial to predicting beak sizes, and in these finches is set by seed hardness (Fig. 2.2). This is a physical rather than purely behavioural limit. Seeds whose hardness falls above the limit are rarely crushed successfully despite repeated (if infrequent) attempts (Schluter and Grant, 1984). The lower limit to the range of seeds is set by seed size. Seeds below a certain size are ignored. This is a behavioural limit but is presumably the result of the declining relative profitability of small seeds with increasing beak and body size. The precise limit did not greatly affect the predicted beak sizes.

112 • *Divergent natural selection between environments*

The total density of seeds within the feasible range for a finch of given mean beak size was then converted to expected finch density. The conversion used was based on empirical relationships between seed density, finch density, and finch body mass (which is positively related to mean beak size). Alternate conversions based on general metabolic requirements give similar results (D. Schluter, unpublished observations). Adaptive landscapes were constructed by calculating expected population density for hypothetical solitary finches, with beak sizes ranging from the smallest to the largest found in finches anywhere on earth (Fig. 5.10).

The resulting landscapes are rugged with many peaks and valleys (Fig. 5.10). The ruggedness arises directly from the discrete nature of seed resources in the environment. Plant diversity on the islands is low, especially on the smallest islands, and the gradients of seed size and hardness are discontinuous. For example, most seeds on Isla Daphne Major (top left panel of Fig. 5.10) are small and soft, and these determine the low peak at the smallest possible beak size. Two seed species having medium hardness are also abundant on the island and these are responsible for the adaptive peak at medium beak size. Other islands also have very hard seeds in abundance, leading in most cases to a third adaptive peak at very large beak size (e.g. Islas Genovesa and Marchena; Fig. 5.10).

Mean beak sizes of finch populations on Galápagos islands correspond relatively well to peaks in the landscapes (Fig. 5.10), supporting both the hypothesis of divergent natural selection and the idea that seed size and hardness distributions are the cause of this selection. Mean beak size tends to lie to the right of the nearest high peak (Fig. 5.10), which may reflect the simplicity of the conversion between seed and finch density or perhaps some additional advantage to large size.

5.6.2 Other examples

Armbruster (1990) predicted pollination performance (rate of pollen arrival and dispersal) for a group of resin-producing neotropical vines (*Dalechampia*) as a function of mean size of the resin-producing gland and mean distance between gland and anthers. A single peak at high gland area and small gland–anther distance was predicted well outside the range of flower morphologies exhibited by species in nature. Adding an assumed cost to resin production converted the surface to a ridge whose backbone extended from one extreme where both trait values are low to the other extreme where both are high (Fig. 5.11). Most of the species reside along this ridge (Armbruster, 1990).

This outcome suggests that drift along an adaptive ridge, rather than divergent natural selection, is responsible for the variety of flower structures seen among species. However, the analysis assumed that all size classes of bees are equally abundant and that this size distribution prevails at all sites. Subtle effects of variation in resource distributions would therefore not be detected (Armbruster, 1990). Armbruster hypothesized that the ridge was not actually a ridge because of competition among species

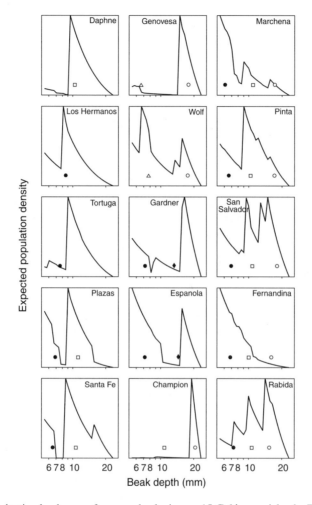

Fig. 5.10 Adaptive landscapes for mean beak size on 15 Galápagos islands. The height of the curve is expected population density for a solitary granivorous finch species. The expected population density in each panel is scaled to the maximum. Mean beak depths of male ground finches on each island are indicated by the position of the symbols: *Geospiza fuliginosa* (●), *G. difficilis* (△), *G. fortis* (□), *G. conirostris* (♦), and *G. magnirostris* (○). The large form of *G. magnirostris* on Champion and nearby islands is extinct. Modified from Schluter and Grant (1984), with permission of University of Chicago Press.

stemming from the negative impacts of sharing pollinators with other plant species including other *Dalechampia*. This hypothesis receives some support from later comparative studies (Hansen et al., 2000). In Chapter 6, I return to the effects of interspecies interactions on the shape of adaptive landscapes.

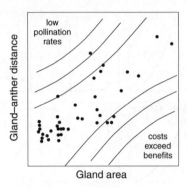

Fig. 5.11 Adaptive landscape for mean resin gland size and mean gland–anther distance in flowers of the neotropical vines, *Dalechampia*. This estimate is based on measures of pollen removal but incorporates an assumed cost to resin production. Contours describe a mean fitness ridge running from lower left to upper right. Symbols indicate means for various species of *Dalechampia*. Modified from Armbruster (1990), with permission of University of Chicago Press.

Kingsolver (1988) predicted thermoregulatory performance of Coliadine and Pierine butterflies in cool environments as a function of basal, medial, and distal melanin in the wings. Predicted peaks in performance occur at different combinations in the two butterfly taxa because of differences in reflectance-basking behaviour. These are broad relationships and were not intended to predict interspecific diversity in wing patterns within the taxa as a function of different environments inhabited.

5.6.3 Remarks

A virtue of empirical adaptive landscapes is that they enable prediction of mean phenotypes from first principles. In the Galápagos finches the landscapes exhibit multiple optima and often steep fitness valleys between peaks. They show that landscapes differ from island to island and that finch beaks vary in concert. This supports the hypothesis of divergent natural selection as the cause of phenotypic differentiation in this group. The correspondence between beaks and peaks gives further confidence that the principal agents of divergent selection, namely seeds, have been correctly identified. In contrast, the fitness landscape for floral shape in *Dalechampia* revealed no peaks. Divergence in this group remains consistent with drift along an adaptive ridge.

But the finch landscapes were built for only one morphological trait, beak size. Might the peaks disappear if we added more traits, revealing the adaptive ridges connecting apparent peaks (Whitlock *et al.*, 1995)? The main argument against this possibility is that while multiple traits distinguish the species, the traits strongly covary. Variation among populations and species in the generalist granivore group occurs largely in a single dimension involving a combination of traits that includes body size and mandibular muscle mass. Beak depth is merely a sensible indicator of the suite of traits. Given this restriction, even though we might imagine combinations of traits that would be candidate intermediate steps along a hypothetical adaptive ridge, it is unlikely that the Galápagos ground finches have ever travelled there.

The evidence for divergent selection could be made even stronger if the shapes of the landscapes calculated from seeds to fitness were corroborated by independent data on natural selection on natural populations.

5.7 How do adaptive peak shifts occur?

The evidence reviewed above suggests that divergent natural selection is a major cause of phenotypic differentiation between populations exploiting different environments, and adaptive landscapes often have multiple peaks and valleys. Differentiated populations and species appear to lie within the domains of separate peaks. This conclusion raises a new issue. Peaks are by definition separated by valleys of low fitness whose crossing is resisted by selection. Therefore, how do peak shifts come about in the first place? This question concerns mechanisms at work in the initial stages of phenotypic differentiation of populations and species. Not much information is available from natural populations, and current thinking is based largely on the plausibility of different mechanisms. Below I summarize briefly the main ideas.

5.7.1 Two types of peak shift

Proposed causes of peak shifts fall into two fundamentally distinct categories. In the first, adaptive valleys are forded by selection. A low-fitness barrier separating domains of adaptive peaks is crossed when the landscape itself is transformed such that the valley temporarily disappears, or the valley may appear for the first time after differentiation has begun. The twin tenets underpinning this idea, both attributed to Fisher (1930), are that the landscape is always changing and that evolution occurs only in directions of increasing mean fitness (i.e. uphill on adaptive landscapes). Nevertheless, a variety of mechanisms are included in this first category of explanations. The adaptive landscape may change because of a fluctuation in the underlying fitness function, arising from changes in the external environment. Or, the landscape may change if the frequency distribution of phenotypes (e.g. population variance) changes. Transitions of the former type are purely adaptive, fuelled by effects that are commonplace and inevitable in nature. Transitions of the latter type may result from either deterministic (e.g. selection, migration) or random forces (e.g. genetic drift, mutation or other contingency) less well documented.

In the second category of peak shift, adaptive valleys are crossed when populations briefly undergo excursions downhill on the adaptive landscape. The plunge begins with genetic drift and occurs despite opposing natural selection (Wright, 1931, 1932). Theoretical models for this type of peak shift assume that the surface is unchanging.

A constant landscape is not necessary for genetic drift to play a role in divergence, but if fluctuations occur frequently then the valley separating peaks will be swiftly crossed via natural selection, and drift's potential role is then much reduced.

5.7.2 Peak shifts by natural selection

Spatial variation in environments—Peak shifts are straightforward if environments and locations of adaptive peaks vary spatially as in a simple two-island model (Fig. 5.12). The first of two episodes of change is initiated when a few individuals from a source population are stranded on a new island with a different landscape. A climb toward the new adaptive peak yields a daughter population phenotypically differentiated from that of the source. Backcolonization of the first island and a shorter climb completes the peak shift on the initial island. The trail of occupied peaks is marked by the populations left behind at intermediate stages, if extinctions do not remove them. At this point the two forms on island 'A' may persist as two sympatric species or as a single variable species balanced between hybridization and selection.

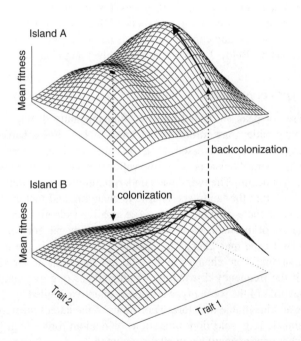

Fig. 5.12 A spatial model of adaptive peak shift. Two islands differ in the mean fitness landscape for two phenotypic traits, particularly the location of adaptive peaks. Colonization of the second island from the first, and eventual backcolonization, circumvents a direct crossing of the valley. Points indicate population means. Solid lines and arrows indicate the uphill direction of evolution on each island.

How do adaptive peak shifts occur? • 117

Alternatively, one of the two sympatric forms may be lost completely (not necessarily the one on the lower peak) if gene flow between the forms is strong.

This model of spatial variation is the only mechanism of peak shift with much supporting evidence from nature. Different environments have different adaptive landscapes. For example, different trajectories of evolution in experimental hybrid populations transplanted to the distinct environments of their parents (e.g. Fig. 5.9) is difficult to explain in any other way. Variations among Galápagos islands in the availability of seeds of different size and hardness sometimes lead to strikingly different adaptive landscapes for finch mean beak size (Fig. 5.10). It is easy to imagine the process illustrated in Fig. 5.12 at work in the Galápagos finches, and cycles of divergence via island colonization and backcolonization is the preferred explanation for the initial stages of speciation and phenotypic diversification in the finches there (Lack, 1947; Grant, 1986). (However, character displacement is thought to have finished the job; see Chapter 6.) The historical sequences of island (and adaptive peak) occupation are not known in the finches but molecular studies may someday help to elucidate this.

In a modification of this model, Fear and Price (1998) suggested that peak shift may be sped up if phenotypic plasticity results in a shift in mean phenotype in the new environment toward the novel peak (see also West-Eberhard, 1989). Such plasticity may make shifts possible between widely separated peaks if it enhances population persistence in the new environment.

Temporal variation in environments—Environments also vary in time, causing changes in adaptive landscapes for traits involved in their exploitation. This process can also facilitate peak shifts. To visualize this process one needs only to imagine that the two landscapes in Fig. 5.12 represent stages in time on a single island.

Information from nature is scant on temporal fluctuations in adaptive landscapes, although temporal variation in selection coefficients is frequently described (Price *et al.*, 1984*a*; Gibbs and Grant, 1987*a*; Schluter *et al.*, 1991). Repeated estimates of landscapes for beak size on Galápagos islands show considerable variation in the absolute heights of peaks and even in their locations (Fig. 5.13). These estimates represent snapshots from wet season to dry season in a single whereas variation between dry seasons is probably more important (Price *et al.*, 1984*a*; Gibbs and Grant, 1987*a*). It is reasonable to imagine that temporal variation facilitates adaptive peak shifts on islands, but direct evidence is lacking.

Resource depletion—Depletion of resources through consumption can radically alter fitnesses of different phenotypes. Theoretical literature on the evolutionary consequences of intra- and interspecific competition abounds with graphical illustrations of resulting changes in fitness functions (e.g. Wilson and Turelli, 1986; Taper and Case, 1992*b*; Dieckmann and Doebeli, 1999). The basic process is intuitive. Resource depletion may create a dimple in the adaptive landscape above the species mean. A peak so occupied may be trimmed to the height of its neighbouring valley, a drop that

118 • *Divergent natural selection between environments*

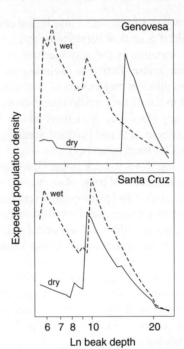

Fig. 5.13 Seasonal variation in estimates of adaptive landscapes for ground finches on 2 Galápagos islands. Curves are from the middle of the wet season (dashed) and the middle of the dry season (solid) of the same year.

could spark a transition of the population to a higher peak, or cause a shift instead in a second species dependent upon the same resources. Unfortunately the best examples are theoretical. Changes in fitness surfaces wrought by resource depletion are poorly documented in nature (see Chapter 6).

Peak fission and the evolution of specialization— The shapes of fitness surfaces are not determined entirely by external environments. They depend also on how resources are utilized. For example, a resource gradient with gaps (e.g. between available seed size categories) may yield a smooth unimodal fitness function if each phenotype is able to utilize a broad range of resources along the gradient. However, discontinuities in the gradient are more noticeable to species able to utilize only a narrow range of seed sizes, with increased ruggedness of the fitness surfaces (and corresponding adaptive landscapes) the result. Separate species occupying the erstwhile solitary peak might evolve increasing specialization and gradually find themselves separated by a fitness depression that was never crossed by either species.

Correlated response— A trait whose mean resides under an optimum in the adaptive landscape may be dragged down a neighbouring valley to a new optimum if the trait is strongly correlated genetically with a second trait under strong directional selection (Price *et al.*, 1993). Mean fitness overall does not decline in this scenario, because the temporary loss in fitness resulting from the shift in the first trait is compensated by the rapid rise in fitness accompanying the change in the selected trait.

Changes in level of phenotypic variance—Variation around the mean of a trait leads to an adaptive landscape that is smoother than the underlying fitness surface because mean fitness is the average of individual fitness over a range of phenotypes (Fig. 5.1). For the same reason an increase in phenotypic variance, whether by an increase of genetic variance or of environmentally induced variance, will cause the adaptive landscape to become even smoother. The result will be the loss of small peaks (Fig. 5.1) precipitating transitions to higher ones (Kirkpatrick, 1982*b*; Whitlock, 1995).

Any number of processes can increase phenotypic variance including introgression, mutation accumulation, and phenotypic plasticity. Stochastic increases in genetic or environmental components of population variance may also occur, such as may occasionally happen after a bottleneck in population size (Whitlock, 1995). This is the first of several mechanisms in which genetic drift may play a role in peak shifts on phenotypic adaptive landscapes.

Divergent response to uniform selection—The shapes of adaptive landscapes are potentially complex even when a small number of traits are considered (e.g. Fig. 5.10) and they must be more so when many traits are subject to change. If a population in a novel environment has many nearby peaks to choose from, the peak actually climbed may sometimes be a matter of chance. Populations starting from the same mean phenotype need not end up at the same adaptive peak (see Chapter 4; Fig. 4.4).

Populations are also expected to diverge despite experiencing the same selection regime if genetic covariances among traits diverge between populations. This is because response to multivariate selection is strongly affected by these parameters (Chapter 9). Chance processes that contribute to change in trait variance and covariance include a different history of accumulation of new mutations (Fisher, 1930) and genetic drift in genetic and environmental components of phenotypic covariance among traits (Whitlock, 1995). Differences in selection regime might also contribute to changes in covariance, directly or via effects on the level of assortative mating or recombination (Williams and Sarkar, 1994).

5.7.3 Peak shifts and genetic drift

Peak shifts via drift in mean phenotype—Genetic drift is ubiquitous, but the extent to which it contributes to phenotypic differentiation in adaptive radiation is not known. Drift along adaptive ridges is one way in which populations may diverge in mean phenotype and use of environment. Drift may do even more: perhaps it can take populations through adaptive valleys and into the domains of new adaptive peaks. A considerable body of theory has examined the plausibility of peak shift by genetic drift of mean phenotype, either in populations of fixed size or during catastrophic reductions in population size (reviewed in Barton, 1989; Coyne et al., 1997). The upshot is that a shift between peaks is plausible when the adaptive valley is very shallow and when population size is very small. If these conditions hold often, drift-induced shifts are indeed a force to consider. (Note, however, that adaptive valleys are often deep; Fig. 5.10.)

The theory for peak shifts by drift was developed for adaptive landscapes that do not vary but such landscapes probably do not exist, at least not for phenotypic traits involved in exploitation of environments. If even a small amount of variation in the topography of adaptive surfaces is allowed, then the shallow adaptive valley would be among the least permanent of its features (Whitlock, 1997). A shallow adaptive valley will frequently disappear, at which point a peak shift would occur much more readily by natural selection. The impermanence of shallow valleys also affects the plausibility of other mechanisms of peak shift involving drift, one of which is discussed next.

The shifting balance—Wright (1931, 1932) envisioned a complex process of adaptive evolution in which transitions between adaptive peaks (always from lower to higher) involve a special combination of drift, migration, and selection he called the 'shifting balance' (reviewed by Provine, 1986; Coyne et al., 1997; Wade and Goodnight, 1998). The process requires that large populations are structured as networks of local demes connected by small amounts of gene flow. In the first of the three phases of the shifting balance, genetic drift within local demes eventually carries one of them down an adaptive valley and into the domain of a higher peak. In the second phase, directional selection brings the mean value of the trait in the affected deme to the top of its new peak. Third, the higher mean fitness so acquired causes the affected deme to export a disproportionate number of migrants to other demes, eventually dragging them up the same high peak.

Arguments over the importance of this process are still being heard (Coyne et al., 1997; Wade and Goodnight, 1998; Coyne et al., 2000) and a review is beyond the scope of this book. As there is virtually no data from nature, most of the debate has centred on the plausibility of specific elements. Two of the arguments against it are ecological and carry as much weight as any of the genetic difficulties that the process also seems to encounter (Coyne et al., 1997). One of these is the shortage of evidence that real populations possess the demic structure required by the shifting balance process. The other argument is that adaptive landscapes, particularly for phenotypic traits involved in extracting resources from environments, are unlikely to be stable enough to warrant the theory in the first place (Whitlock, 1997). The first and third phases of the process are most likely when valleys between adaptive peaks are shallow; yet such valleys are likely to fluctuate in and out of existence and precipitate peak shifts by selection under a wide range of realistic population structures.

5.8 Discussion

Under the ecological theory of adaptive radiation, divergent natural selection arising from differences between environments is the major cause of phenotypic differentiation. This claim goes beyond the proposition that different species are adapted to exploiting different environments. Rather, under divergent selection trait means of populations and species are pulled apart and intermediate phenotypes have lower fitness.

Divergent natural selection is not the only possible explanation for phenotypic differentiation of functionally significant traits. An alternate hypothesis is that species diverge via mutation and drift along adaptive ridges lacking distinct peaks. In this alternate scenario, differences between species are neutral even while the traits are not. Adaptive ridges probably exist in nature and many of the cases summarized here are unable to rule it out. Part of the problem is that much of the data have been collected without the distinction between divergent selection and drift along ridges in mind. Most tests are designed to distinguish 'selection' from 'no selection', the latter depicted by a completely flat adaptive landscape. The 'no selection' alternative is unrealistic especially for taxa already meeting the phenotypic criteria for adaptive radiation (Chapter 2). Therefore future tests of divergent natural selection must focus more directly on evaluating the fitness of intermediate phenotypes. Such an inquiry would be aided by redoubled efforts to elucidate the mechanisms of selection, for this would pinpoint the relevant intermediate environments in which the fitness of intermediate phenotypes ought to be judged and their prevalence in nature.

The data represent a wide diversity of taxa, most of which have not yet been shown to fulfil all the criteria of adaptive radiation. Therefore it will be important to ask how the results are different when only the adaptive radiations are considered. At the very least, effort should focus on those traits which show divergence in association with environment, because an explanation of such associations is among the foremost of our goals.

While these limitations of the data cannot be dismissed, on balance the evidence so far weighs in favour of divergent natural selection. Population differentiation in at least some phenotypic traits is faster than neutral genetic markers, at least in the early stages of divergence. Drift along ridges would predict slower divergence than the pure neutral expectation. In transplant experiments the native phenotype almost always has higher fitness than the immigrant phenotype. The magnitude of such trade-offs between environments is typically large. The main uncertainty concerns the fitness of intermediate phenotypes in intermediate environments if such environments exist. Direct measurements of natural selection and evolution also tend to support the idea that landscapes have multiple peaks. This is particularly true in perturbation experiments that attempt to recreate an earlier stage of divergence. In at least one case, estimates of adaptive surfaces for phenotypic traits based on measurements of resource availability provide compelling evidence that population and species means lie in the vicinity of adaptive peaks separated by valleys of lower mean fitness.

In combination these results yield a picture of divergence more consistent overall with the ecological theory than any other. Empirical study over the past 50 years has by and large proved the utility of Simpson's multiple-peaks model of phenotypic divergence. Adaptive peaks corresponding to niches in the environment remain a compelling depiction of selection on phenotypes in nature. Populations and species often appear to lie on opposite sides of adaptive valleys. The most glaring omission from this view is an accepted explanation of the routes by which two populations arising from a common ancestor managed to arrive at this state. We know little about the mechanisms of peak shift in nature.

The evidence for spatial differences in adaptive landscapes is strong (e.g. Fig. 5.10). Temporal variation in adaptive peaks is less well documented. The main argument in their favour is their plausibility. It is easy to see how such temporal and spatial variability in location and height of adaptive peaks might facilitate a crossing from one peak to another, but we are unable to point to examples from nature. Peak shifts involving genetic drift, in which mean fitness temporarily declines, are less plausible than other mechanisms but have not been ruled out by data. The influence of stochastic changes in genetic and phenotypic variance and covariance is potentially great, but evidence from nature is scarce.

6

Divergence and species interactions

> *The meeting of two forms in the same region to form new species must ... result in subdivision of the food or habitat, and so to increased specialization. The repetition of this process has produced the adaptive radiation of Darwin's finches.*
> —Lack (1947)

6.1 Introduction

Interspecific competition for resources was regarded by the early naturalists as the second major cause of phenotypic differentiation in adaptive radiation. Under this view depletion of shared resources drives species to exploit new types, where they come under the influence of contrasting selection pressures. Lack (1947) was a strong proponent of this view. Interspecific competition was an important part of his overall interpretation of the extraordinary morphological variety of the Galápagos finches. His demonstration that the small and medium ground finches, which differ greatly in beak size where they occur together but converge in phenotype when present alone, remains among the most frequently cited examples of competition-induced divergence in phenotype. It pursuaded most naturalists of the time.

The recent situation could hardly have been more different. Skepticism became commonplace, the result of a series of critical studies that highlighted the doubtful quality and completeness of the evidence for competition's role in phenotypic divergence (Grant, 1975; Strong *et al.*, 1979; Connell, 1980; Simberloff and Boecklen, 1981; Arthur, 1982; Gotelli and Graves, 1996). The storm has settled somewhat over the past few years, in large part because much new information has accumulated. In the present chapter, I weigh the observational, predictive, and experimental evidence from closely related species for competition's role in phenotypic divergence.

A challenge of a different sort ultimately requires as much attention, concerning not whether competition is important in phenotypic divergence but whether other interactions among close relatives are any less so. The most spectacular adaptive radiations, at least those in vertebrates (e.g. Figs 1.1, 1.3), are celebrated in part because the descendant species have in a short period of time come not only to utilize different foods but to occupy completely different trophic levels in community food webs. A great complexity of interactions evidently connects species, and a major question for the future is how much role these have played in ecological differentiation. Ecologists modelling nature have largely moved on from equations of pairwise competition to more comprehensive models of webs of species that interact not only via depletion of shared resources but also via shared predators and as predator and prey. Should students of adaptive radiation now do likewise? Our ability to answer this question is presently very limited. Below, I summarize a small amount of theory and a few case studies which reveal the potential role of alternative interactions in adaptive radiation.

6.2 Divergence between competitors

A rich body of theory has explored the conditions that favour divergent 'ecological character displacement', defined as the process of phenotypic divergence caused or maintained by interspecific resource competition. In what follows, I summarize aspects of the theory that help to illuminate mechanisms driving divergence and that help interpret patterns in data presented in subsequent sections. A more thorough review of the theory can be found in Abrams (1986) and Taper and Case (1992*b*).

6.2.1 Conceptual framework

The outstanding feature of natural selection resulting from resource competition is that fitness of phenotypes depends on the total *density* of individuals and on the *frequency* of different phenotype classes (e.g. Slatkin, 1979, 1980; Taper and Case, 1992*b*). This process is most simply envisioned with two species and a single, continuously varying phenotypic trait z (e.g. body size) that determines ability of individuals to exploit prey distributed along an underlying resource gradient (e.g. prey size).

The conventional fitness function incorporating these elements extends the usual Lotka–Volterra equations for competition:

$$f(z) = 1 + r - \frac{r}{K(z)} \sum_{i=1}^{2} \int p_i(x) \alpha(z, x) N_i \, dx$$

(Slatkin, 1980). N_i is population density of species i, and $p_i(x)$ is its frequency of phenotype x. $K(z)$ is the 'resource function', indirectly describing prey available to each phenotype z by the equilibrium number of identical individuals that can be supported ('carrying capacity'; Roughgarden, 1972, 1976; Slatkin, 1980). $K(z)$ is often assumed to be normal (gaussian), meaning that it is symmetric and bell-shaped with a single peak at an intermediate phenotype. Width of the gaussian bell indirectly reflects the breadth of resources available. The constant r is the instantaneous rate of population growth in both species, assumed to be independent of z.

$\alpha(z, x)$ is the intensity of competition between an individual of phenotype z and another individual whose value for the same trait is x. This competition function mirrors overlap in resource use between phenotypes (Roughgarden, 1972, 1976; Taper and Case, 1985). $\alpha(z, x)$ is usually assumed to attain its maximum value at $x = z$, corresponding to complete resource sharing, and declines gradually for larger or smaller values of x. This decline is described by another gaussian curve whose width symbolizes the breadth of resources utilized by each phenotype. When utilization is broad, $\alpha(z, x)$ declines slowly with increasing distance between z and x, implying a high degree of resource sharing. Low resource sharing and narrow utilization are implied when $\alpha(z, x)$ falls steeply with increasing distance between phenotypes.

The summation in the above equation describes the total competition an individual of phenotype z experiences from members of its own and the other species. This total competition term depends in turn on the densities and frequencies of all phenotypes. Fitness $f(z)$ declines with increasing total competition. It is high when densities N_i are low and when most individuals in the community are unlike it in phenotype.

The above fitness function has been the basis of a number of computer simulations of the character displacement process (e.g. Fig. 6.1). A simulation begins with a choice of starting values for the mean and variance of z in each of the two species. Selection differentials on each mean are computed from the distribution of fitnesses of

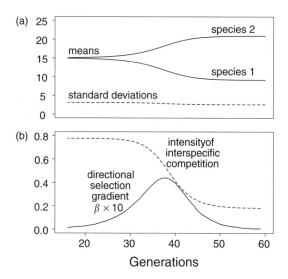

Fig. 6.1 Divergent character displacement in an additive, quantitative trait (Doebeli, 1996). (a) Means and standard deviations in two species through time, beginning from negligible initial differences. Divergence is slow at first but picks up speed as the species begin to diverge, and finally slows as equilibrium values are approached. The first 15 generations of virtually no evolution are not shown. The rate of evolution is unnaturally high because the heritability of the trait was set to 1. (b) The strength of selection favouring divergence compared with the average strength of interspecific competition through time. Competition intensity is the average $\alpha(z, z')$ between pairs of phenotypes in the two species. Selection intensity is the directional selection gradient (Chapter 5). It peaks at an intermediate intensity of interspecific competition. Based on Fig. 1(b) in Doebeli (1996, personal communication), with permission of the Ecological Society of America.

all members of the population. These differentials are then used to compute changes in means in the next generation (see Chapter 9). Population sizes in the next generation are computed from the average fitnesses of phenotypes in the current generation. This series of calculations is then iterated through many generations until an equilibrium is reached. Most variation in z is assumed to be genetic to speed up the divergence process. Displacement is much slower when a reasonable part of this variation is nonheritable but final results are otherwise unaffected. Variation is assumed to be continuous and determined by the summed effects of a large number of loci. Upper and lower bounds are typically placed on the genetic variance within species. The simplest assumption is that the variance is small and remains constant. This accomodates the fact that variances evolve more slowly than means and do not increase indefinitely (see references in Grant and Price, 1981; Taper and Case, 1992b; Bell, 1997). In the example shown in Fig. 6.1, the variances evolve but assumptions about the underlying genetic basis of trait variation constrain their values (Doebeli, 1996). The constancy of the function $\alpha(z, x)$ implies that breadth of resources utilized is also fixed. Models in which utilization breadth also evolves are presented in Taper and Case (1985).

6.2.2 Multiple adaptive peaks

Divergent character displacement occurs easily under the above conditions, the magnitude depending on starting values for the mean, details of $K(z)$, $\alpha(z, x)$, and the limits placed on the evolution of genetic variance. The theory therefore confirms the intuition that interspecific competition may be an agent of divergent natural selection, generating differences between sympatric species beyond those expected from resource distributions alone. Resource depletion is the mechanism driving divergence in species' mean phenotypes. Change occurs via the 'positive response' to resource density, in the absence of complicating factors: organisms evolve greater ability to utilize more abundant resources and reduced ability to utilize rarer (including depleted) ones. Resource density is not explicitly included in the above type of model but is represented by total population size.

Under these simple assumptions, mean phenotype of a single population evolves toward a peak in the resource distribution (represented here by the carrying capacity curve $K(z)$). When more than one species is present, this tendency in any one of them is counteracted by resource depletion by other species, which favours the use of otherwise less abundant resources toward the extremes of the resource distribution. At evolutionary and demographic equilibrium, two species share the commonest resources at the centre of the resource gradient but use opposite tails of the gradient (Fig. 6.2).

In effect, competition between two species generates two 'adaptive peaks' in mean phenotype even though the resource function itself is assumed to have only one peak. The approximate location of these peaks is illuminated by plotting attainable population size of each species against mean phenotype. Consider two species at equilibrium at the end of the character displacement process. If we hold fixed the

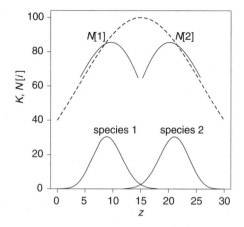

Fig. 6.2 Phenotype distributions and conditional population sizes of the two species at equilibrium (cf. Fig. 6.1). The dashed curve is the carrying capacity function $K(z)$. $N[1]$ indicates 'conditional population size' of species 1, its population size as a function of its mean phenotype when the mean phenotype and population size of the second species are held constant at their ecological and evolutionary equilibrium values (Roughgarden, 1976). $N[2]$ is the equivalent variable in the second species.

equilibrium population size and mean phenotype of species 1 and plot the population size of species 2 at different values for its mean, a peak in population size for species 2 is revealed on the right flank of the resource function $K(z)$ (Fig. 6.2). An equivalent peak for species 1 sits on the left flank of $K(z)$. Each of these two curves depicts what Roughgarden (1976) called 'conditional population size,' which is population size stemming from resources after resource competition from other species is accounted for.

Although the concept of 'adaptive peak' helps intuition, interspecific competition muddies the very notion (Fear and Price, 1998). The 'peaks' move whenever the species means change, over and above those changes induced by external environment. More annoying to the researcher hoping to apply this theory is that evolution will not as a rule bring population means to lie directly under the peaks when selection is frequency-dependent (Wright, 1959; Lande, 1976; Roughgarden, 1976; Abrams, 1989; Taper and Case, 1985, 1992*b*). The 'positive response' breaks down if we depart from the simple assumptions used above. For example, if competition is asymmetric, with larger individuals exerting a stronger effect on smaller individuals than vice versa, then the mean size at equilibrium is expected to be larger than that corresponding to peak resources even though the result is a reduction in population size. At best, peaks in mean fitness serve as reference points from which frequency-dependent departures may be calculated (e.g. Day and Taylor, 1996). These departures might be large if competition is strongly asymmetric or if agents of selection other than resources intervene. Occasionally, peaks in conditional population size will predict mean phenotypes of competing species in nature with considerable accuracy (Section 6.4.2), but it cannot always be expected to do so.

6.2.3 Initial differences

Under these models, divergent natural selection is weak, and the rate of phenotypic divergence slow, when initial differences between the species are small even though interspecific competition is then most intense (Slatkin, 1980; Milligan, 1985; Doebeli, 1996; Fig. 6.1(b)). Weak divergent selection when competition is strong may seem paradoxical. The relevant quantity, however, is not the intensity of interspecific competition but the rate at which interspecific competition is reduced per increment of divergence. This rate is low when species lie next to one another and overlap broadly in their utilization of resources. Selection reaches maximum intensity at an intermediate distance between the species before declining again as differences between the species become large. The details of this result may be model-dependent and its robustness is unknown.

The significance of this result is that character displacement may occur most readily when species are already somewhat differentiated. The implication is that divergence between competitors is aided if other factors initiate divergence. For example, phenotypic divergence between closely related species may begin in allopatry through the action of drift or spatially divergent natural selection (Chapter 5), and these initial differences are secondarily built upon in sympatry by interspecific competition. This was essentially Lack's view of the origin of differences between Galápagos finches (Lack, 1947; Grant, 1986). These ideas focus on differences in the single trait z. However, initial differences between species in other traits also facilitate character displacement in z (Milligan, 1985). A second reason for thinking that initial differences aid character displacement is that without them competitive exclusion may occur instead (Milligan, 1985).

6.2.4 Resource functions

At least two characteristics of the resource function significantly alter the outcome of character displacement. The first is resource breadth: the greatest divergence is expected when resource distributions are widest (Slatkin, 1980; Taper and Case, 1985, 1992b; Doebeli, 1996; Drossel and McKane, 1999). Breadth in the above theory may be modelled by a wide gaussian $K(z)$ curve, by a resource function more flat-topped than the gaussian (e.g. Slatkin, 1980), or by a function having more than one peak (e.g. Fig. 5.10). The effective width of K may also be less than the full breadth of resources if other taxa usurp portions of the resource gradient.

The outcome of character displacement is also affected by skew or asymmetry in the resource distribution: greater divergence is expected when resource functions are asymmetric, all else being equal. A skewed resource distribution is one in which the peak in $K(z)$ (or the highest peak in a many-peaked resource function) sits close to one end of the permissible range of z values rather than in the middle (Slatkin, 1980). Divergence in this case should also be asymmetric. The mean of one species should lie close to the resource peak in both sympatry and allopatry, whereas the other species should shift markedly away from the resource peak in sympatry.

6.2.5 Remarks

The process of character displacement encounters few theoretical obstacles. The main challenge is deciding whether this theory has a basis in fact, particularly in adaptive radiations at low taxonomic level. Does character displacement occur frequently? When it occurs, do its properties match theoretical expectations? Most attention has been paid to the first of these questions, surprisingly little to the second. I have listed only a few of these expectations but others may be equally useful. For example, the amount of displacement should increase with population size (Abrams, 1986).

Other models of character displacement not summarized here incorporate explicit resource dynamics (Lawlor and Maynard Smith, 1976; Rosenzweig, 1981; Taper and Case, 1985; and especially Abrams, 1986, 1987, 1990). The qualitative predictions of these models for divergent character displacement are similar to those above as long as species respond positively to the more abundant and less depleted resources. However, more complicated responses to resource depletion are also possible when additional features are added. For example, convergence and parallel shifts in sympatry may occur instead of divergence when resources are not nutritionally substitutable or when populations competing for a given set of resources are limited by alternative resources or predation (Abrams, 1986, 1987, 1990, 1996). Other variations include models with more than two species (Taper and Case, 1985), models incorporating environmental variability (Gotelli and Bossert, 1991), and models in which deleterious mutations accumulate in genes that underlie traits determining exploitation of little-used resouces (Kawecki and Abrams, 1999).

Although convergence is theoretically possible, divergence may be the most common outcome of competition when species are closely related and morphologically similar (indeed, there are no examples of convergent character displacement, but perhaps nobody has looked). Competition is therefore an agent of divergent natural selection. The process generates an adaptive valley between species means that drives them apart. A compelling demonstration of divergent character displacement is therefore virtual proof that adaptive landscapes are not flat ridges along which divergence occurs by drift and mutation alone. Consideration must therefore be given to competition when testing alternatives to the adaptive ridge hypothesis (cf. Chapter 5).

6.3 Observational evidence

The evidence for character displacement is observational, predictive, and experimental. By observational I mean that competition's effects are inferred from patterns of species differences in existing assemblages. The predictive approach uses optimality models to generate quantitative expectations of mean phenotypes under character displacement, and compares these expectations with predictions of alternative models lacking competition. Experiments directly measure changes in competition, natural selection, and evolution upon addition or deletion of related species differing in phenotype. Fossil sequences are observational because the information they yield about mechanism is similar to that of spatial sequences of contemporary populations (except that the direction of change is laid bare). Although observational evidence is indirect, it can be compelling. It is also available in greatest abundance because it is the easiest to gather.

The present section reviews observational evidence for divergent character displacement. The following two sections cover predictive and experimental evidence. I limit these surveys to cases of displacement between close relatives because here I am interested in the outcome of interactions within lineages. In practice, this means congeneric species. Evidence for displacement between more distantly related species is covered in Chapter 7 in the context of impacts of other lineages on adaptive radiation.

6.3.1 Three types of patterns in nature

Three kinds of patterns make up the observational evidence for divergent character displacement. The most common is *exaggerated divergence in sympatry* (Table 6.1) in which phenotypic differences between species are found to be greater where they coexist (sympatry) than where they occur separately (allopatry) (Lack, 1947; Brown and Wilson, 1956). Two examples, *Anolis* lizards and three-spine sticklebacks, are illustrated in Fig. 6.3.

The second most common type of pattern is *trait over-dispersion* (also called 'constant size ratios') in which the mean phenotypes of species belonging to a guild are more evenly spaced along a size or other trait axis than a random collection (Table 6.2). For example, mean diameters of canine teeth of the small native cats of Israel exhibit an astonishing regularity in the sizes of intervals separating morphological neighbours (Fig. 6.4). In this example, sexes are dimorphic and behave as distinct 'morphospecies' in the sequence. The smallest difference between any two morphospecies (0.62 mm) is unusually large: four means thrown randomly and uniformly onto the interval bracketed by the smallest and the largest species would yield a minimum difference as large as 0.62 mm only one in a thousand times. Constant size-ratios are not a necessary or even a likely outcome of character displacement. Constancy is expected only under special conditions that include: competition is confined to a single resource gradient; the frequency distribution of resources and resource utilizations along this gradient are highly symmetric; and the relationship between morphology and resource utilization is linear (Taper and Case, 1985). The hope is that character displacement would still leave a signature even if these assumptions are violated, perhaps not by the complete absence of close size-neighbours but at least by a deficit of them (e.g. *Accipiter* hawks and *Stylidium* triggerplants; Table 6.2).

The third pattern is *species-for-species matching*, defined as replicated guild structure between sets of species that evolved independently (Table 6.3). Matching does not require that trait means within a community follow any particular distribution predictable *a priori* (e.g. constant size ratios). It requires only that trait distributions tend to be repeated between communities evolving in similar environments. Significant matching implies that species are nonrandomly apportioned to niche categories, which implies in turn that species traits within a community are partly the outcome of interspecific interactions (Schluter, 1990). Two possible cases were discussed in Chapter 3: the largely independent origin of the same *Anolis* ecomorphs on different large islands of the Greater Antilles (Table 3.3); and the repetitive origin of planktivorous and benthic fish species pairs in postglacial lakes (Table 3.2). Matching is not the same as convergence, which only requires that the communities being compared are more similar than was the case between their ancestors (Schluter, 1986, 1990; Schluter and Ricklefs, 1993).

6.3.2 Criteria for character displacement

I scored the observational evidence for character displacement by the extent to which individual cases satisfy six criteria (Tables 6.1–6.3). These six are taken from a compilation and reworking by Schluter and McPhail (1992) of the most important ones discussed by earlier researchers (e.g. Grant, 1972; Arthur, 1982; Strong *et al.*, 1979; see also Taper and Case, 1992*b*). Cases that fulfilled none of these criteria were discarded, as were cases that failed a test of any criterion no matter how many of the others were satisfied. The six criteria are:

(1) *Phenotypic differences between populations and species should have a genetic basis.*

This criterion applies mainly to shifts in means between sites of sympatry and allopatry (Table 6.1), which must be shown to be genetically based. This is not the same as showing that the trait in question has significant heritability, which only indicates within-population genetic variation. Demonstrating that a trait is highly susceptible to environmental induction within a site does not rule out a genetic basis to shifts between sites.

Differences between populations often include a genetic component and it is likely that most of the cases in Table 6.1 will eventually survive this test. The *Hydrobia* snails may eventually fail it. Direct competition between *H. ulvae* and *H. ventrosa* induces transitions in growth rate that lead to a larger shell in the former and a smaller shell in the latter (Gorbushin, 1996). A full common garden experiment using the original sympatric and allopatric populations has not yet been carried out (but see Grudemo and Johannesson (1999) for an experiment using other populations), and I have retained the case in Table 6.1 pending the outcome. I assume that the genetic criterion is met in all cases of trait overdispersion (Table 6.2) and species-for-species matching (Table 6.3). These patterns are based on means of species living in sympatry, and whose differences must therefore be partly genetic.

Table 6.1 Examples of character displacement between close relatives: exaggerated divergence in sympatry. Criteria 1–6 are those listed in the text. 'Ratio' is the ratio of means of the two species in allopatry (a) and sympatry (s). Symmetry ('Symm') is the ratio of the displacements of the two species, the smaller divided by the larger; each displacement is the logarithm of the ratio of trait means in sympatry and allopatry. Symmetry ranges from 0 (only one of two species shifts from allopatry to sympatry) to 1 (both species shift equal amounts)

Species	Trait	Ratio (a)	Ratio (s)	Symm	Criteria 1[a] Genetic	2[b] Chance	3[c] Diverge	4 Resource	5 Controls	6[d] Compete	Source
Vertebrates											
• *Gasterosteus aculeatus* complex	Body size	1	1.55	0.89	L	M	R	Diet, habitat	Predators	E	Schluter and McPhail (1992); Schluter (1994); Pritchard and Schluter (manuscript)
	Body shape	1	1.98	0.57					Lake size		
• *Catostomus discobolus* C. *platyrhynchus*	# Gill rakers	—	—	—	—	M	I,R	—	Physical[e]	—	Dunham et al. (1979)
• *Coregonus sardinella* C. *clupeaformis*	# Gill rakers	—	1.91	—	—	M[f]	I	—	—	—	Lindsey (1981)
• C. *clupeaformis* Atlantic C. *clupeaformis* Acadian	Body size	—	1.82	—	—	M	I,R	—	—	—	Fenderson (1964); Bernatchez and Dodson (1990)
	# Gill rakers	—	1.15								
• *Lepomis gibbosus* L. *macrochirus*	Body shape	—	—	—	—	M	I,R	Diet, habitat	Predators Lake size	E	Robinson et al. (1993); Werner and Hall (1976, 1977, 1979); Robinson and Wilson (1996, 2000)
• *Paragalaxias dissimilis* P. *eleotroides*	# Gill rakers	1.10	1.75	—	—	—	R,P	Diet, habitat	—	—	McDowall (1998)
	Body shape	—	—								
• *Raja erinacea* R. *ocellata*	Number of tooth rows	—	1.68	—	—	—	—	Diet	—	—	McEachran and Martin (1977)
• *Plethodon cinereus* P. *hoffmani*	Squamosal/ dentary ratio	1.04	1.35	0.70	—	M	R	Diet	—	—	Adams and Rohlf (2000)
• *Spea bombifrons* S. *Multiplicata*	Polyphenism	—	—	—	L	M	R	Diet	—	E	Pfennig and Murphy (2000)
• *Anolis wattsi* group A. *bimaculatus* group	Jaw length	1	1.81	0.64	—	M,P	R,P	Micro-habitat	—	E	Schoener (1970); Losos (1990c); Pacala and Roughgarden (1982, 1985); Rummel and Roughgarden (1985)

Species	Trait									Reference
• Cnemidophorus tigris C. hyperythurus	Body length	1.04	1.52	0.82	—	M,P	I,R,P	Prey size	—	—	Case (1979); Radtkey et al. (manuscript)
• Typhlosaurus lineatus T. gariepensis	Head length	—	1.24	—	—	S	I,R	Diet	—	—	Huey et al. (1974)
• Phelsuma sundbergi[h] P. astriata	Body length	1.12	1.45	0.19	—	M	I,R,P	Diet	—	D	Radtkey (1996)
• Geospiza fortis G. fuliginosa	Beak depth	1.19	1.56	0.72	F	M	I,R	Diet	Seeds	F	Grant and Schluter (1984); Schluter et al. (1985); Grant et al. (1985)
• Rollandia microptera R. rolland	Beak length	—	1.71	—	—	S	I,R	Diet	—	—	Fjeldså (1983)
• Podiceps taczanowskii P. occipitalis	Beak length	—	1.66	—	—	S	I,R	Diet	—	—	Fjeldså (1983)
• Podiceps gallardoi P. occipitalis	Beak length	—	1.15	—	—	S	I,R	—	—	—	Fjeldså (1983)
• Podiceps griseigena P. cristatus	Beak length	1	1.24	0	—	S	I,R	Diet	—	—	Fjeldså (1983)
• Tachybaptus ruficollis T. novaehollandiae	Beak length	1.09	1.32	0.91	—	S	I,R	—	—	—	Fjeldså (1983)
• Myzomela pammelaena M. sclateri	Body mass$^{1/3}$	1.06	1.17	0.07	—	M	I,R	—	—	—	Diamond et al. (1989)
• Dendroica dominica D. pinus	Beak length	1.01	1.22	0.21	—	—	I,R	Microhabitat	—	—	Ficken et al. (1968)
• Parus ater Two other Parus	Body mass$^{1/3}$	—	1.06	—	F	—	I,R	Microhabitat	—	D	Alatalo et al. (1986); Gustafsson (1988); Alatalo and Gustafsson (1988)
• Talpa europea T. romana	Skull size	1.05	1.03	—	—	—	I	—	—	—	Loy and Capanna (1999)
• Neomys fodiens N. anomalus	Jaw size Jaw shape	1.02 —	1.04 —	0 0	—	—	I	—	—	—	Racz and Demeter (1999)

Table 6.1 (*Continued*)

					Criteria						
Species	Trait	Ratio (a)	Ratio (s)	Symm	1[a] Genetic	2[b] Chance	3[c] Diverge	4 Resource	5 Controls	6[d] Compete	Source
● *Sorex minutus* *S. araneus*	Jaw size	1.28	1.41	0.20	—	—	I,R	—	—	—	Malmquist (1985)
● *Apodemus sylvaticus* *A. flavicollis*	Body length	—	1.08	—	—	M	I	—	Climate	—	Angerbjörn (1986)
Invertebrates											
● *Hydrobia ulvae* *H. ventrosa*	Shell length	1.03	1.53	0.72	—	M	I,R	Diet	Habitat[i]	E	Fenchel (1975); Fenchel and Kofoed (1976); Saloniemi (1993); Gorbushin (1996)
● *Littorina saxatilis* *L. arcana*	Whorl width	—	1.03	—	—	M[j]	I	—	—	—	Grahame and Mill (1989)
● *Poecilozonites discrepens* *P. circumfirmatus*	Spire height Shell shape	1.08	1.25 1.25	0.45 0	—	M	I	—	—	—	Schindel and Gould (1997)
● *Mandarina hahajimana* other *Mandarina*	Shell length, height, and colour	—	—	—	—	—	I,R	Habitat	—	—	Chiba (1996, 1999*b*)
● *Chalcosoma atlas* *C. caucasus*	Body length	1.09	1.27	0.88	—	M	I,R	Habitat	—	—	Kawano (1995)

Protozoa

Species	Trait										Reference
• *Eucyrtidium matuyamai* / *E. calvertense*	Body width	1.09	1.59	0.86	—	S	I	—	—	—	Kellogg (1975)

Plants

Species	Trait										Reference
• *Stylidium diuroides*[k] / other *Stylidium*	Flower column reach	—	1.77	—	—	M	I	—	—	—	Armbruster *et al.* (1994)
• *Dalechampia scandens* and other *Dalechampia*	Glandarea Gland-stigma distance	—	—	—	—	M	I	Pollinators	—	—	Hansen *et al.* (2000)
• *Erodium cicutarium*[l] / *E. obtusiplicatum*	Competitive ability	1.11	1.28	0.19	L	—	I	—	All	E	Martin and Harding (1981)

[a] L, laboratory test; F, field test.

[b] M, significant shift in distribution of population means between sympatry and allopatry; P, shifts implied by phylogeny; S, abrupt changes in space or time at sympatry–allopatry boundary.

[c] I, intraspecific variation; R, species means in sympatry exceed allopatric extremes; P, shifts inferred along a phylogeny.

[d] E, field experiments; D, negative correlations in density; F, food limitation and depletion.

[e] Physical controls include elevation, discharge, gradient, latitude, and longitude.

[f] Modal counts compared between lakes having cisco and those without: $F_{1,19} = 38.67$, $P < 0.0001$.

[g] Symmetry is the ratio of competition coefficients in narrow sympatry. Authors argue competition is entirely interference, but that it evolved in response to exploitative competition.

[h] Contrast is between old and recent sympatry between species.

[i] Controlling for habitat (four types) did not alter the statistical effect of *H. ventrosa* on the size of *H. ulvae*; controlling for shelter index did reduce the effect, but not significantly.

[j] Comparison of means of four sympatric and six allopatric populations: $F_{1,8} = 7.95$, $P = 0.022$.

[k] At least some of the displacement is in pollen placement on the same pollinators, suggesting reproductive interference is at least part of the story.

[l] Competitive ability was measured by seed output when each species was equally frequent. Competition experiments were done in the lab, not in wild. Authors controlled for all environmental differences during test of competition, not for environmental differences between sites of sympatry and allopatry.

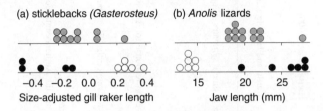

Fig. 6.3 Exaggerated differences in sympatry between sticklebacks (*Gasterosteus spp.*) in small lakes of coastal British Columbia (a) and between *Anolis* lizards on islands of the northern Lesser Antilles (b). Symbols are population means. Each species indicated by an open circle co-occurs with a second species indicated by a filled circle. Shaded circles represent allopatric populations. Body shape in (a) is a composite variable representing the major axis of variation between populations in size-adjusted traits. Populations on the right have many, long gill rakers, narrow gapes, and thin bodies compared with those on the left. Points are averages of male and female means. *Gasterosteus* data are from Schluter and McPhail (1992). *Anolis* jaw length measurements are from Schoener (1970) and Losos (1990c), males only.

Nongenetic shifts in sympatry are interesting in their own right (Robinson and Wilson, 1994). They might facilitate character displacement by generating an initial difference in sympatry. Plasticity itself may even evolve in response to competition. The sole example is the spadefoot toad, *Spea multiplicata*, whose tadpoles show a genetically based reduction in expression of a carnivore phenotype when sympatric with *S. bombifrons* (Pfennig and Murphy, 2000).

(2) *Chance should be ruled out as an explanation of the pattern.*

Two species occurring together on one island and living separately on two other islands are expected to exhibit greater divergence in sympatry one time in four if population means are assigned randomly to islands. The possibility of being misled at such a high rate stresses the need to rule out chance associations. This requires a number of replicate populations and a demonstration that the frequency distribution of population means differs between sympatry and allopatry. Tests of this criterion are currently tentative because they do not correct for nonindependence of populations arising from phylogeny or from gene flow between nearby populations. Hansen *et al.* (2000) have developed a test to overcome these problems but it is not yet in wide use.

Trait overdispersion is tested by comparing the pattern of size differences between species in a community with that in a 'null' distribution of phenotypes generated by random sampling from formal probability distributions (e.g. uniform or lognormal) or from an empirical distribution obtained by randomization of existing means. Choice of the null distribution is critical, and much of the debate has centred on the appropriateness of alternatives (Gotelli and Graves, 1996). Species-for-species matching is tested by showing that differences between two or more communities is significantly smaller than that expected in random assemblages (Schluter, 1990).

(3) *Population and species differences must represent evolutionary shifts and not just species sorting.*

Resource competition may give rise to the patterns in Tables 6.1–6.3 by either of two mechanisms. One is evolutionary divergence in sympatry, which is character displacement when resource competition is the cause. The other is species sorting driven by extinction, which is not an evolutionary process (Strong *et al.*, 1979; Case and Sidell, 1983; Waser, 1983). The two processes are related because character displacement involves 'sorting' of genotypes rather than species. Ecological circumstances favouring one of the processes likely also favour the other, and sorting may usually occur alongside true character displacement (e.g. Grant and Schluter, 1984; Schluter and Grant, 1984; Losos, 1990c; Armbruster *et al.*, 1994). I assumed this criterion to be met if one or more of the following conditions were met: differences between sympatry and allopatry were intraspecific; difference between species means in sympatry exceeded range of means in allopatric populations; or shifts were implied by character changes on a phylogeny.

(4) *Shifts in resource use should match changes in morphology or other phenotypic traits.*

Competition is assumed to arise from depletion of shared resources, hence a link between trait means and similarity in resource use must be shown. The appropriate test depends on the type of evidence. Tests of exaggerated divergence in sympatry (Table 6.1) require a demonstration of a shift in resource use between sympatry and allopatry. Tests of trait overdispersion and species matching (Tables 6.2, 6.3) require evidence that spacing of resource utilization curves correlates with the pattern in morphology. In Table 6.2 we require a demonstration that spacing in resource utilization underlies the pattern in morphology.

(5) *Environmental differences between sites of sympatry and allopatry must be controlled.*

Uncertainty over hidden causes is the curse of observational study. It is never possible to rule out all environmental agents that might cause the pattern except competition, but progress should be made in ruling out the most obvious alternative agents (e.g. resource availability). Resource differences between sites of sympatry and allopatry are known in several of the cases listed in Table 6.1 (food types in *Sorex*, habitat in *Typhlosaurus*) but are thought not to be responsible for greater differentiation in sympatry. I scored cases by whether or not the hypothesis of character displacement survived at least one attempt to incorporate these differences. For example, the case for character displacement in Galápagos ground finches has withstood incorporation of differences between islands in seeds, their main resource (Schluter and Grant, 1984; Schluter *et al.*, 1985). Variation in a specific habitat variable (shelter index) accounts for most of the observed shifts in *Hydrobia* mud snails (Saloniemi, 1993) but displacement did not disappear entirely. I assume this criterion is met in all

Table 6.2 Examples of character displacement between close relatives: trait overdispersion. The number of sympatric congeneric species in the comparison is given in parentheses. A congeneric set may represent a portion of a more diverse assemblage that also includes noncongeners (see Table 7.1 in the following chapter). Multiple sets of congeners are listed under a single case if the test combined their information; they are listed separately if they represent independent communities. 'Ratio' is the mean average ratio of adjacent congeners (larger divided by smaller). Criteria 1 and 5 were considered fulfilled in all cases (see text for justification)

			Criteria					
			1[a] Genetic	2[b] Chance	3[c] Diverge	4 Resource	5 Controls	6[d] Compete	
Species	Trait	Ratio							Source
Vertebrates									
• *Calidris* (9)	Beak length	1.08	F	M	—	—	All	—	Eldridge and Johnson (1988)
Tringa (3)	Beak length	1.22							
• *Accipiter* 12×(2)[e]	Wing length	1.38	F	M	—	Prey size	All	—	Schoener (1984)
Micrastur (2)	Wing length	1.52							
• *Cathartes* (3)	Skull length	—	F	M	—	—	All	—	Hertel (1994)
• *Gyps* (3)	Skull length	—	F	M	—	—	All	—	Hertel (1994)
• *Gyps* (4)	Skull length	—	F	M	—	—	All	—	Hertel (1994)
• *Anas* (2–5)	Body size	—	F	M	—	Feeding depth	All	—	Pöysä *et al.* (1994)
• *Geospiza* (2–5)	Beak depth	1.43	F	M	I	Seed size	All	F	Case and Sidall (1983); Grant and Schluter (1984); Grant *et al.* (1985)
	Beak length	1.37							
• *Camarhynchus* (2–4)	Beak length	1.37	F	M	—	Prey size	All	—	Case and Sidall (1983);
Many bird genera	Beak and body shape	—	F	—[f]	—	Diet, foraging	All	—	Karr and James (1975)
• *Felis* (3)	Gape	1.37	F	M	—	—	All	—	Kiltie (1984)
• *Felis* (6)[g]	Canine diameter	1.13	F	M	—	—	All	—	Dayan *et al.* (1990)

Taxon	Character	Ratio							References
• *Vulpes* (3)	Skull length	1.17	F	M	I	—	All	—	Dayan et al. (1989b); Dayan et al. (1992)
	Carnassial length	1.19							
Canis (2)	Skull length	1.31							
	Carnassial length	1.33							
• *Mustela* (3)[g]	Skull length	1.13	F	M	I	Prey size	All	—	Dayan et al. (1989a); Dayan and Simberloff (1994a)
	Canine diameter	1.20							
• *Gerbillus* (3)[h]	Tooth row length	1.21	F	M	—	Diet	All	E	Yom-Tov (1991); Abramsky et al. (1990)
	Skull length	1.15							
Meriones (2)	Tooth row length	1.31							
	Skull length	1.12							
• *Dipodomys* (3–4)	Incisor width	1.18	F	M	I	—	All	E	Dayan and Simberloff (1994b); Brown and Munger (1985); Heske et al. (1994)
	Pouch volume$^{1/3}$	1.15							
Perognathus (3)	Incisor width	1.19							
	Pouch volume$^{1/3}$	1.12							
Chaetodipus (4)	Incisor width	1.09							
• *Gerbillus* (2)	Incisor width	1.06	F	M	—	—	All	—	Parra et al. (1999)
• *Hylomyscus* (2)	Incisor width	1.24	F	M	—	—	All	—	Parra et al. (1999); Gautier-Hion et al. (1985)
	Incisor radius	1.20							
• *Funisciurus* (2)	Incisor width	1.11	F	M	—	—	All	—	Parra et al. (1999); Emmons (1980)
• *Calomys* (2)	Incisor width	1.15	F	M	—	—	All	—	Parra et al. (1999)
Akodon (2)	Incisor width	1.09							
• *Pipistrellus* (2)	Tooth row length	1.32	F	M	—	—	All	—	Yom-Tov (1993a)
	Skull length	1.20							
Rhinolophus (2)	Tooth row length	1.45							
	Skull length	1.32							
Taphozous (2)	Tooth row length	1.39							
	Skull length	1.36							

Table 6.2 (*Continued*)

			Criteria						
Species	Trait	Ratio	1[a] Genetic	2[b] Chance	3[c] Diverge	4 Resource	5 Controls	6[d] Compete	Source
• *Dasyurus* (4)[g]	Canine strength	1.51	F	M	I	Prey size	All	—	Jones (1997)
	Muscle size	1.19							
	Prey size	—							
Invertebrates									
• *Odontochila* (2–5)	Mandible length	1.37	F	M	—	—	All	—	Pearson (1980)
• *Therates* (4)	Mandible length	1.27	F	M	—	—	All	—	Pearson (1980)
Plants									
• *Stylidium* (2–5)[i]	Flower column reach	1.72	F	M	I	—	All	—	Armbruster *et al.* (1994)
• *Acacia* (4)	Flowering time	—	F	M	—	Pollinators	All	—	Stone *et al.* (1996)
• *Heliconia* (5)	Flowering time	—	F	M	—	Pollinators	All	—	Stiles (1977); Cole (1981)

[a]F, Field test (differences evident in sympatry).
[b]M, significant overdispersion.
[c]I, intraspecific variation present.
[d]E, field experiments; F, food limitation and depletion.
[e]Pairs only analysed here, but Schoener (1984) also shows effects in trios and quartets.
[f]Congeners in seven families are nearest morphological neighbours less often than expected, but statistical significance not provided.
[g]Treats sexes as separate 'species'.
[h]Interspecific interference competition demonstrated in experiment.
[i]Ratio is the mean for sympatric populations having the same flower column position. Niche separation detected mainly involves differences in where pollen deposited on bees, suggesting that the effect may be due at least partly to reproductive interference.

Observational evidence • 141

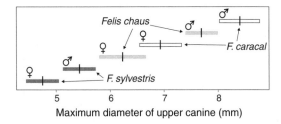

Fig. 6.4 Unusually constant size differences between adjacent cat species and sexes in the maximum diameter of the upper canine teeth. Means and standard deviations are shown. Modified from Dayan *et al.* (1990), with permission of University of Chicago Press.

cases of trait overdispersion and species matching because the species being compared are all sympatric and presumably have access to the same resource base.

(6) *Independent evidence should be gained that similar phenotypes compete for resources.*

Interspecific resource competition is the proposed cause of displacement and a case is greatly strengthened when species are shown to compete. This is important because alternative interactions may also produce exaggerated divergence in sympatry, trait overdispersion, and species matching including aggressive interference, reproductive interference, 'apparent' competition (competition for enemy free space) and intraguild predation. Divergence driven by aggressive interference might still be regarded as character displacement if the behaviour evolved in response to resource competition, but this too would need to be tested. Other interactions may also favour the evolution of interference (e.g. apparent competition and intraguild predation). I scored both experimental and observational evidence for competition.

Dytiscid beetles are an example of a case that failed this criterion. These beetles show significant overdispersion in body size but experimental study suggested that food limitation was absent at natural beetle densities (Juliano and Lawton, 1990*a*, *b*). If so, resource competition is not the mechanism behind trait overdispersion. The case was excluded from my tables only because here I am focussing on competition. A most interesting possibility is that another unidentified interaction has produced the divergence.

By using these six criteria I do not mean to imply that we should adhere to them exclusively. For now, they provide a useful measure of the completeness of the observational evidence for character displacement. They remind us that character displacement is not the only explanation for exaggerated divergence in sympatry, trait overdispersion, and species-for-species matching. Fulfilling a criterion means that the corresponding alternative hypothesis has been tested and rejected at least once. Fulfilling all criteria does not prove character displacement (we are dealing only with

Table 6.3 Examples of character displacement between close relatives: species-for-species matching. Multiple sets of congeners are listed under a single case if the test combined their information. Criteria are considered met if they are fulfilled in a subset of cases tested. Criteria 1 and 5 were considered fulfilled in all cases (see text for justification)

		Criteria						
		1[a]	2[b]	3[c]	4	5	6[d]	
Species	Trait	Genetic	Chance	Diverge	Resource	Controls	Compete	Source
• Pairs of sister fish species in postglacial lakes *Gasterosteus* (2) *Coregonus* E. Canada (2) *Coregonus* Yukon (2) *Salmo* (2) *Salvelinus* Scotland (2) *Salvelinus* Iceland (2) *Osmerus* (2)	Benthivore and planktivore	L,F	M[e]	I,R,P	Habitat	All	E	Table 3.2
• *Anolis* of Greater Antilles 3×(4–6)	Ecomorphs	F	M[f]	P	Perching habitat	All	E	Table 3.3
• Mediterranean climate finches *Pipilo* (2) *Carpodacus* (2) *Zonotrichia* (2) *Carduelis* (2) *Serinus* (2)	Body size	F	M[g]	—	—	All	—	Schluter (1986, 1990)

[a]L, laboratory test; F, field test.
[b]M, significant similarity (Schluter 1990).
[c]I, intraspecific geographic variation; R, species means in sympatry exceed allopatric extremes; P, shifts inferred along a phylogeny.
[d]E, field experiments.
[e]A preliminary test based on eight independent cases of species pairs in postglacial lakes (listed in Table 3.2) finds the distribution of species into ecotypes too even to have occurred by chance (contingency $\chi_5^2 = 0$, $P < 0.001$). The correct P is somewhat larger, pending inclusion of cases that violate the pattern (frequency unknown).
[f]The proportion of all species found in each ecomorph category is too similar between islands (first four ecomorphs only, $\chi_6^2 = 0.87$, $P = 0.01$).
[g]Differences among community mean body sizes are too small to be explained by chance (Schluter 1990).

observational evidence here) but a case is increasingly compelling as each of the most likely alternative hypotheses fails.

Robinson and Wilson (1994) regarded the need to fulfil all six criteria in every putative case as overly rigid and suggested that fulfilling a criterion in a subsample of cases would bolster the evidence in general. Their idea makes sense when applied to a body of data sufficiently homogeneous to constitute essentially a single large case (e.g. replicated divergence in fishes of postglacial lakes; Table 6.3). It applies less well to collections of cases that otherwise have little in common. The converse also holds: rejection of the case for character displacement in *Sitta* nuthatches (Grant, 1975) did not require that we downgrade other examples from birds. For this reason I adopt a case-by-case approach while striving to extract a general message.

6.3.3 *The evidence*

A search of the literature turned up 64 cases that include at least two congeneric species and that fulfil one or more of the six standard criteria (Tables 6.1–6.3). By 'case' I mean a unique pair of species (Table 6.1), a unique assemblage of species (Table 6.2), or a unique set of assemblages (Table 6.3). The summaries below are based on 61 cases only, as it does not include three new cases added in proofs (*Plethodon, Spea,* and *Dalechampia*; Table 6.1).

Many of these 61 cases are compelling. Twenty-three satisfy at least four of the six criteria and can be regarded as 'strong candidates.' Five cases fulfil all the criteria. More than 200 species are represented in these tables. Not all have undergone shifts, as divergence is sometimes highly asymmetric. Data in Table 6.1 indicate that more than three of every four species are shifted in sympatry. Both species are displaced in over half the documented pairs (displacement symmetry > 0.5) whereas shifts are mainly restricted to one of two species in the remainder (symmetry < 0.5). If this is a general rule, and character displacement is the cause, then the documented number of species affected is over 150. This count is itself probably conservative because it does not take account of multiple independent shifts in several of the cases (e.g. *Gasterosteus* (Schluter and McPhail, 1992) and *Cnemidophorus* (Radtkey et al., 1997)).

None of the putative cases is unassailable, as the evidence is all observational. The lists nevertheless indicate that patterns in support of divergent character displacement are not hard to find. They also suggest that alternative hypotheses to account for these patterns have not fared well when tested. Most of this supporting evidence is recent. Ten years ago only four cases would have made the 'strong candidate' grade.

What criteria are most often missing (Table 6.4)? The rarest evidence overall is that the species compete for resources. Much of the existing evidence for competition is observational, not experimental. While I counted all experiments in which a negative interaction between species was detected, in at least some of these cases resource competition was not confirmed as the basis of the result (other negative interactions could not be ruled out). The frequency of other missing criteria depends on the type of pattern. Tests of exaggerated divergence in sympatry frequently lack

Table 6.4 The number of cases of character displacement from Tables 6.1–6.3 fulfilling each of the six criteria for observational data. The total number of cases is 61

Criterion	Number of cases
Chance ruled out	51
Divergence, not sorting	39
Environmental controls	36
Genetic basis	33
Resource use	30
Species compete	13

Table 6.5 The number of cases of character displacement from Tables 6.1–6.3 according to the trophic status of species

Trophic category	Number of cases
Carnivore	35
Herbivore, granivore, one omnivore	14
Primary producer	5
Scavenger and detritivore	4
Microbivore	2

evidence of a genetic basis to divergence (Table 6.1), and are often missing controls for environmental differences between sites of sympatry and allopatry (controls for resource availability have been implemented in only a single case). Examples of trait overdispersion usually lack evidence that the pattern represents evolutionary divergence rather than species sorting and that the pattern in morphology matches differences in resource use (Table 6.2).

6.3.4 Trophic level bias

The most striking pattern in the observational data is the vast overrepresentation of carnivores compared with other trophic groups (Table 6.5). Herbivores (mainly granivores) are next followed by primary producers (plants). Examples of character displacement in microbivores and detritivores are very few. At least five explanations may account for this trophic level bias.

(a) Character displacement may be most frequent at trophic levels experiencing the strongest interspecific competition. Carnivores experience stronger competition than other tropic levels because theirs is the only level not affected by predation. In contrast, lower trophic levels are limited by both resources and predation.

This explanation has several counterarguments. Even top predators may be limited by parasites and disease (Marcogliese and Cone, 1997), not just resources. Many carnivores in the study are not top carnivores but intermediate-level predators that are themselves preyed upon. Many species in lower trophic levels are highly resistant to predation and experience mainly resource limitation (e.g. Osenberg and Mittelbach, 1996). The explanation also does not account for differences among the lower trophic levels in the frequency of cases (Table 6.5). There is no reason to think that resource limitation should be more prevalent in herbivores than their plants.

(b) The main cause of phenotypic and ecological differentiation is direct interference and intraguild predation rather than resource competition (cf. Polis *et al.*, 1989). Intraguild predation is likely to be most prevalent among the carnivores for obvious reasons.

Interspecific killing is common among mammalian carnivores, where cases involving similar-sized species predominate (Palomares and Caro, 1999). Perhaps this is also true of carnivores in other taxa. It is not clear how this explanation could account for variability in the number of cases of character displacement among the lower trophic levels.

(c) Carnivores may diverge more often than other trophic levels in response to competition because their resources are nutritionally substitutable. When resources differ in essential nutrients, as might be true for herbivores and plants, character displacement is more complex and may include convergence (Abrams, 1986, 1987, 1990).

Unfortunately, there are no good examples of convergent character displacement at any trophic level. Abrams (1996) argues that this is because nobody has looked.

(d) Greater species richness at the lower trophic levels dilutes the outcome of competition between pairs of species. Character displacement, if it occurs, is weaker and less detectable.

This possibility is contradicted by observations that closely related species of carnivores frequently exhibit character displacement even when they are part of a more diverse carnivore assemblage (see Table 6.2 and Chapter 7).

(e) Variation between trophic levels results from simple detection bias. Character displacement is most easily detected when species differ in easily measured morphological traits that map linearly onto resource utilization. This is more likely to be the case in carnivores than any other trophic level.

Detection problems are almost certainly partly to blame for trophic level bias even if they are not the full explanation. Simple relationships between morphology and resource use in carnivores are more abundant in the literature than relationships involving any other trophic level. Simple linear relationships are also common in granivores, which make up the majority of the herbivore cases (Table 6.5). Straightforward relationships between phenotypes and resource use are scarce in plants and detritivores. It is likely that physiological differences frequently distinguish closely related species of plants (especially vegetative differences) and herbivores (especially insect folivores) and these differences are less easily measured than beak, body, or tooth size.

6.3.5 Symmetry

Among cases of exaggerated divergence in sympatry, symmetry of divergence is related to the magnitude of the difference between trait means in sympatry (Fig. 6.5), but in a way opposite to that predicted by some theory. Average size-ratio in sympatry is greatest when displacement is relatively symmetric (mean 1.55, $n = 8$) and least when displacement is asymmetric (mean 1.29, $n = 7$) (Mann–Whitney $U = 51$, $P = 0.006$).

Asymmetry in divergence may reflect asymmetry in resource gradients. Yet, resource asymmetry is thought to facilitate divergence, not hinder it (Slatkin, 1980). This discrepancy is resolved if the most symmetric resource gradients in nature are also the broadest, since high resource breadth also favours large displacement (e.g. Slatkin, 1980; Doebeli, 1996). Resource gradients may have high effective breadth when potentially competing taxa are absent: some of the largest and most symmetric shifts are from depauperate environments (e.g. *Gasterosteus* sticklebacks, *Anolis* and *Cnemidophorus* lizards, and Galápagos ground finches *Geospiza*; Table 6.1).

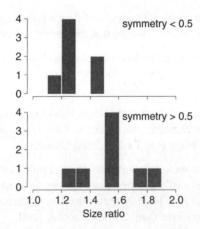

Fig. 6.5 Character ratios in sympatry according to the symmetry of character displacement. Data are from the subset of cases in Table 6.1 providing full sets of measurements in sympatry and allopatry. Symmetry is the ratio of the shifts of the two species between sympatry and allopatry, the smaller shift divided by the larger shift (see Table 6.1). Symmetry ranges from 0 (only one of two species shifts from allopatry to sympatry) to 1 (both species shift equal amounts).

6.3.6 Character ratios

Character ratios after displacement (i.e. in sympatry) are variable and not always large (range: 1.03–1.98; Tables 6.1, 6.2), with a mean ratio over all the data of 1.33. The range is similar to that in random assemblages (Eadie *et al.*, 1987). Although a formal analysis of observed ratios is not presented, it would not be reckless to conclude that the magnitude of character ratios provides little information about whether size differences in sympatry are the result of character displacement.

6.3.7 Remarks

The prevalence of any pattern in nature is difficult to assess from a literature that tends to present only the positive cases. Might the list not exaggerate the prevalence of character displacement? After all, many of the most familiar and spectacular adaptive radiations lack table entries. Not a single Hawaiian taxon and only one Galápagos taxon is represented. Character displacement has not been described for any of the cichlid radiations. Do the tables therefore overstate the case?

Possibly, but a countervailing influence also needs to be weighed: that character displacement, when present, infrequently leaves a clear trace in the form of exaggerated divergence in sympatry, trait overdispersion, or species-for-species matching. The argument for this view is that observational evidence for displacement requires special conditions that are infrequently met, and whose absence blurs the picture. Detecting exaggerated divergence in sympatry requires an effect sufficiently strong to overwhelm any environmental differences between allopatric and sympatric sites, which may be substantial. Statistical tests of trait overdispersion are weak (Losos *et al.*, 1989). They usually assume that displacement is confined to single traits that map linearly onto resource utilization. Finally, tests of overdispersion take the morphological diversity exhibited by all species as given, and ask only how it is arranged within these fixed extremes. They cannot test whether competition has played a role in generating this diversity in the first place. Tests of matching suffer similar problems.

These are among the reasons why observational evidence for character displacement will never allow us to fully assess the role of competition between close relatives in phenotypic diversification. Some of these weaknesses are overcome in two additional types of evidence, to which I now turn.

6.4 Evidence from prediction

A second type of evidence for divergent character displacement comes from successful prediction of mean phenotypes using models of adaptive evolution that incorporate interspecific competition. An estimate of natural selection pressures stemming from resources—the resource function or adaptive landscape—is a requirement for prediction. Character displacement is tested by determining whether the incorporation of competition significantly improves the prediction over that expected from resources alone.

6.4.1 The positive response to resources

The theory of divergent character displacement relies on the 'positive response' to resource density, which is the tendency of organisms to evolve a greater utilization of more abundant resources and a lower utilization of rarer ones (Abrams, 1986, 1987). Existence of such a response has rarely been tested on its own.

Aquatic isopods show the positive response (Fig. 6.6). Individuals of two species from allopatric populations exhibit enhanced survival on those foods (fungi) that are the most abundant in the rivers they inhabit. High survival reflects an evolved capacity to tolerate and metabolize fungus-specific toxins and nutrients (Rossi *et al.*, 1983). In contrast, isopods do not survive well when raised on fungi that are rare in the wild. This pattern weakens in sympatry because each species specializes on a subset of the fungi it uses in allopatry. Nevertheless, the very abundant fungi continue to be used by both species (Rossi *et al.*, 1983; this case does not involve congeners and, therefore, is listed in Chapter 7 instead).

Peaks in adaptive landscapes for Galápagos ground finches (Fig. 5.10) reflect peaks in the underlying distribution of seed resource. A match between peaks and beaks

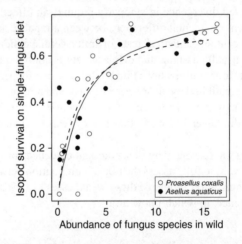

Fig. 6.6 The 'positive response' to resource abundance in two allopatric detritivorous isopods. The X-axis measures the abundance of 16 different fungus species in each of two rivers, measured as number of colonies/30 petri dishes. The Y-axis measures survival in the laboratory of isopods restricted to a diet of exactly one species of fungus. Survival was measured on F1 offspring of wild-caught parents. Curves are Michaelis–Menten equations fit by least squares (dashed line *P. coxalis*, $R^2 = 0.73$, solid line *A. aquaticus*, $R^2 = 0.60$). Data from Rossi *et al.* (1983).

implies that the finches have evolved to exploit the most abundant seed types available on islands. Peaks predict beak sizes well for solitary species: each beak is close to an adaptive peak, usually the highest (Fig. 5.10).

6.4.2 Evidence

The predictive approach has been tried only with *Cnemidophorus* lizards from Baja California (Case, 1979) and with Galápagos ground finches *Geospiza* (Schluter and Grant, 1984; Schluter *et al.*, 1985). Both studies used 'conditional population size' to predict mean phenotypes (Roughgarden, 1976; cf. Fig. 6.2). Case (1979) estimated the carrying capacity K for a range of body sizes of *Cnemidophorus* lizards using measures of insect abundance, diet overlap, and lizard population size. The estimated K curve had a single peak at 80 mm snout–vent length. Predicted optimal body sizes in sympatry were 55 and 90 mm. This agrees well with the observed means of the two *Cnemidophorus* species present at the site, 57 and 85 mm. The two species were not displaced equally to either side of 80 mm, the peak in K, possibly because larger individuals were competitively superior to smaller individuals (Case, 1979). A second pair of optimal body sizes in sympatry was found at 85 and 100 mm, which do not match the observed sizes.

Schluter and Grant (1984; Schluter *et al.*, 1985) used a similar approach, but the discrete nature of the resource base meant that a continuous competition function would not work well. Furthermore, the episodic nature of seed production on Galápagos islands does not satisfy the population dynamic assumptions of standard character displacement theory. Instead, abundance of resources available to a finch population of given beak size was simply discounted if other species also used it. Results were fairly robust to variations in the method of discounting. Simulation was then used to generate predicted beak sizes island by island (cf. Fig. 5.10). Predicted mean beak sizes were closely correlated with observed means.

6.4.3 Remarks

Predictive evidence is stronger than pattern evidence because it requires one to develop a specific alternative to the statistical 'null' hypothesis of no character displacement. The value of prediction might be usefully gauged by the amount of extra information needed to generate it. Prediction also addresses the origin of diversity not just its rearrangement. The predictions of Case (1979) and Schluter and Grant (1984) were not determined by the ranges of species sizes already present on the islands, and hence they are less prone to the problems of circular dependence that afflict some types of pattern evidence for character displacement (see Gotelli and Graves, 1996 for a review).

I have already discussed potential problems associated with using conditional population size to predict mean morphology during character displacement (e.g. Taper and Case, 1992*b*; Brown and Vincent, 1987). The prediction is therefore not expected to be robust. Nevertheless, the predictions were moderately successful in the lizard and finch studies, suggesting that asymmetric competition and other complications had little impact there.

6.5 Evidence from field experiments

Whereas field experiments on competition abound (Schoener, 1983; Connell, 1983; Gurevitch *et al.*, 1992), species thought to have undergone character displacement are not frequently represented. Nevertheless, experiments have confirmed a negative interspecific interaction in at least some putative cases of character displacement (Tables 6.1–6.3) where resource depletion is known or strongly suspected to be the underlying mechanism (e.g. *Lepomis, Dipodomys*, and *Erodium*).

Field experiments on character displacement, those testing more specific predictions of the hypothesis, are even rarer. Such experiments go further than classic competition experiments by exploring how interaction strength has changed during the putative character displacement sequence, and how natural selection pressures on species are altered when a second, closely related species is present in the environment. Below, I summarize work to date in this area. I focus on experiments on cases of character displacement in the wild. Taper and Case (1992*b*) summarize results of a laboratory experiment on character displacement in *Callosobruchus* beetles.

6.5.1 *Kinds of experiments*

Two kinds of experiments have been carried out to date. The first tests whether competition between species declined through time as divergence proceeded, yielding a level of interaction between present-day species that is a mere 'ghost' of its former strength (cf. Connell, 1980). The prediction is tested by contrasting the strength of competition between species under simulated 'pre-displacement' and 'post-displacement' conditions. For example, using field enclosures Pacala and Roughgarden (1985) showed that competition was stronger between the two sympatric species of *Anolis* lizards on the Caribbean island of St. Maarten, whose morphological separation is small, than between the morphologically more divergent species on St. Eustatius.

Pritchard and Schluter (manuscript) tested the prediction with three-spine sticklebacks, *Gasterosteus spp*. In nature, sympatric stickleback species in small lakes always consist of a zooplanktivore (the limnetic) and a benthivore (the benthic), whereas solitary species tend to be morphologically and ecologically intermediate (Fig. 6.3). Indirect evidence suggested that the solitary form represents the first stage in the evolution of a species pair, and that this form was displaced toward the benthic phenotype when the lake was later colonized by the ancestral marine species, a zooplanktivore (Schluter and McPhail, 1992). Using experimental ponds, Pritchard and Schluter (manuscript) showed that competition experienced by the ancestral marine three-spine stickleback was greater when it was paired with the intermediate species (pre-displacement) than when with the benthic species (post-displacement).

The second kind of experiment tests whether natural selection favours divergence between sympatric species because of competition for resources. In an experiment with three-spine sticklebacks, selection on the intermediate species when present alone was contrasted with that when a zooplanktivorous stickleback species (the limnetic) was added (Schluter, 1994). The latter treatment simulated the 'pre-displacement' conditions thought to have prevailed when certain lakes were colonized a second time by the marine ancestor. As predicted by the character displacement hypothesis, addition of the zooplanktivore generated selection in favour of the more benthic phenotypes within the intermediate species (Fig. 6.7). The effect diminished gradually with increasing morphological distance from the zooplanktivore such that no growth depression was detected in the most benthic phenotypes within the target species. The growth differential was steepest in the pond where final limnetic density was highest. This experiment was carried out within a single generation (fish were introduced as larvae in spring and retrieved in autumn as subadults). Selection, not evolution, was the measured feature. In a later experiment Schluter (manuscript) showed that divergent natural selection arising from interspecific competition was frequency-dependent. One treatment paired the intermediate species with a zooplanktivore

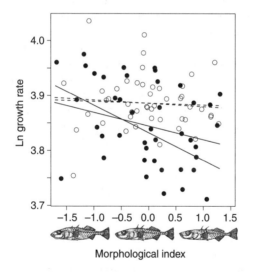

Fig. 6.7 Natural selection for divergence in a competition experiment. Symbols indicate growth rates of different phenotypes within an intermediate species in the presence (filled) and absence (open) of a zooplanktivorous competitor (a limnetic species). Data are pooled observations from experimental and control sides of two divided ponds. Lines are regressions of ln(growth rate) on morphology for each pond in the presence (solid) and absence (dashed) of the planktivore. The morphological axis distinguishes more benthic-like individuals on the left from more limnetic-like individuals on the right, in arbitrary units. Each symbol is an average of three adjacent points. Reproduced from Schluter (1996), with permission of University of Chicago Press.

(the limnetic) as before, whereas the second treatment paired the intermediate species with the benthic species. As predicted, selection pressures on the intermediate species differed between the two treatments, with the phenotypes most different from the added competitor being favoured in each case.

6.5.2 Remarks

Field experiments are a final frontier of evolutionary study and are still rare. To a large extent we continue to rely on indirect evidence when assessing the role of divergent character displacement in adaptive radiation. This state of affairs should change dramatically in the next decade as further experiments are attempted.

6.6 Other interactions promoting divergence

Interspecific competition for resources is the main interaction driving divergence, according to the ecological theory of adaptive radiation. The above sections present substantial evidence for competition's role in divergence. But, is this traditional view of divergence complete? Closely related species interact in a great variety of ways many of which may also promote divergence. Indeed, many of the putative examples of character displacement summarized in Tables 6.1–6.3 might have causes other than resource competition. Sharing a predator, for example, can yield an antagonism between prey species whose ecological and evolutionary effects qualitatively mimic those of interspecific competition (Holt, 1977; Doebeli, 1996; Abrams, 2000). Competition and 'apparent competition' are interactions in which neither party benefits. However, exploitative interactions which benefit one or more of a set of closely related species (e.g. predation, mutualism, or indirect facilitation) may also enhance their divergence. This section briefly reviews the possibilities.

The literature on alternative interactions and divergence does not permit as complete a treatment as that for resource competition, for the simple reason that compelling examples are so few. Juliano and Lawton (1990a,b) have provided the only known case in which criterion 6 for character displacement, when tested, has failed. They suggested that mechanisms other than competition might be responsible for trait overdispersion in dytiscid beetles but no candidates have come forth. The shortage of positive evidence might mean that interactions other than competition are insignificant by comparison. However, the blame is just as likely to rest on the scant effort that has yet been dedicated to measuring their evolutionary consequences (Abrams, 1996, 2000). The theory of divergence under alternative interactions is also poorly developed (but see Abrams, 2000; Doebeli and Dieckmann, 2000). In the meantime, I am able to highlight only a small number of possible cases, in the hope that this area will receive more attention in future.

6.6.1 Divergence via apparent competition

'Apparent competition' (also known as 'competition for enemy-free space') occurs when increased numbers of one prey species intensify the predation rate on another prey species, reducing the latter's population size, sometimes to the point of elimination (Holt, 1977; Ricklefs and O'Rourke, 1975; reviewed in Jeffries and Lawton, 1984; Holt and Lawton, 1994). Short-term apparent competition may result if greater numbers of the first prey enhance general search efficiency by predators, but a longer-term effect requires an increase in predator population size (Holt and Lawton, 1994). Apparent competition can act on a relatively broad scale if predators move freely between patches containing prey; for example, it may occur between specialized prey species that never encounter one another (e.g. because they occur on different host plants). In reviewing this topic, I use the term 'predation' to refer generally to whole or partial consumption by predators, parasites, diseases, and herbivores.

A way to intuit the evolutionary consequences of apparent competition is to momentarily contemplate its inverse. When prey species are noxious or harmful to predators and use conspicuous colouration or behaviour to advertise this fact, interspecific *convergence* of signals is favoured (Müllerian mimicry: Turner, 1976; Gilbert, 1983; Mallet and Barton, 1989). Convergence is favoured because a lower diversity of signals improves predator learning and reduces the mortality rate. However, when prey are palatable and rely on crypsis, evasion, or other mode of escape then the presence of similar prey species may intensify total predation rate. If so, then prey species effectively 'compete' for a given escape mode and divergence may be favoured. Several examples are known in which this mechanism has led to diversification of colour patterns and behaviours within single species (Clarke, 1962; Owen, 1963; Owen and Whiteley, 1989). However, there are no strongly supported cases of divergence between species via apparent competition.

Apparent competition may promote divergence under a variety of scenarios according to the structure of environments, the types of traits under selection, the degree of specialization of predators, and the type of initial differences between prey species. Under many realistic circumstances, apparent competition favours parallel change in prey species rather than divergence (Abrams, 2000). A simple scenario for divergence is that of a prey species evicted from those of its habitats colonized by a close relative, as a result of increases in the densities of their common predators. This process would require an initial difference between prey species to account for their contrasting vulnerabilities to predation, but can precipitate substantial further differentiation. Perhaps many cases of divergence of closely related species along gradients of overall predation risk (e.g. Wellborn *et al.*, 1996) or between environments containing different predators (e.g. McPeek, 1990*a*) are the result of this type of mechanism. Such examples are not of themselves evidence for divergence via apparent competition however, because the same pattern may result from simple divergent natural selection between environments varying in predation risk, without predator-mediated interactions between prey.

Holt (1977) envisioned two prey species varying in a continuous trait z, where the rate of prey consumption by predators depends on prey position along this dimension. Increases in predator abundance intensify predation on phenotypes in the region of z where frequency distributions of prey species overlap, since total prey abundance is elevated in this region, possibly favouring divergence between prey. In a more elaborate setting, prey species compete for resources too and predators themselves evolve in response to divergence of prey (Brown and Vincent, 1992; see Fig. 4.5). Here the strength of both resource competition and apparent competition between prey species depends on their resemblance in phenotype, with the result that apparent competition complements resource competition and may yield divergence under conditions when competition alone would not. Under completely symmetric assumptions, numbers of each prey species at equilibrium are limited by both predation and resource competition. Less symmetric assumptions have not been tried but might lead instead to an evolutionary equilibrium at which one of the species is better at avoiding predation and is limited mainly by food, and the second species is a more efficient resource competitor but is susceptible to predation (Holt *et al.*, 1994; Abrams, 2000).

Once prey begin to diverge under apparent competition, competing predators may now persist and diverge from one another in turn, placing further pressure on prey that leads to yet another round of divergence (Brown and Vincent, 1992). This theoretical cycle of differentiation and diversity buildup in predators and prey is reminiscent of Whittaker's (1977) verbal argument for the 'mutual facilitation of diversity growth by interacting trophic groups' (p. 40). Whittaker imagined the process increasing indefinitely, but in Brown and Vincent's model each new round of diversity buildup and divergence is increasingly difficult, bringing the process to an eventual halt.

Promising examples from nature of divergence via apparent competition include: consistent separation between cryptic moths in appearance and behaviour (Ricklefs and O'Rourke, 1975); high diversity of nesting microhabitats in open-nesting bird assemblages (Martin, 1988); extraordinary leaf shape variation among *Passiflora* species attacked by egg-laying female *Heliconius* butterflies (Gilbert, 1975; Fig. 6.8); and the variety of gall shapes in oak gall wasps attacked by parasitoids (Askew, 1961). The majority of these examples invoke apparent competition to explain unusually high levels of interspecific differentiation in appearance. However, divergence in traits affecting handling efficiency by predators (e.g. body size) and in habitat use are also likely to be affected by apparent competition.

The problem with the pattern evidence so far is that little of it would satisfy criteria analogous to the standard six routinely applied to cases of character displacement arising from resource competition (Tables 6.1–6.3). These six criteria, suitably modified for apparent competition, would be as follows:

(1) Phenotypic differences between populations and species should have a genetic basis.
(2) Chance should be ruled out as an explanation of the pattern.
(3) Population and species differences must represent evolutionary shifts not just species sorting.

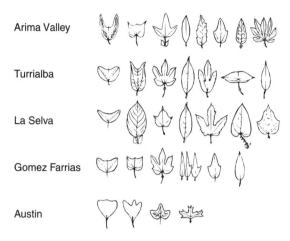

Fig. 6.8 The shapes of leaves of different species of passion flower vine (*Passiflora*). The hypothesis is that a different leaf shape in each *Passiflora* species minimizes detection by visually hunting female *Heliconius* butterflies searching for leaves on which to lay their eggs. Figure adapted from Coevolution of Animals and Plants edited by Lawrence E. Gilbert and Peter H. Raven, Copyright (c) 1975, 1980. By permission of the University of Texas Press.

(4) Shifts in susceptibility to predators should match changes in morphology or other phenotypic traits.
(5) Environmental differences between sites of sympatry and allopatry must be controlled.
(6) Independent evidence should be gained that similar phenotypes interact via apparent competition.

Most of the cases mentioned above satisfy criterion 1 but few would meet any of the others. Additionally, there have been no predictive or experimental tests of divergence via apparent competition. This seems meagre progress on an idea nearly 40 years old (Jeffries and Lawton, 1984) and which gained prominence over 20 years ago (Holt, 1977). On the other hand, most of the best evidence for character displacement is less than a decade old, having accumulated well after Brown and Wilson's (1956) seminal paper. This gives grounds for optimism that divergence via apparent competition will soon experience a focussed effort so that its importance can be better gauged.

6.6.2 *Mechanisms of divergence in food chains*

Competition and apparent competition are interactions mediated through different trophic levels. As such, the relative importance of each to divergence between species should depend on the position of lineages in food chains, if the importance of limitation by resources and predators also differs among food chain levels. In particular, divergence among species at the highest trophic level (e.g. carnivores or their parasites, but sometimes herbivores) should be influenced mainly by character displacement, whereas apparent competition should come into play at lower levels. Hairston, Smith, and Slobodkin (1960) suggested that species at the highest level in a food chain (e.g. carnivores) should grow in abundance until they are limited only by their prey, causing their prey (e.g. herbivores) to be mainly predator-limited, which should cause the next level down (e.g. primary producers) to be resource limited. Oksanen *et al.* (1981) extended this 'HSS hypothesis' to longer food chains.

An evolutionary extension of HSS to the problem of divergence via species interactions might predict classic character displacement to predominate at odd-numbered levels in the food chain (highest, third-highest, etc.), and divergence via apparent competition to prevail at the even-numbered levels. This evolutionary HSS suffers at least two counterarguments, however. The first is that the evidence for alternating control by predators and resources at different levels is not overwhelming (DeAngelis *et al.*, 1996; Osenberg and Mittelbach, 1996). The most significant objection is that every trophic level includes species highly resistant to predation, whose presence partly compensates for reductions in the numbers of more vulnerable species (Hunter and Price, 1992; Strong, 1992; Osenberg and Mittelbach, 1996; Leibold *et al.*, 1997). Also, the simple model ignores the fact that many species are omnivorous, taking prey from more than one trophic level (Spiller and Schoener, 1996). Consequently, most trophic levels may be limited by both resources and predation.

It might be more realistic to expect character displacement at the highest trophic level and mixtures of this and divergence via apparent competition at all lower levels. This outcome prevails in the models of Brown and Vincent (1992), although they did not include evolution in the resources exploited by the prey. A case for an evolutionary HSS would decidedly be easier to make with a more sizeable list of cases and their breakdown by trophic level. Such a list is not yet available.

A second counterargument is that trophic interactions engender a far more complex array of indirect interactions than simply competition and apparent competition. For example, sharing a predator does not automatically lead to apparent competition. It can simply lead to weaker competition or even apparent mutualism (Abrams and Matsuda, 1996). In a later section, I briefly consider interspecific facilitation and divergence.

6.6.3 Directly harmful interactions

Directly harmful interactions are underrated causes of ecological and morphological divergence between closely related species. By 'direct' I mean that individuals from two species consume one another or engage in reproductive and/or antagonistic interference. (Exploitative competition is indirect since it is mediated via other species at lower trophic levels or via abiotic nutrients.)

Antagonistic interference is widespread. Robinson and Terborgh (1995) attribute habitat partitioning between closely related Amazonian bird species primarily to interspecific aggression. Aggression is prevalent in at least some of the species sets listed in the character displacement tables (e.g. *Anolis*, *Parus*, and *Accipiter*), and it may occur in many of the other less well-studied cases. Interspecific aggression might favour convergence in behavioural signals (Cody, 1973) but it enforces spatial separation and hence may promote divergent natural selection on other traits, to the extent that it confines species to different environments.

Aggression itself may be an evolved response to resource competition, in which case the ensuing divergence is an extension of standard ecological character displacement. However, there are several reasons for exploring antagonistic interference further. It may promote divergence when resource competition by itself does not, or inhibit divergence despite resource competition. Interspecific aggression may evolve in response to processes other than resource competition, including 'apparent competition'. It may also occur as an incidental by-product of selection for aggressive behaviour in intraspecific interactions.

Interference may additionally arise in the context of reproduction and lead to reproductive character displacement or reinforcement, which is the evolution of traits that reduce hybridization frequency (Butlin, 1989; Howard, 1993; Liou and Price, 1994; Noor, 1995; Sætre *et al.*, 1997; Rundle and Schluter, 1998; Kirkpatrick and Servedio, 1999). Often the traits reducing hybridization frequency are ecological, such as habitat preference or pollinator type (Waser, 1983; Armbruster *et al.*, 1994), with the result that reproductive character displacement enhances niche differentiation between species. Some putative cases of ecological character displacement may instead be the result of this alternative mechanism (e.g. *Stylidium*; Armbruster *et al.*, 1994).

6.6.4 Exploitative interactions

The literature on interactions favouring divergence between close relatives pays most attention to mutual antagonisms, interactions leading to a decline of fitness in both parties. But species frequently (perhaps always) open up novel opportunities for other species which may evolve to exploit them, with diversification as the result (Whittaker, 1977). The question is whether exploitative interactions evolve between close relatives and whether they facilitate divergence in adaptive radiation.

Yucca moths (*Tegeticula*) provide a clear example (Pellmyr *et al.*, 1996; Pellmyr and Leebens-Mack, 2000). Most *Tegeticula* species actively pollinate flowers of a *Yucca* host shortly after depositing a small number of eggs in the unfertilized ovaries. Destruction of too many ovaries within a fruit provokes fruit abortion by the *Yucca* within a few days. But at least two species oviposit past the period of fruit abortion in fruits of flowers pollinated earlier by actively pollinating species. The number of seeds destroyed by these 'cheaters' has evolved to high levels. These seed predators are the result of two secondary losses of pollination behaviour within the yucca moth clade (Fig. 6.9). Eight of 13 pollinator lineages had matching nonpollinators of the same *Yucca* host (Fig. 6.9). Because the seed predator relies on prior pollination by other *Tegeticula*, its evolution is among the best evidence yet for divergence promoted by a facilitative interaction within a lineage.

158 • *Divergence and species interactions*

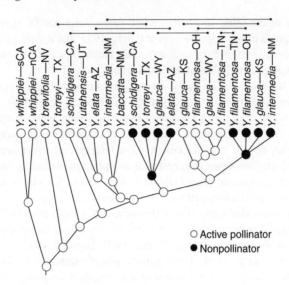

Fig. 6.9 Phylogeny of yucca moth populations (*Tegeticula*) classified by *Yucca* hosts exploited and geographical location. Most *Tegeticula* actively pollinate *Yucca* flowers (○) but two species have independently lost this behaviour ('cheaters'; ●). The nonpollinators oviposit into the fruits of flowers fertilized by actively pollinating yucca moths and their young eat the seeds. Ancestor state estimates assume that loss of active pollination is irreversible (Pellmyr *et al.*, 1996). Horizontal lines join pairs of active pollinators and seed predators exploiting the same *Yucca* host. Based on Fig. 1 in Pellmyr *et al.* (1996). The tree was altered so that each cheater species is monophyletic (cf. Pellmyr and Leebens-Mack, 2000).

In other cases, mutualists may facilitate the origin of other mutualists exploiting the same resource. For example, increased numbers of one flowering plant species may lead to a larger pool of pollinators which are then available for other species (Rathcke, 1983; Waser and Real, 1979; Stone *et al.*, 1996). Flowers of some species have taken the same route as nonpollinating yucca moths, by becoming nectarless 'cheaters' (e.g. Chase and Hills, 1992).

Other exploitative interactions that may evolve between close relatives include various forms of direct predation and parasitism. A rich variety of types of social parasitism is found in the social insects, especially the Hymenoptera (Wilson, 1971). These range from kleptoparasitism (theft of food) and temporary parasitism (queen substitution with eventual worker replacement) to slavery (theft of pupae) and inquilinism (permanent queen substitution and loss of the worker caste). As many as 10% of wasp species in certain genera are permanent parasites of closely related species (Wilson, 1971). Over a hundred parasitic ant species are known, and in the majority of cases the host is a close relative. The routes to parasitism are varied. In ants parasitism may have arisen from initially predatory, nearly neutral, or even mutualistic associations involving nest-sharing (Wilson, 1971).

Predation or parasitism of close relatives is relatively common in fishes. Interspecific predation occurs in the endemic cichlids of lakes Malawi, Tanganyika, and Victoria. About 40% of the cichlid species of Lake Victoria are fish predators, primarily of eggs and juveniles of other cichlids (Fryer and Iles, 1972). A total of nine or so species of cichlid fishes in these same lakes also eat scales of other fishes including other cichlids. Fryer and Iles (1972) suggest that scale-eating developed from herbivory (rock-scraping) in one case and from direct predation in other cases.

Predation between close relatives often involves species otherwise occupying the same trophic level ('intraguild predation'; Polis et al., 1989; Holt and Polis, 1997). This interaction apparently underlies many examples of behavioural niche shift (Polis et al., 1989). The resulting change in niche-specific selection pressures could precipitate evolutionary divergence in traits involved in niche exploitation, including morphology and physiology. The evolutionary consequences of intraguild predation have not been modelled or described.

6.6.5 The evolution of food webs

The above discussions focus mostly on interactions that promote divergence between closely related species inhabiting the same trophic level. Nevertheless, some of the most spectacular adaptive radiations in animals and plants have occurred where extant food webs are sparse, and which have resulted in clades of species distributed over several trophic levels. The significance of this process lies in its implications for the types of interactions that cause divergence. A theory that encompasses only competition for resources will clearly not go far toward explaining radiations that build food webs from scratch.

New trophic levels seem to evolve most rapidly in vertebrates. Transitions between herbivory (of seeds or leaves) and carnivory (especially of invertebrates) and between different levels of carnivory (consuming herbivores vs consuming carnivores) are the most common. The Galápagos finches have experienced one or two transitions between diets of primarily seeds or insects, as well as one switch from insectivory to folivory (Fig. 1.1). All of the herbivorous finches are omnivorous in the breeding season, when they also consume insects (Grant, 1986), and this omnivory may facilitate trophic level transitions.

Cichlid fishes are the supreme switchers. The monophyletic cichlid assemblage in the West African crater lake of Barombi Mbo has only 11 species but includes detritivores/microbivores, phytoplankton and macrophyte feeders, insectivores, and a predator of other fishes (Fig. 6.10). A new trophic level has emerged roughly every two or three speciation events, all within about one million years (Schliewen et al., 1994). The cichlids of the large East African lakes also include a large diversity of trophic levels (Fryer and Iles, 1972). Trophic level transitions are not uncommon in other fishes too, especially between intermediate carnivore and top carnivore levels (e.g. Westneat, 1995; Hynes et al., 1996).

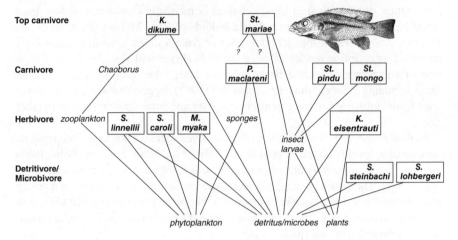

Fig. 6.10 Partial food web illustrating trophic relationships among the 11 endemic, monophyletic cichlids of Barombi Mbo, a crater lake in western Cameroon (Trewevas *et al.*, 1972; Schliewen *et al.*, 1994). The cichlids form an endemic monophyletic group that originated in sympatry within the past million years (Schliewen *et al.*, 1994). Omnivores are placed at the highest trophic levels to which they belong. Only predominant foods are indicated. *Stomatepia mariae* is a piscivore but its diet is unknown. Five other fish species inhabit the lake (two piscivores and three insectivores, the latter taking insects mainly from the water surface). This figure is a reduced version of a larger web in Trewevas *et al.* (1972). Image reproduced with permission of Cambridge University Press.

The rate of transition in invertebrates may not be as high, as indicated by the high taxonomic position at which trophic level variation is exhibited. The members of entire insect orders may be restricted to a single trophic level (Southwood, 1972). Phytophagy—feeding on living tissues of higher plants—has arisen 50 times or more in the insects from predaceous or detritivorous ancestry (Mitter *et al.*, 1988). While this sounds like a great many times, it is small next to the number of insect species, and is not among the most common ecological transitions between close relatives. Phytophagy has arisen more frequently in the Diptera than in any other insect order (Mitter *et al.*, 1988), yet almost all the Hawaiian *Drosophila* are saprophages of plants (larvae in one clade are parasites of spider eggs; Carson and Kaneshiro, 1976).

Trophic level transitions appear rare in plants, though there are singular exceptions. Carnivory is rare, but has apparently been gained once and lost once in *Brocchinia* bromeliads (Givnish *et al.*, 1997). Noncarnivorous species in this genus include a nitrogen-fixer and species that obtain nutrients from diverse sources of dead and decaying plant and insect debris.

The extent to which trophic level switches, and the interspecific interactions that may give rise to them, need to be considered in adaptive radiation may thus differ between taxa. When common, it is unlikely that the same interactions involved in transitions between levels (food web 'height') are the same as those within a level (food web 'width'). There is as yet little indication as to the typical direction of trophic level transitions. A transition to a higher trophic level might involve facilitation in at least one direction (e.g. prey on predator) but what interactions would favour a descent to a lower level? Increased focus on these questions would better illuminate the types of interactions that govern divergence generally.

6.7 Discussion

On balance, the data support a role for competition as envisioned in the ecological theory. The evidence is similar in quality to that for divergent natural selection discussed in the previous chapter: plenty of interesting and suggestive examples, most of them incomplete in some way, very little evidence to the contrary, plus a few cases in which the evidence is very strong. Though uncertainties remain, there is no longer any reason to dismiss the concept of competitive divergence in adaptive radiation as lacking support. The view still widespread in the ecological literature, that little or no evidence exists for character displacement, is becoming increasingly difficult to sustain. The results summarized here indicate that the view is outdated and should be abandoned.

Nevertheless, we still know far too little about the dynamics of competition and selection, the conditions that favour character displacement and the details of the process. We rely too heavily on studies of pattern for our evidence. The causes of variation in the form of divergence (symmetry, magnitude) are still somewhat obscure. There is also a strong trophic bias to the literature: most examples of character displacement are from carnivores. This raises the possibility that trophic position strongly influences the probability of divergence via character displacement, but detection bias probably plays some role in producing the pattern.

Beyond showing that character displacement is easy under varied circumstances, theory has not yet contributed much to empirical study of the process. Again this stems from the descriptive nature of most of the evidence, a state of affairs that should change as experimental studies become more common. Slatkin's (1980) quantitative genetic models initially appeared to show that character displacement rarely led to significant divergence, but further examination of the theory showed this preliminary result to be the outcome of overly restrictive assumptions about the shape and symmetry of resource functions on the one hand, and overly liberal assumptions about how easily trait variances change on the other (Slatkin, 1980; Milligan, 1985; Taper and Case, 1985; Doebeli, 1996). An important objection by Wiens (1977), that environments are too variable to favour divergence via character displacement, has also been shown to be unfounded. Theory shows that divergence can occur even in variable environments (Gotelli and Bossert, 1991). One of the most strongly supported cases of character displacement, in the Galápagos ground finches *Geospiza* (Table 6.1), has an extraordinarily variable environment as its setting (Grant, 1986; Gibbs and

Grant, 1987*a*, *b*). Connell (1980) suggested that competition would be too infrequent and inconsistent to effect much displacement in diverse communities. This is more an argument for expecting 'diffuse' displacement rather than none at all. In any case, it is a minor concern in adaptive radiations for two reasons: divergence often takes place in species-poor settings; and young species begin from the same ancestral phenotype, ensuring a high initial ecological similarity (and high potential for competition) between them. Abrams (1986, 1987, 1990) has outlined conditions under which one should expect outcomes other than divergence (e.g. parallel shifts and convergence) but tests of his ideas await the discovery of examples.

The most serious gap in our understanding of interactions in adaptive radiation is the dearth of information on the contribution to divergence of interactions other than resource competition. A role for apparent competition seems likely but examples are few and further tests are wanting. Consideration of the mechanisms of competition and apparent competition hint that the likelihood of each is related to trophic position of lineages in food webs, but no evidence yet bears on this. Divergence may also be aided by other direct and indirect interactions between close relatives, but here again the evidence is scant.

7
Ecological opportunity

> [S]ome time after a rather distinctive new adaptive type has developed it often becomes highly diversified. ... The same sort of diversification follows, and this case begins almost immediately, when a group spreads to a new and, for it, ecologically open territory.
>
> —Simpson (1953)

7.1 Introduction

In the most remarkable living examples of adaptive radiation, the span of resources utilized by the taxon surpasses typical bounds to include items normally taken by unrelated taxa missing or poorly represented in the region. For example, in addition to ordinary seeds finches on remote archipelagoes take foods otherwise exploited by warblers, woodpeckers, and honeyeaters in more diverse mainland faunas. The marsupial mammals attain their highest diversity in Australia where placental mammals are underrepresented, and the vast array of phenotypes to have evolved there include likenesses of wolves, shrews, moles, and other types of placental mammals that roam the other continents.

From such observations Lack, Mayr, Simpson, and others reasoned that competition from other taxa must hinder adaptive radiation whereas absence of competing taxa promotes it. They viewed phenotypic divergence and speciation in depauperate environments as a kind of release from the burden of competition imposed by other taxa. These ideas make up the bulk of the hypothesis of 'ecological opportunity'. Adaptive radiation requires more than an adaptive zone: the zone must also be empty or, at the very least, underutilized.

Colonizing a remote archipelago, surviving a mass extinction, or being in the right place when environments change are among the ways lineages may stumble upon an array of unexploited niches. Simpson (1953), however, attached great importance to an additional mechanism: the acquisition of evolutionary novelties ('key innovations'). This idea explained the apparent association between the first appearance of an ecologically useful trait and large-scale adaptive radiation. 'Acquisition, by any process, of such prospective adaptation may and frequently does lead to occupation of a new zone without involving physical movement of the group or ecological change around it. ... The horses that became grazers did not go anywhere to do so ...'

(p. 207). His key innovations hypothesis, then, is a special case of a general idea linking adaptive radiation with ecological opportunity.

In this chapter I evaluate evidence for the hypothesis of ecological opportunity and its putative mechanisms, mainly using adaptive radiations at low taxonomic level. My goal is to determine whether the processes hypothesized, largely inspired by large-scale patterns, leave their signatures on contemporary species where their mechanisms may be better studied.

7.2 Ecological opportunity and morphological divergence

As originally formulated, the concept of ecological opportunity involved both a wealth of resources and freedom from competition. Resources themselves are rarely measured and so most tests focus on the presence/absence of competitors. For example, Benton (1996*a*, *b*) asked whether the rise and expansion of individual families of tetrapods (amphibians, reptiles, birds, and mammals) was biased toward resources that were unused by other tetrapods extant at the time. By examining the coarse ecological characteristics (e.g. diet (carnivore, omnivore, or herbivore) and habitat (terrestrial, freshwater, marine, arboreal, arial, and subterranean)) of all 840 families having a fossil record and more than one species, he showed that over the long history of the tetrapods a clear majority of families (74–87%) arose in vacant niche categories, those not utilized by another tetrapod family at the same place and time.

Below I focus on several patterns in contemporary populations to address the question: do competitors in other lineages affect the direction and breadth of expansion in morphology and use of environments?

7.2.1 Divergent character displacement between distant taxa

Demonstrations of character displacement between distant taxa provide strong indications that members of one lineage constrain phenotypic evolution in members of another lineage. Observational study has uncovered at least 25 instances of divergent character displacement between noncongeners (Table 7.1). Fourteen of these cases were also listed in Tables 6.2 and 6.3 because the data sets also include congeners. As with the cases involving congeners, examples may be scored by the number of standard criteria met, which gives an indication of how well each case has fared when tested against simple alternatives. More than a third (11) satisfy four or more of the six standard criteria (i.e. are 'strong' candidates). Evidence that competition is the mechanism of interaction is again the criterion least often met.

The number of cases involving members of different taxa (genera or higher) is small compared with the list of examples involving only congeners. The frequency of cases drops still further with increasing taxonomic distance (Table 7.2). The majority of cases of character displacement between noncongeneric species fall into the nearest category of taxonomic similarity, species in different genera within the same family. Fewer cases are known from greater taxonomic distances. Perhaps the likelihood of competition or the strength of selection resulting from competition also falls with increasing taxonomic distance. If so, then ecological opportunity is principally freedom from competition from members of not-too-distant lineages. Alternatively, the decline may have more to do with biased research effort (evolutionary studies tend to focus on close relatives) and increased problems of detection with increasing taxonomic distance (character displacement between species in distant taxa likely involves dissimilar traits).

Often when unrelated species compete for a common resource one of the lineages, the one with a long evolutionary history of association with the resource, may be designated the incumbent (e.g. nectar-feeding bees) and the other may be designated the newcomer (e.g. nectar-feeding finches). In such cases the incumbent should show a smaller shift in sympatry than the newcomer (Schluter, 1990). Data to test this possibility are few. Nevertheless character displacement is asymmetric in the two instances listed in Table 7.1 in which shifts have been measured in both parties (*Asellus* vs *Proasellus*, and *Trifolium* vs *Lolium*).

7.2.2 Morphological divergence on islands and mainlands

Is morphological divergence accelerated in depauperate environments where competitors are fewer? I address the question using island–mainland comparisons. Finches are among the best-known taxa from this standpoint. When set against the Galápagos and Hawaiian finches, divergence in mainland finches seems slow and unremarkable whether calculated per unit time (Fig. 7.1) or per speciation event (Schluter, 1988c). This is so even when comparison is restricted to fully sympatric sets of species, which should equalize the opportunity for divergent character displacement. The pattern also holds when we confine attention to just the seed-eaters, a role fulfilled on Galápagos by the ground finches *Geospiza* (Fig. 7.1).

Seeds from just about every species of angiosperm on the Galápagos islands are eaten by one or another species of *Geospiza*. The few seed types not consumed are probably toxic (Schluter, 1988c). Similar data are unavailable for Hawaii because most finch species, including many of the granivores in *Psittirostra*, have been extirpated within historical times (Olson and James, 1982). However, the range of beak sizes on Hawaii was even greater than in Galápagos *Geospiza* (Schluter, 1998). In contrast, many mainland seeds go untouched by finches, particularly the largest but also often the smallest (e.g. Fig. 7.2). When a fairly broad range of seed sizes is exploited on mainlands, the species taking the larger seeds are in a different subfamily or even a different family than those taking the smaller seeds. Nevertheless the resource base is often similar between islands and mainlands in terms of size-frequency distribution (e.g. Fig. 7.2) and taxonomic composition.

The finch patterns are consistent with the hypothesis that adaptive radiation on islands represents release from competition. The most likely competitors for large and small seeds are other avian lineages (pigeons, parrots, quail) plus granivorous

Table 7.1 Examples of character displacement between noncongeners and more distantly related taxa. Some of the congeners included here are also listed in Table 6.2. Genera and families are listed multiple times under a single case if the test combined their information or if they have species or genera in common. Criteria 1 and 5 were considered fulfilled in all cases of trait over-dispersion (see text for justification)

Species	Trait	Greatest taxonomic distance	Criteria						Source
			1[a] Genetic	2[b] Chance	3[c] Diverge	4 Resource	5 Controls	6[d] Compete	
Exaggerated divergence in sympatry									
● *Coregonus hoyi*	Gill raker number	Order	—	—	I	Habitat, diet	—	D	Crowder (1984, 1986)
Alosa pseudoharengus	—								
● *Coregonus sp.*	Gill raker number	Order	—	M	—	Diet	Several[a]	D	Westman et al. (manuscript)
Several planktivorous fish	—								
● *Salvelinus fontinalis*	Gill raker length	Order	—	M	I	Habitat, diet	—	D,E	Magnan (1988); Lachance and Magnan (1990)
Catostomus commersoni	—								
● *Geospiza fuliginosa*	Tarsus length	Phylum	—	M	I	Diet	—	—	Schluter (1990)
Xylocopa darwini	—								
● *Loxia curvirostra*	Beak size	Class	—	—	I	Diet	Food availability	D,F	Benkman (1989, 1999)
Tamiasciurus hudsonicus	—								
● *Rhabdomys pumilio*	Skull length	Genus	—	S	I	—	Temperature	—	Yom-Tov (1993b)
Lemniscomys griselda	—								
● *Asellus aquaticus*	Diet	Genus	L	—	I	Diet	Food abundance	—	Rossi et al. (1983)
Proasellus coxalis	Diet								
● *Trifolium repens*	Competitive ability	Class	L	M	I	—	—	E	Evans et al. (1985, 1989); Turkington (1989); Lüscher et al. (1992)
Lolium perenne	Competitive ability								
Trait over dispersion									
● Shorebirds, 13 spp. in 3 genera	Beak length	Genus	F	M	—	—	All	—	Eldridge and Johnson (1988)

Taxa	Character	Level	F	M		Diet	All		Reference
Hawks, 7 noncongeneric pairs in 5 genera	Wing length	Genus	F	M	—	Diet	All	—	Schoener (1984)
Old World vultures 9 spp. in 7 genera 8 spp. in 5 genera	Skull length Skull length	Genus	F	M	—	—	All	—	Hertel (1994)
New World vultures 7 spp. in 5 genera 6 spp. in 6 genera	Skull length Skull length	Genus	F	M	—	—	All	—	Hertel (1994)
Felids, 4 spp. in 2 genera	Gape	Genus	F	M	—	—	All	—	Kiltie (1984)
Canids, 4 spp. in 4 genera Canids, 5 spp. in 2 genera	Carnassial length Skull length Carnassial length	Genus	F	M	—	—	All	—	Kieser (1995) Dayan et al. (1992)
Mustelids and viverrids[e] 6 spp. in 6 genera	Canine diameter	Family	F	M	—	—	All	—	Dayan et al. (1989a)
Hyaenas, 7 spp. in 7 genera	Carnassial length	Genus	F	M	—	—	All	—	Werdelin (1996)
Bats, 9 spp. in 6 genera	Tooth row length Skull length	Order	F	M	—	—	All	—	Yom-Tov (1993a)
Gerbils, 6 spp. in 3 genera	Tooth row length Skull length	Genus	F	M	—	Diet	All	E[e]	Yom-Tov (1991); Abramsky et al. (1990)
Heteromyids, 8 spp. in 3 genera Heteromyids, 7 spp. in 3 genera	Incisor width Pouch volume Incisor width	Genus	F	M	I	—	All	E	Dayan and Simberloff (1994a); Brown and Munger (1985); Heske et al. (1994)
Muridae, 6 spp. in 5 genera	Incisor width Incisor radius	Genus	F	M	—	Diet	All	—	Parra et al. (1999); Gautier-Hion et al. (1985)
Muridae, 7 spp. in 5 genera	Incisor radius	Genus	F	M	—	—	All	—	Parra et al. (1999)
Sciuridae, 7 spp. in 6 genera	Incisor width	Genus	F	M	—	Diet	All	—	Parra et al. (1999); Emmons (1980)
Rodents, 6 spp. in 5 genera	Incisor width	Family	F	M	—	—	All	—	Parra et al. (1999)

Table 7.1 (Continued)

Species	Trait	Greatest taxonomic distance	Criteria						Source
			1[a] Genetic	2[b] Chance	3[c] Diverge	4 Resource	5 Controls	6[d] Compete	
• Angiosperms, 11 spp. in 7 genera	Flowering time	Family	F	M	—	Time	All	—	Stiles (1977); Cole (1981)
Species-for-species matching									
• Mediterranean finches 20 spp. in 12 genera	Body size	Family	F	M	—	—	All	—	Schluter (1986, 1990)

[a] L, laboratory test; F, field test.
[b] M, significant shift in distributions of population means; P, phylogenetic test; S, spatial or temporal test, not rigorous.
[c] I, intraspecific geographic variation; R, species means in sympatry exceed allopatric extremes; P, shifts inferred along a phylogeny.
[d] E, field experiments; D, negative correlations in density; F, food limitation and depletion.
[e] Interference competition demonstrated in the experiment by Abramsky et al. (1990).

Table 7.2 The number of cases of character displacement according to the greatest taxonomic distance between the species included in the study. The category 'species' includes 61 cases involving congeners (i.e. those in Tables 6.1–6.3) minus 17 cases also listed in Table 7.1 because they include more distantly related species as well

Taxonomic distance	Total no. cases
Species	44
Genus	17
Family	4
Order	4
Class	2
Phylum	1
Kingdom	0

rodents and ants. Perhaps reduced predation also plays a role in morphological release on islands. The largest beaks in finches of Galápagos and Hawaii are larger per gram of body mass than is found on mainlands (Benkman, 1991). Yet the large beaks are cumbersome, and possibly the greater investment in body mass on mainlands that would be required to lift and rapidly accelerate such an unwieldy beak may reduce the value of the large beak itself when predators are present (Benkman, 1991). This could slow divergence at the top end of the beak size spectrum.

Greater morphological and ecological diversity on islands than mainlands is seen in many of the Hawaiian radiations, such as the Hawaiian *Drosophila* and their relatives (Carson and Kaneshiro, 1976), but the patterns are poorly quantified. For example, the variance of maximum heights is greater within the silversword alliance (Fig. 7.3(a)) and within the Hawaiian *Bidens* (Fig. 7.3(b)) than is seen in their mainland relatives.

The same pattern does not hold in *Anolis* lizards on large islands of the Greater Antilles and nearby mainland areas. Central America was colonized from the islands at least once, and one of these colonists went on to produce about 200 species. The island ancestor of this mainland clade is also the ancestor to the *Anolis* lineage that subsequently radiated on Jamaica (Irschick et al., 1997; Jackman et al., 1997). The mainland and Jamaican clades may even be sister taxa. A small fraction of the mainland species has been measured. Unexpectedly, the mainland clade is morphologically no less diverse than the Jamaican species (Fig. 7.4). Nor is the diversity of perch types used any less on the mainland (Irschick et al., 1997). The greater number of competing taxa (and predators) on the mainland seems not to have damped the rate of morphological evolution nor the resulting span of phenotypes.

7.2.3 Intraspecific release in northern lakes

Theoretical studies of character displacement also predict increases in trait variance when a competitor is removed, the amount depending on genetic assumptions and the

170 • Ecological opportunity

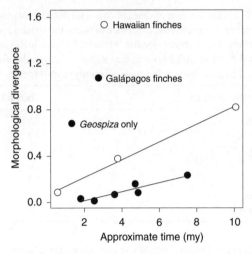

Fig. 7.1 Divergence in beak and body dimensions in the Galápagos and Hawaiian finches compared with mainland relatives, and with time. Symbols indicate emberizine (●) and cardueline (○) finch taxa. Time for each clade is twice the distance from the tip of the phylogeny to its deepest node (all phylogenies had contemporaneous tips), assuming that a million years is roughly 1/18 units of Nei's or Rogers' allozyme distance, and 0.5% sequence divergence of mtDNA (Zink, 1991; Klinka and Zink, 1997). Morphological divergence is the rate of morphological evolution multiplied by time. Rate is measured as the diffusion coefficient of a Brownian motion process fitted to the data, calculated using the method of Schluter *et al.* (1997) (Box 5.2). Rate is the sum of rates calculated for the first two principal components based on ln-transformed species means of four traits: wing length, tarsus length, beak length, and beak depth (separate principal components were carried out on each clade). Mainland emberizine points, from left to right, are: *Melospiza*, *Spizella*, *Zonotrichia*, *Ammodramus* (mtDNA), the larger '*Zonotrichia*' clade (including *Zonotrichia* proper, *Melospiza*, *Junco* and *Passerella*), and *Ammodramus* (allozymes). Mainland cardueline points, from left to right, are *Loxia curvirostris* complex, *Carduelis*, and the cardualine clade consisting of *Carduelis*, *Carpodacus*, *Coccothraustes* and *Loxia*. Phylogenies are from Yang and Patton (1981), Marten and Johnson (1986), Groth (1993), Zink and Avise (1990), and Johnson *et al.* (1991). Measurements are from Amadon (1950), Grant *et al.* (1985), Schluter (1988c, unpublished data) and C.W. Benkman (unpublished data).

breadth of resources that a given phenotype can efficiently consume (Roughgarden, 1972; Van Valen, 1965; Slatkin, 1980; Taper and Case, 1985; Doebeli, 1996). This result suggests another broad test of ecological opportunity: trait variance should be highest in depauperate environments where competitors are fewest.

The literature on changes in trait variance during character displacement has yielded mixed results. Variances are sometimes higher in depauperate environments

Ecological opportunity and morphological divergence • **171**

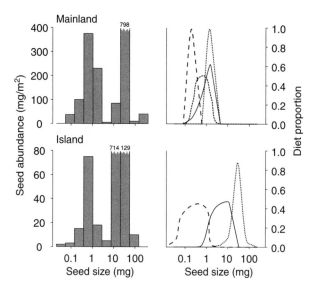

Fig. 7.2 Availability and use of seeds by finches on a Galápagos island and in a similar environment in Kenya. Galápagos finches from left to right are *Geospiza fuliginosa, G. fortis*, and *G. magniriostris*. Kenya species are *Estrilda rhodopyga, Plocepasser mahali, Passer griseus*, and *Dinemellia dinemelli*. Modified from Schluter (1988c).

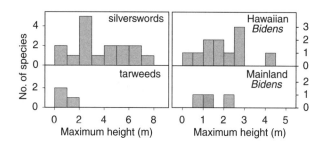

Fig. 7.3 Maximum heights of the Hawaiian silverswords (a) and Hawaiian *Bidens* (b) compared with mainland relatives. Maximum heights of Hawaiian plants were obtained from Wagner *et al.* (1990). Heights of three California tarweeds, which include the sister species to the silverswords, are from Munz and Keck (1970). Heights of three mainland *Bidens* (Helenurm and Ganders, 1985) are from Wagner *et al.* (1990) and Peterson and McKenny (1968).

and sometimes not. Few of the examples of character displacement listed in the tables (6.1–6.3 and 7.1) recorded greater variances in allopatry than sympatry. For example, solitary sticklebacks (Fig. 6.3(a)) are not as a rule more variable morphologically than paired species (D. Schluter, unpublished observations).

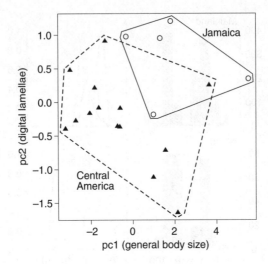

Fig. 7.4 Morphology of island (○) and mainland (▲) *Anolis* lizards. Each point represents a species mean. The island points are from Jamaica, the mainland ones are from Central America. Both clades share a comparatively recent common ancestor and may be sister taxa (Jackman et al., 1997). Axes are principal components extracted from the correlation matrix based on log-transformed species means for six morphological traits. pc1 is a general body size variable, whereas pc2 mainly reflects the number of lamellae found on toe pads. The two axes account for 86% and 8% of the total morphological variance. Data are from Losos (1990a) and Irschick et al. (1997).

Nevertheless, instances of unusually high trait variance, when found, tend to occur in depauperate environments. The most systematic survey has been carried out on 'trophic polymorphisms' in freshwater fishes of northern North America (Robinson and Schluter, 1999). A trophic polymorphism was defined as excessive phenotypic variation associated with intraspecific niche differentiation, where 'excessive' was judged by the authors of the articles describing each case. The distribution of phenotypes within each population was not necessarily bimodal. Instances of trophic polymorphism were found to be inversely related to total fish species diversity (Fig. 7.5). The highest incidence of these polymorphisms occurred in lakes of previously glaciated areas. Unfortunately, the genetic basis of enhanced variation has been little-explored. At least some of the variation was found to be genetic in one case investigated using common garden experiments (Robinson and Wilson, 1996).

To summarize, not all or even most fishes in depauperate lakes exhibit trophic polymorphisms. But polymorphisms, when they occur, are associated with low fish diversity. Why they occur only in some lakes and not others is unknown in general. However, its incidence in pumpkinseed sunfishes is correlated with both environmental heterogeneity and the diversity of competitors and predators (Robinson and Wilson, 2000).

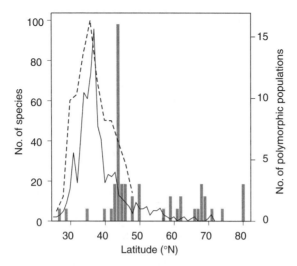

Fig. 7.5 Instances of intraspecific 'trophic polymorphisms' in fishes at different latitudes of northern North America (histogram, right axis). Fish species diversity is indicated for comparison (lines, left axis). The solid line indicates the number of species having their range centres at given latitudes. The dashed line indicates the number of fish species whose ranges overlap each latitude, up to 48°N only (read as the number indicated on the left axis × 5). Intraspecific polymorphisms are most frequent at high latitudes where fish species diversity is low. The tally does not include undescribed sympatric species pairs which also tend to occur at relatively high latitudes (e.g. Table 3.2). Modified from Robinson and Schluter (1999), with permission of Oxford University Press.

7.2.4 Other trends in depauperate faunas

Several other patterns in contemporary lineages can be interpreted as competitive release in depauperate faunas but cannot at this point be regarded as tests of the hypothesis. Two trends warrant mention.

Body size convergence in island mammals—Mammals range in mass from about 10 grams to several tons, and the diets and life histories of species are strongly tied to body size. Perhaps this huge range of sizes is partly the result of interactions between mammal species within and/or between lineages. Change in this distribution in depauperate faunas provides a potential test. On islands, which tend to have fewer mammals than mainlands, individual species show consistent shifts to the middle of the size distribution (Fig. 7.6). On average, small mainland mammals are larger on islands whereas large mainland mammals are smaller on islands. One interpretation is that body sizes relax toward a common optimum in the absence of other taxa. An alternative hypothesis is that resource distributions are contracted on islands.

Increased woodiness in plants—A second multispecies trend, arborescence in island plants, can be interpreted in a similar light. Plants that colonized Hawaii

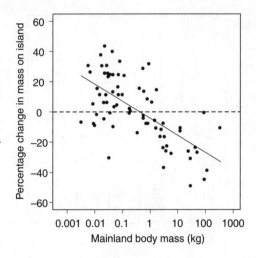

Fig. 7.6 Body sizes of mammal species found on both mainlands and nearby islands. Modified from Lomolino (1985), with permission of University of Chicago Press.

from distant mainlands were predominantly herbaceous, possibly because of a dispersal advantage (Carlquist, 1974, 1980). On the mainlands they exploited chiefly open habitats (Givnish, 1998). On Hawaii, however, the descendants of many of the colonists evolved taller, woody forms especially after entering more mesic habitats. A similar trend is evident on Galápagos in *Opuntia* and several lineages of Asteraceae (Wiggins and Porter, 1971), and in other Asteraceae of many oceanic islands (Givnish, 1998). One explanation of these trends is that intraspecific competition for light favours greater height ('competitive overtopping') particularly in mesic habitats having high vegetation cover. Presumably, the same trends were not favoured on mainlands because of interspecific competition from other woody taxa. Arborescence may therefore represent large-scale release from interspecific competition. Carlquist (1974) proposed an alternative explanation: that greater height is favoured in the moderate climates of oceanic islands.

7.2.5 Remarks

Evidence that ecological opportunity spurs morphological divergence consists in part of observations that divergence in taxa on Hawaii and Galápagos reaches extraordinary levels compared to mainlands. Quantitative evidence is surprisingly scarce, however, at least from patterns at low taxonomic level. Morphological diversity among closely related species is indeed greater in finches on islands than mainlands. The same is true of some Hawaiian plants compared to mainland relatives, but the same is not true in *Anolis* lizards.

Cases of character displacement are known that involve species more distantly related than congeners, but the frequency of cases drops off quickly with increasing taxonomic distance. As a result we know much less about how competition leads to divergence between species in distant taxa than between congeners. Morphological variance within species is not consistently higher in depauperate environments than in species-rich areas, but in postglacial fishes instances of high trait variance tend to be concentrated in the depauperate areas. This pattern is consistent with the hypothesis of ecological opportunity but we have little idea how widespread and significant the phenomenon is because of the shortage of quantitative measures of divergence in depauperate and species-rich environments. But since none of the patterns appears universal and there are many exceptions, additional unknown factors probably exert a strong influence on morphological divergence.

Resource heterogeneity may be one such factor. Depauperate environments may have fewer competitors but they should tend to have fewer resources too, and effects of the latter may counteract or even supercede effects of the former. Resource heterogeneity has been shown to have a strong influence on the evolution of polymorphism in experimental microcosms of bacteria (Rainey and Travisano, 1998), but tests of ecological opportunity in nature have not typically accounted for it.

Also scarce is independent evidence that reduced competition from other taxa is the mechanism behind greater morphological divergence in depauperate environments. Such evidence may be easier to acquire for some patterns than others. For example, the magnitude of character displacement in *Salvelinus* and *Coregonus* (Table 7.1) is correlated not just with the presence or absence of putative competitors but also with the density and diversity of species in other taxa, which is more consistent with competition (or perhaps intraguild predation) than other factors (Magnan, 1988; Lachance and Magnan, 1990; Westman *et al.*, manuscript). Testing whether competition is indeed the interaction producing greater morphological divergence in depauperate environments may be more difficult to test. The role played by predation, which may also be reduced in depauperate environments, is obscure.

7.3 Ecological opportunity and speciation rate

Three large-scale patterns in the fossil record support the notion that ecological opportunity promotes speciation. The total diversity of skeletalized marine animals (number of genera and families) was comparatively stable through much of the Paleozoic, a stretch of over 200 million years, despite constant turnover in species composition (Bambach, 1977; Stanley, 1979; Sepkoski, 1979, 1984, 1988, 1996; Jablonski and Sepkoski, 1996). Diversity rebounded predictably after mass extinctions, mainly by increases in the per capita rates of speciation (or origination rates of higher-level taxa) not reductions in extinction probability. Finally, diversity also seems to rise after nutrient inputs to the biosphere (Vermeij, 1995).

In this section I focus instead on smaller-scale patterns consistent with the hypothesis that ecological opportunity promotes speciation. I focus on comparisons of speciation rates between depauperate and species-rich environments. A high species count by itself is not adaptive radiation, and so I focus on cases in which phenotypic and ecological diversification also seem to take place around the same time. I am more interested in speciation than extinction, but species counts inevitably represent the difference between the two.

7.3.1 Speciation rates on islands and mainlands

Isolated archipelagoes may present greater ecological opportunities to colonists than mainland source areas. This is supported by the greater morphological differentiation of many taxa of Hawaii and Galápagos (Figs 7.1, 7.3) though there are striking exceptions (Fig. 7.4). Are speciation rates also higher in such environments?

Figure 7.7 plots species richnesses of clades that have radiated on isolated archipelagoes (mainly Hawaii and Galápagos) and compares them to their mainland sister clades (or to two or more comparable mainland taxa if the sister clade is unavailable). The majority of points fall above the $Y = X$ line, indicating that diversities indeed tend to be higher on the archipelagoes, but the trend is not overwhelming. The island clade is the more diverse in just 9 of 12 comparisons (one-tailed binomial test, $P = 0.073$; 11 out of 14 if we include comparisons in which the sister clade occurred on other islands not mainlands, $P = 0.029$). The full 800+ species of Hawaiian Drosophilidae are not added because the sister group is not known (DeSalle, 1995). One candidate is the continental subgenus *Drosophila* which has 759 species (Carson *et al.*, 1981), in which case diversity in the island group is again marginally higher.

One potential problem with this comparison is that the available sample of island clades is biased toward larger groups. More exhaustive sampling of island–mainland sister pairs would be needed to overcome this bias. Another aspect to take into account is that mainland species counts are often accumulated over huge geographic areas. For example, the Galápagos *Geospiza* are barely more diverse than seed-eating clades of emberizine finches on mainlands (lower left pair of symbols in Fig. 7.7) but the latter are spread over the whole of the continents of North and South America. The extraordinary aspect of the Galápagos and Hawaiian radiations is the extremely confined geographical areas in which they have occurred.

The exceptions to the main trend are worth contemplating as they might give clues to the other factors influencing speciation rate. In two island clades in particular (*Anolis* lizards of Jamaica and the endemic asters *Hesperomannia* of Hawaii), diversity is substantially greater in their putative mainland sister taxa. One interpretation is that we have underestimated ecological opportunities on mainlands. Alternatively, many other factors such as opportunities for geographical isolation override the influence of vacant niches in these cases.

Note that lower diversity in an island clade relative to its mainland sister group does not mean that the island clade flunks the rapid-speciation criterion for adaptive radiation. Sister-group comparison is just one of several ways to judge rapid speciation. For example, the *Anolis* of large islands of the Greater Antilles (including Jamaica) pass the test in other ways (Chapter 2).

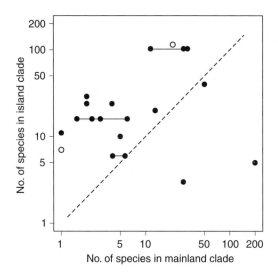

Fig. 7.7 Species numbers in clades on isolated archipelagoes and mainlands (●). Connected filled circles represent comparisons between one island clade and two or more comparable but nonsister mainland clades. Open symbols (○) are Hawaiian clades whose outgroups are on other Pacific islands rather than on mainlands. The dashed line represents $Y = X$. Mainland diversities were corrected for age by estimating speciation rate in the clade, and then calculating number of species expected for a clade with age equal to the corresponding island clade. Eleven island clades fall above the line. From top to bottom these are: (1) Hawaiian lobeliads + lobelioids vs its sister clade *Sclerotheca* + one *Lobelia* (Givnish *et al.*, manuscript); (2) Hawaiian *Drosophila* of the picture wing clade (Ayala, 1975) vs three mainland *Drosophila* clades, *D. melanogaster* subgroup (Eisses *et al.*, 1979), *D. obscura* subgroup (Lakovaara *et al.*, 1972; Cabrera *et al.*, 1983), and *D. willstoni* group (Ayala *et al.*, 1974); (3) Hawaiian *Schiedea* + *Alsinidendron* vs its sister clade, tentatively one or two *Minuartia* (Wagner *et al.*, 1995); (4) Hawaiian silversword alliance vs its sister clade *Raillardiopsis* (Baldwin, 1997); (5) *Argyranthemum* from Macaronesia vs its sister clade *Ismelia* + *Heteranthemis* + *Chrysanthemum* (Francisco-Ortega *et al.*, 1996, J. Francisco-Ortega, personal communication); (6) Hawaiian *Bidens* vs probable mainland *Bidens* sister clade (Helenurm and Ganders, 1985; F. Ganders, personal communication); (7) Hawaiian *Psittirostra*, including subfossil species (Olson and James, 1982; Johnson *et al.*, 1991), vs four cardualine clades, *Coccothraustes* (Marten and Johnson, 1986), *Carpodacus* (Marten and Johnson, 1986), *Carduelis* + *Acanthis* + *Serinus* (Marten and Johnson, 1986); *Loxia* (Marten and Johnson, 1986; Groth, 1993; C.W. Benkman, personal communication); (8) Hawaiian *Tetramolopium* vs its sister clade, *T. alinae* (Lowrey, 1995); (9) Galápagos *Galapaganus* vs its sister clade, *G. howdenae* (Sequeira *et al.*, manuscript); (10) Hawaiian *Scaevola* excluding *S. glabra* vs its sister, *S. sericea* (Patterson, 1995); (11) Galápagos *Geospiza* (Yang and Patton, 1981) vs two ecologically similar emberizine clades, 'greater *Zonotrichia*' clade (Zink, 1982), *Ammodramus* (Zink and Avise, 1990). Three island clades fall below the dashed line. From top to bottom these are: (12) Hawaiian *Sarona* vs its sister clade, *Slaterocoris* + *Scalponotatus* (Asquith, 1995); (13) Jamaican *Anolis* vs its possible sister, the Central American *Anolis* clade (Irschick *et al.*, 1997); (14) Hawaiian *Hesperomannia* vs its sister, *Veronia* subsection *Strobocalyx* (Kim *et al.*, 1998).

7.3.2 Latitudinal gradient in fish speciation

Freshwater fish species diversity in northern lakes shows a steep decline north of 50°N despite the multitude of lakes there (Fig. 7.5). This low diversity must partly reflect the outcome of glaciation, which periodically defaunated the region. The last major cycle ended only 10 000–15 000 years ago. Once the ice disappeared, the colonization of newly formed lakes from glacial refugia was impeded by a limited number and duration of passage routes. Marine access to coastal lakes and rivers was limited to salt-tolerant species, and changes in sea level meant that many lakes were accessible for only a short period (McPhail and Lindsey, 1986). Inland species dispersed via infrequent changes in the drainage patterns of lakes and rivers during deglaciation (McPhail and Lindsey, 1986). Many lakes and watersheds of previously glaciated areas are therefore depauperate and represent aquatic islands in a sea of land.

Two measures reveal the implications of a depauperate northern fish fauna for fish speciation. First, closely related fish species living at high latitudes (mainly sister species) exhibit reduced levels of mtDNA genetic divergence (Fig. 7.8). On average, northern fishes sit on the tips of shorter branches on phylogenetic trees than do fishes of unglaciated areas. The average branch length shortens with increasing latitude, but only above the maximum southern extension of the ice sheets. Extinction alone would eliminate species at the tips of both long and short branches so it cannot be solely to blame for the pattern. Rather, lakes of previously glaciated areas appear to

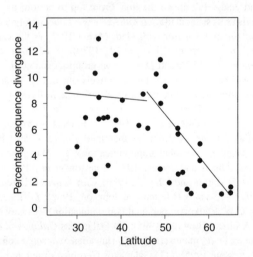

Fig. 7.8 Mitochondrial DNA divergence between closely related species of northern fishes (mainly sister species) in relation to their median latitude of distribution. Separate regressions are plotted above and below 46°, which is close to the median latitude of maximum glacial advance in the Pleistocene. The tally does not include undescribed sympatric species pairs of fishes, which also tend to occur at relatively high latitudes (e.g. Table 3.2). Modified from Bernatchez and Wilson (1998), with permission of Blackwell Science Ltd.

have experienced a recent rise in speciation rate not seen in fish at the lower latitudes. Range contraction and isolation may have contributed to the higher rate, but it is not clear that these ingredients are in short supply at the lower latitudes.

Second, a slew of very young sympatric pairs of fish species has recently come to light, and these too are concentrated in lakes and rivers of previously glaciated areas (Schluter, 1996; Bernatchez and Wilson, 1998). Examples from lakes were noted earlier in Table 3.2. The level of mtDNA differentiation between the sympatric species is typically less than 1%, which is much less than that between the species discussed above. Genetic differentiation and reproductive isolation has been confirmed only recently in most cases, and as yet none of the species has been formally described. These examples are not included in Fig. 7.8. I am not aware of the existence of similar cases at comparable frequency at lower latitudes.

Both these indications are consistent with the idea that ecological opportunity stemming from low fish species diversity has contributed to an elevated rate of speciation in recent fishes. In Chapter 8 I review evidence that selection pressures stemming from the environment have contributed to the evolution of reproductive isolation in some of these cases.

7.3.3 Saturation and speciation

Another prediction is that as niches become used up the rate of speciation should slow down. Few tests of this pattern have been carried out using recent taxa. Nee *et al.* (1992) developed a statistical test of saturation using branching rates of phylogenetic trees and applied it to a molecular phylogeny of extant birds. Attention was limited to the first 121 nodes of the 9700-species phylogeny, covering the time from the origin of birds to about the origin of the modern bird families. Rates of species accumulation over this interval indeed slowed through time as diversity built up but never reached an asymptote. An extension to taxa of the more recent past would be profitable but such a test would probably require combining patterns from many taxa. Zink and Slowinski (1995) suggested that speciation rates in North American birds slowed over a period lasting from before the Pleistocene to the present time. However, the pattern is at least partly an artefact of the lag time necessary for bird speciation rather than a reduced speciation rate in the very recent past (Avise and Walker, 1998).

7.3.4 Speciation in the presence and absence of a competitor

Observational tests for ecological character displacement contrast morphological differences between two species where they occur together (sympatry) with that when they occur separately (allopatry) (Tables 6.1, 7.1). The analogous test for competition and speciation would compare number of speciation events in each of two lineages where they occur together with that where the lineages occur separately. I know of no studies of this type but analysis of crossbill divergence in the presence and absence of squirrels is a step in that direction (Benkman, 1999).

180 • Ecological opportunity

The case concerns divergent natural selection on a species of the red crossbill complex (*Loxia curvirostra*) stemming from spatial variation in the characteristics of lodgepole pine cones (*Pinus contorta*) that are its major food. Cone characteristics are strongly affected by the presence or absence of red squirrels, *Tamiasciurus hudsonicus*. Where squirrels are present they eat most of the cones. Crossbills are rare and are not an important selective force on cone traits. In response to predominantly squirrel predation the cones are small, wide, and have thick basal scales and few seeds. In two regions where squirrels are absent, however, crossbills are abundant and a dominant selective force, and the cones there have the opposite features. The crossbills in these two sites have deeper, shorter beaks than elsewhere and are more efficient at extracting seeds from the larger, wider cones. Thus the presence or absence of squirrels leads to beak differences between crossbill populations. The only missing step is speciation. Within the red crossbill complex, beak differences are almost always associated with differences in song and some degree of reproductive isolation. We can imagine the same happening here between beak morphs in the presence or absence of squirrels, although this part is still speculative.

7.3.5 Remarks

Evidence exists from contemporary lineages that ecological opportunity influences speciation rate, but it is not abundant. Adaptive radiations on depauperate islands provide evidence in favour of the hypothesis but the pattern is not as consistent as expected. It is weaker than that suggested earlier from a smaller set of cases (Schluter, 1998). Further comparisons from other systems come to mind that may be added, such as between new lakes and adjacent river systems. The cichlid radiations of the East African Great Lakes have been attributed in part to the wealth of resources available when the lakes formed and the dearth of competing lineages (Fryer and Iles, 1972). A systematic exploration of this effect is needed. Finally, a latitudinal gradient in speciation rate of fishes supports the hypothesis.

Tests of alternative hypotheses are needed. Remote archipelagoes may have lots of vacant niches but they may also have heightened opportunities for geographic isolation, and the contributions of both need to be weighed. The solitary Cocos island was colonized long ago by a finch from Galápagos (Petren *et al.*, 1999) but no speciation ensued, presumably because of the lack of barriers to gene flow (Lack, 1947). On the other hand, opportunities for geographic isolation are surely no greater in Hawaii and Galápagos than on the whole continents and their satellite islands where their relatives have diversified (Fig. 7.7). The East African lakes are not more subdivided spatially than adjacent rivers. In general it would be useful to know the extent to which the bursts of speciation that characterize adaptive radiation are explained by expansion to new environments free of competitors or by the range contractions and isolated refugia that may have prevailed at the same time, or both.

7.4 Key evolutionary innovations

Simpson (1953) proposed that adaptive radiations may be initiated when a novel trait is acquired that influences ability to exploit resources hitherto little used. A lineage may achieve access to ecological opportunity, then, not just by being the first to stumble upon a new location having abundant untapped resources but alternatively by acquiring the right characters to make effective use of resources already present in the region. The kinds of traits Simpson had in mind were not just ordinary preadaptations to particular kinds of environments. Rather, a 'key innovation' was the ultimate novelty that, once added to the phenotype, made all the difference in the world and triggered adaptive radiation. It was not necessary that the radiation should follow immediately, but rather diversification 'may follow soon after the origin of such a type or it may be long delayed...' (Simpson, 1953, p. 223) depending in part on continued modification of the key trait. In this section I summarize a few case studies of putative key innovations and our understanding of their mechanisms.

Most tests of key innovation hypotheses attempt to correlate appearance of a novelty with change in the net rate of speciation rather than with adaptive radiation, of which speciation is only part. The lack of attention to effects of novel traits on ecological and phenotypic expansion is an outstanding gap in the study of key innovations. The problem with focussing only on speciation is that evolutionary novelty may influence speciation (and even adaptive radiation) without increasing access to new resources. Traits such as lower dispersal or a high rate of evolution of preferences for secondary sexual characters may be examples (e.g. Price, 1998). Such traits are not key innovations in the classic sense and I consider them no further here. Sexual selection is dealt with instead in Chapter 8.

I focus on those few candidate key innovations that have reasonable statistical support. My main objective is to evaluate whether the mechanism linking appearance of these traits with adaptive radiation is enhanced ecological opportunity rather than some other mechanism. The large number of putative key innovations as yet without statistical support are not covered here (see Heard and Hauser (1995) for a list up to 1995).

7.4.1 A comment on statistical tests

Several methods exist for testing statistically whether the first appearance of a trait in a lineage coincides with an increase in speciation rate (or speciation minus extinction). All of them use sister-group comparison: an effect on speciation rate is determined by whether taxa possessing the novelty are more diverse than their sister taxa which lack it. The approach is not foolproof. A true key innovation may not leave its mark in a difference between sister taxa if other factors are also at work. As well, speciation by itself is a poor indicator of adaptive radiation. Finally, sister taxa are not necessarily the ideal or only groups worth comparing when testing key innovation hypotheses (Hunter, 1998).

182 • Ecological opportunity

The method of Slowinski and Guyer (1993) tests whether the number of species in the taxon possessing the trait is significantly greater than that in its sister taxon lacking the trait. Sister taxa are the same age and are likely to be similar in many other respects except the one of interest. The null hypothesis is that the rates of speciation and extinction are the same in the two lineages, and that observed differences in diversity are merely stochastic. Sanderson and Donoghue (1994) go further by testing whether the precise location in the phylogenetic tree at which an increase in speciation rate occurs is the same as that at which the putative key innovation first appears.

Neither of these methods requires more than a single appearance of a putative key trait in history. The independent trials on which calculation of probability depends are the number of speciation events, not the number of replications of the evolutionary events 'gain/loss of trait' and 'rise/fall in speciation rate'. Consequently the methods of Slowinski and Guyer (1993) and Sanderson and Donoghue (1994) are not strictly tests of whether trait appearance and speciation rate are correlated. Treating them as tests of correlation is analogous to pseudoreplication in experimental designs (Hurlbert, 1984). This problem is not diminished when probabilities from different tests are combined (Slowinski and Guyer, 1993; Goudet, 1999) since the probabilities themselves are not based on replicated events. For this reason I restrict attention below to the small number of cases in which the evolutionary events have been repeated multiple times.

7.4.2 Herbivory

The extraordinary diversity of insects that feed on vascular plants, and the apparently contemporaneous rise of insect diversity with diversity of flowering plants, has fuelled speculations that each group has influenced diversification in the other. In their 'escape and radiation coevolution' model of mutual diversification, Ehrlich and Raven (1964) proposed that lineages of herbivorous insects and plants are locked in a perpetual arms race in which traits that facilitate herbivory accumulate in insect lineages whereas traits that deter herbivory spread in lineages of plants. Diversification in each is sped whenever it temporarily gains the upper hand. Insect innovations that overcame plant defences opened up a wealth of new resources for exploitation. Conversely, new ecological opportunities were afforded plant lineages acquiring novel defences that helped shed their insect burden. Two elements of this hypothesis, which is explicitly about novel traits and ecological opportunity, have now been tested. The first is summarized here and the second is reviewed in the section immediately following.

Mitter et al. (1988) examined species number in 13 insect taxa that have independently acquired plant-feeding (phytophagy) with that in their saprophagous or predatory sister taxa. The phytophagous clade was more diverse in 11 of the 13 sister pairs (e.g. Fig. 7.9). The increase in species diversity associated with plant-feeding was about 20-fold. Farrell (1998) followed this up with a comparison of diversities between sister taxa within the Phytophaga, the largest and oldest radiation of

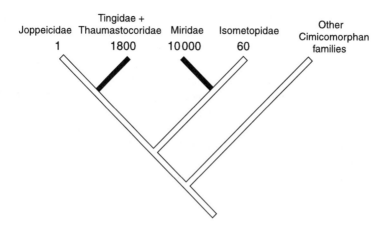

Fig. 7.9 Example comparison used in testing whether plant-feeders have elevated speciation rate. The tree is a section of the phylogeny for the insect order Heteroptera, which is primitively predaceous (pale branches). Plant-feeding has evolved at least twice in this section, and these clades have many more species than predaceous sister taxa. After Mitter *et al.* (1988).

phytophagous beetles. In all five pairs of sister taxa available, in each of which one sister consumed mainly angiosperms and the other used mainly gymnosperms (the ancestral state, according to fossils), the angiosperm-eating lineage was the more diverse. The average increase in diversity associated with a shift to angiosperms was 170-fold. Together these studies show that increased diversity is associated with both *de novo* evolution of plant-feeding by nonherbivorous taxa and shifts to angiosperms within phytophagous taxa. The increases probably represent many adaptive radiations not just many new species, because of the diversity of feeding modes and life histories that have accompanied the transitions. Although no phytophagous insects were included in my list of examples of adaptive radiation (Chapter 2), reciprocal transplants of closely related species within several beetle lineages have indicated divergent adaptations to distinct host plants (e.g. Rausher, 1984; Futuyma *et al.*, 1995; Funk, 1998).

Phytophagy is better thought of as the environment not the 'key innovation', whereas the innovations themselves are general traits that permit plant-feeding (or angiosperm-feeding). Combining these into a single test seems reasonable even though the specific morphological and physiological adaptations involved may be different each time. Whether these traits are key innovations then rests on elucidation of the mechanism generating new species at an accelerated pace. Plant-feeding undoubtedly represents an 'adaptive zone' with a huge diversity of potential niches. What remains to be determined is whether ecological opportunities are indeed greater for phytophages than their sister taxa, and whether these greater ecological opportunities are the reason behind the greater diversity of phytophages. Understanding the mechanisms of speciation would help establish the latter requirement. In support, there is indication from leaf beetles and phytophagous flies that adaptation to different hosts is partly responsible for the evolution of barriers to gene exchange (speciation) between populations on different hosts plants (Funk, 1998; Feder, 1998; see Chapter 8).

Herbivory and diversity are also correlated in mammals, and in this case a candidate evolutionary novelty, the hypocone, has been suggested (Hunter and Jernvall, 1995). The hypocone is a cusp on the upper molar teeth of mammals that has appeared in over 20 mammalian lineages during the Cenozoic. The hypocone may be an adaptation to a generalized diet that assisted the maceration of fibrous plant tissue and so permitted the evolution of full herbivory. Species diversity of mammalian groups possessing the hypocone increased through the Cenozoic compared with groups lacking the hypocone.

7.4.3 Defence against herbivores

The second line of evidence relevant to the escape and radiation hypothesis is the higher diversities of plant lineages that evolved resin canals, a highly effective deterrent against herbivorous insects (Farrell *et al.*, 1991). This defence has evolved independently multiple times over the past 10–180 million years, allowing a test of the key innovation hypothesis using comparison across multiple pairs of sister taxa. The canal-bearing clade was the more diverse in 13 of 16 sister pairs tabulated, with a median six-fold excess of species richness. These findings are among the strongest evidence to date that ecological interactions govern speciation and/or extinction. Information on morphological and ecological diversity in the clades needs to be provided to ensure that the increases in diversity are associated with adaptive radiation.

The next problem is to determine the mechanism underlying enhanced speciation rates (or lowered extinction). Possibly the reduced insect burden stemming from canal defences has given defended lineages the edge in competition for resources against species in other taxa. Or, the trait has enhanced persistence in novel environments, facilitating ecological divergence and speciation. Both mechanisms would be compatible with 'ecological opportunity'. The only observation relevant to these ideas is that tree species with latex or resin canals are numerically more abundant in tropical forests than trees lacking the defence (Farrell *et al.*, 1991).

7.4.4 Nectar spurs and pollination

The columbines (*Aquilegia*) are one of several angiosperm taxa that have independently evolved nectar spurs (Fig. 2.5). Nectar spurs are tubular extensions of flowers. They produce nectar at their base, and a long tongue is needed to extract it. Species with spurs have a reduced diversity of effective animal pollinators in comparison with related species. Taxa having spurs also tend to be diverse. Seven of eight taxa possessing spurs are more diverse than their sister taxa which lack spurs (Hodges and Arnold, 1995; Hodges, 1997; Wilcoxon signed rank test, $N = 8$, $P = 0.016$). Overall, nectar spurs are associated with an average 16-fold increase in diversity, corresponding to about a three-fold increase in the rate of speciation (assuming exponential growth of species diversity and basing the calculation on species richness +1). Nectar spurs are therefore outstanding candidate key innovations, at least in the columbines where a full-blown adaptive radiation is underway (Chapter 2).

Why should spurs have such a positive influence on speciation rate (assuming that spurs and not some correlated trait is responsible)? One possibility is that a change in spur length greatly altered the community of pollinators visiting the flowers, and brought about a degree of reproductive isolation as a by-product. However, experimental shortening of spur length in *Aquilegia pubescens*, a species having long spurs, did not alter hawkmoth visitation rates (Fulton and Hodges, 1999). Shortening only reduced the frequency of physical contact between hawkmoths and anthers, causing a decline in the rate of pollen removal. The manipulation reduced plant fitness but did not induce a switch in major pollinator. Hummingbirds continued to avoid them.

However, if the presence of spurs of a given length means that only a few animal species can effectively pollinate, then colonization of a new habitat having different pollinators inevitably leads to a drop in effective pollination rate. Subsequent strong selection to improve rates of pollen removal might then lead to a shift in spur length and other floral traits, all of which eventually bring about reproductive isolation. Flowers lacking spurs do not experience the same selection pressures upon colonization of a new habitat because the pollinators there are more likely to be similar to those of the old habitat. Under this scenario nectar spurs are a key innovation because they facilitate transitions between different suites of specialized animal pollinators.

7.4.5 *Remarks*

The frequency of large disparities in species diversity between related taxa ensures that interest in key innovations will remain high (Heard and Hauser, 1995). For example, angiosperm plants are over 300 times more diverse than gymnosperms. Might this be attributable to one or a few traits not possessed by gymnosperms? The above summary does not do justice to the literature on the subject of key innovations, which is large but almost wholly untested. Rather, I focussed on a few reasonably well-supported candidates that seem to connect appearance of a novelty with adaptive radiation via increased ecological opportunity. However, the connection to ecological opportunity is not yet proved.

A problem with most putative cases of key innovation is that lack of replication prevents statistical testing. The extraordinary diversity of angiosperms might stem from the evolution of the flower, but the prospects of testing this are grim given that flowers evolved only once in the history of life. A more promising avenue is to focus on diversity differences between families of flowering plants that exhibit repeated transitions in floral and other structures (e.g. Eriksson and Bremer, 1992; Ricklefs and Renner, 1994; Dodd *et al.*, 1999). Plant families whose flowers are pollinated by biotic agents are more diverse than their sister taxa pollinated by abiotic agents.

Families all of whose species are herbaceous are more diverse than sister taxa all of whose species are nonherbaceous. Whether these differences in diversity reflect adaptive radiation remains to be confirmed.

A more general claim that has emerged in many guises is that adaptive radiation and/or speciation is spurred by increased 'evolvability' or 'versatility', the capacity to rapidly produce a variety of different phenotypes exploiting the environment in novel ways. Evolvability may stem from a phenotype made up of a large number of elements that can be modified independently (Vermeij, 1974). For example, the exceptional species richness and diversity of body forms of cichlids in the East African lakes has been attributed to structural modifications in the head that allow independent movement of the upper and lower pharyngeal jaw, and that free the oral jaws from involvement in food processing (Liem, 1973; Galis and Druckner, 1996). The diversity of teleost fishes in general compared to other vertebrate taxa has been attributed to the large number of bony elements in the head region (Galis and Metz, 1998). The success of the vertebrates as a whole may have been facilitated by the increased versatility resulting from one or two complete genome duplications, generated perhaps by rounds of hybridization and polyploidy (Spring, 1997). Hypotheses of versatility may be testable if the property can be identified in many taxa and scored *a priori*.

The statistical tests of key innovations are only correlative and eventually we would like to know the mechanisms underlying associations. The manipulative studies of Fulton and Hodges (1999) show that it is possible to test novel predictions of alternative hypotheses using contemporary populations, and this may be the most promising step toward an understanding of key innovations and their role in adaptive radiation.

7.5 Discussion

The hypothesis of ecological opportunity was invented to account for several patterns in adaptive radiation that seemed difficult to account for in other ways. These include the extraordinary ecological breadth of diversification in several lineages inhabiting environments isolated from other potentially competing taxa, and the fact that diversity of previously minor groups often soars after previously dominant taxa have become decimated in mass extinction. These observations are pursuasive and further exploration of specific cases (e.g. Galápagos ground finches) has not found evidence to reject the hypothesis.

Many tasks nevertheless remain. The relationship between ecological opportunity and phenotypic divergence is poorly quantified in contemporary taxa. As a result we are not able to say much about the size of the contribution made by ecological opportunity, its generality, or even its necessity. The relationship between opportunity and adaptive radiation is noisy, if comparison is any guide. On the other hand, the 'island vs mainland' proxy for 'high vs low ecological opportunity' is probably imperfect. Mainlands have more competitors and predators but probably a wider diversity of resources.

For example, the *Anolis* of Jamaica are an adaptive radiation famous for the ecomorphs that evolved independently there, duplicating those found on other Caribbean islands. Yet, a mainland *Anolis* clade, possibly the sister group to the Jamaican clade, has as much or more phenotypic diversity. Perhaps ecological opportunity for *Anolis* is not less on the mainland than the island, given the more extensive geographic area. While possibly true, such a retrospective interpretation is *ad hoc* and exposes a grave weakness of the hypothesis, namely that it is founded on a feature of environment (ecological opportunity) not easily recognizable *a priori*. The same difficulties apply to tests of a link between ecological opportunity and speciation. Rates of species production may indeed be higher in depauperate environments but island–mainland comparisons suggest that the relationship is noisy.

Unfortunately, we have only a dim grasp of what ecological opportunity is made of. A wide resource spectrum with lots of adaptive peaks, presumably. Freedom from competition against other taxa was emphasized by the early naturalists but there is some indication that this is probably not the only important consideration. The comparative study of diversity and latex/resin canals is the first clear indication that freedom from enemies also promotes species production and/or slows extinction. Competition and predation are very different processes, and in most cases we are unable to say how each of them contributes to, or inhibits, adaptive radiation.

In general, we know less about the impacts on divergence of species in distant taxa than about the forces driving divergence between closely related species. Cases of character displacement between noncongeners are known but their frequency drops steeply with increasing taxonomic distance. Perhaps this decline is a true measure of the influence of taxa on one another, but researcher bias needs to be accounted for. As with character displacement between close taxa, evidence of resource competition is lacking in most examples, which is a worry if other interactions also cause divergence (Chapter 6).

Finally, novel traits almost certainly spur adaptive radiation but we do not yet know why. The bulk of the evidence comes from comparisons of rates of speciation only, which is not the same as adaptive radiation. Phenotypic divergence was an important part of the original key innovation hypothesis but has become neglected. Ecological opportunity of some sort may be involved in the link between traits and speciation, but this has not been confirmed in a single case. Although further comparative study is essential, microevolutionary studies of the process hold promise that predictions of alternative mechanisms may be tested.

8

The ecological basis of speciation

> *The genotype of a species is an integrated system adapted to the ecological niche in which the species lives. Gene recombination in the offspring of species hybrids may lead to formation of discordant gene patterns.*
> —T. Dobzhansky (1951)

8.1 Introduction

The rise of ecological and phenotypic diversity in adaptive radiation is accompanied by one or more episodes of speciation. Determining how the new species form, and why the pace of species formation is higher during some periods than others, is thus crucial to understanding the process. For organisms engaging in sexual reproduction, effort centres on the mechanisms that elevate levels of reproductive isolation between populations.

According to the ecological theory of adaptive radiation, reproductive isolation evolves ultimately as a consequence of divergent natural selection ('ecological speciation'). This idea derives from Mayr (1942) and Dobzhansky (1951), who saw that pre- and postmating isolation could arise incidentally between populations that diverged between distinct selection environments. I interpret their concept as a general principle covering a variety of potential mechanisms, including cases in which reproductive isolation is entirely a by-product of divergent natural selection and those in which speciation is begun by divergent selection but later completed by reinforcement (Dobzhansky, 1951). This concept applies whether speciation occurred in allopatry, parapatry, or sympatry. Many proposed models of speciation invoking sexual selection are also included. Under this ecological view, speciation may be rapid in adaptive radiation because reproductive isolation evolves quickly when divergent natural selection is strong.

The alternative hypothesis is that speciation is driven instead by nonecological mechanisms. In nonecological speciation, reproductive isolation evolves as a consequence of genetic drift in stable populations (Wright, 1940), founder events and population bottlenecks (Mayr, 1954), fixation of alternative advantageous genes in allopatric populations experiencing similar selection pressures (Muller, 1940), or of hybridization and/or polyploidy (Stebbins, 1950; V. Grant, 1981; Rieseberg, 1997; Ramsey and Schemske, 1998). Many such processes may involve sexual selection (e.g. Lande, 1981; Kirkpatrick, 1982a; Holland and Rice, 1998). An explanation for high speciation rates under this alternative view is that high ecological opportunity increases the viability of populations undergoing speciation ('ecological persistence' hypothesis; Schluter, 1998), but does not necessarily hasten the rate at which any one of the populations evolves reproductive isolation from the others. Species accumulate more quickly because more populations avoid extinction long enough to attain the status of biological species. The same factors would presumably also reduce extinction rates of species once formed.

In this chapter I evaluate the evidence for ecological speciation in nature. I begin by briefly summarizing theoretical and experimental models for ecological speciation. I then outline a variety of tests that may help distinguish the hypothesis from alternative, nonecological mechanisms of speciation. I treat sexual selection separately because it can oppose natural selection and may be involved in both ecological and nonecological speciation. As will become evident, research on ecological speciation in nature is still at an early stage. I therefore point to the kinds of tests that might profitably be carried out.

8.2 Models of ecological speciation

This section reviews several ideas on how divergent natural selection might cause reproductive isolation to evolve. The simplest model is the by-product mechanism, whereby reproductive isolation builds as organisms adapt to different environments having contrasting selection pressures. The process may be completed by reinforcement, the strengthening of premating isolation in sympatry in response to reduced fitness of hybrids. Or, selection against intermediate phenotypes may drive the evolution of premating isolation from beginning to end without a by-product phase. Sexual selection may be involved at any stage.

8.2.1 By-product speciation

Divergent natural selection on phenotypes may cause reproductive isolation to evolve incidentally between populations inhabiting different environments (Dobzhansky, 1951; Mayr, 1942, 1963; Endler, 1977; Rice and Hostert, 1993). The mechanism is called 'by-product' because reproductive isolation is not directly favoured by selection. Complete allopatry is probably not essential for this model to work, at least under some genetic models of the process, but speciation is most likely and most rapid when barriers between populations keep gene flow to a minimum.

In one version of by-product speciation, postmating isolation (hybrid sterility or inviability) evolves if the genetic changes favoured in different environments are 'complementary' and incompatible when combined or recombined in hybrids (Muller, 1940; Orr, 1995). Dobzhansky (1951) especially favoured this mechanism: 'physiological isolating mechanisms may be a product of natural selection. . . . Gene recombination in the offspring of species hybrids may lead to formation of discordant gene patterns.'

Premating isolation may evolve as a by-product of divergent natural selection if mating preferences are genetically correlated with morphological or physiological traits that are the targets of divergent selection. Or, mate preferences may be direct targets of selection, for example if mate recognition or mate detection in contrasting environment favours alternative adjustments of sensory structures or neurological pathways (Endler, 1992; Endler and Basolo, 1998). Mate preferences are doubly potent agents of assortative mating because each increment of divergence also contributes to postmating isolation, with hybrids becoming increasingly unattractive as mates (Liou and Price, 1994).

Several laboratory experiments with *Drosophila* have simulated by-product speciation (reviewed in Rice and Hostert, 1993), and they provide a good reference for how the process may work in nature. Kilias *et al.* (1980) raised different lines of *D. melanogaster* for five years in either a cold–dry–dark or a warm–damp–light environment. Dodd (1989) examined mating preferences in replicate lines of *D. pseudoobscura* raised for one year on either starch-based or maltose-based larval medium (Fig. 8.1). In both cases some premating reproductive isolation evolved between lines that had experienced different selective environments, whereas no premating isolation evolved between independent lines raised in the same environment.

In all models of by-product speciation, the populations are presumed not to interact ecologically as interference or resource competitors, intraguild predators, predator and prey, or mutualists, even though they occur together in sympatry and presumably encounter one another. However, reproductive isolation may evolve even more quickly if ecological interactions also occur. For example, divergent natural selection could generate body size differences that incidentally cause individuals of the larger type to prey upon individuals of the smaller type. The evolution of behavioural defences in the smaller type would then likely reduce the frequency of mating between the types as an incidental by-product. This kind of ecological interaction in sympatry has received no treatment in models of speciation. The process is distinct from reinforcement (see below) because the evolution of premating isolation is not driven by the low fitness of hybrids. Rather, premating isolation increases as a by-product of adaptation to the ecological interaction. A possible example in the pied flycatcher is mentioned later (Section 8.4.2).

8.2.2 Competitive speciation

When intermediate genotypes in a population are at a fitness disadvantage, selection may favour a population split leading to speciation. The process is called 'reinforcement' if the intermediates are hybrids between subpopulations that came together after a period of allopatry during which incomplete pre- and postmating isolation built up. The process is 'sympatric speciation' if the allopatric phase is absent altogether and speciation is initiated by disruptive selection in a randomly mating population. Both

Models of ecological speciation • 191

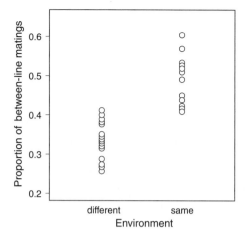

Fig. 8.1 Assortative mating between lines of *Drosophila pseudoobscura* raised separately for one year on either starch-based or maltose-based larval medium. Each point is the result of a multiple choice trial involving a different pair of lines. Each trial placed equal numbers (12) of virgin males and females from each of the two lines in a single chamber. The Y axis reports the proportion of total pairings observed that were between individuals from different lines. Points on the left are of trials between lines from different environments, whereas those on the right are from the same environment. Flies were raised one generation on a standard agar medium before experimentation. Data are from Dodd (1989).

ideas are distinct from the pure by-product model because premating isolation is directly favoured in the sympatric phase. The theoretical plausibility of both processes has been debated extensively, as has the question of whether there is any good evidence to support them (Coyne and Orr, manuscript).

From an ecological perspective, reinforcement and sympatric speciation differ from the simple by-product model in an important respect. The sympatric phase allows ecological interactions between phenotypes to contribute to reduced fitness of intermediates and help drive the speciation process. Rosenzweig (1978) suggested sympatric speciation could happen this way if phenotypes competed for resources. Wilson and Turelli (1986) gave a clear demonstration of how consumer and resource dynamics create the necessary conditions. In their model, resource consumption is governed by a single locus and two alleles. Genotype *AA* is efficient at harvesting the first resource but not the second resource, whereas *aa* is best on the second resource. Heterozygotes *Aa* are intermediate in consumption of both resources, but are slightly worse than the average of *AA* and *aa*. Under random mating, a stable polymorphism results in which *Aa* is the most frequent genotype yet it has lower fitness than the homozygotes. The polymorphism is stable because frequency- and density-dependent selection favour the rare allele. For example, if allele *a* becomes rare (such that *aa* has all but disappeared) then the ensuing increase in the abundance of the second resource increases the fitness of Aa over AA (because Aa is better than AA at consuming it). The homozygote aa reappears as the a allele increases in frequency, and it is favoured over Aa (because it is even better on the second resource). As the second resource gradually becomes depleted, an equilibrium state is reestablished in which AA and aa have equal fitness and Aa is inferior. The process works provided that the consumption disadvantage of Aa is not too great.

Once the equilibrium is established, selection should favour the evolution of assortative mating in the extreme genotypes because this leads to fewer offspring having the less fit intermediate genotype. Whether assortative mating evolves, however, depends on the genetic details of the traits and the strength of selection (Felsenstein, 1981). Consider three possible scenarios. In the first, a mating character codes for the behaviour 'mate with someone having the same phenotype as yourself' or 'mate where you eat'. This character has no trouble increasing in frequency because the same alleles are favoured in both homozygotes (AA and aa). Sympatric speciation in this case is theoretically plausible, as laboratory experiments have confirmed (reviewed in Rice and Hostert, 1993).

The process is more difficult in a second scenario in which assortative mating is based on a separate trait such as colour: 'mate with an individual the same colour as yourself'. In this case, speciation requires that alleles for one colour value (e.g. red) all end up in one of the genotypes favoured by selection (e.g. AA) and alleles for another colour value (e.g. blue) all end up in the other favoured genotype. The principal difficulty is that even though selection favours the buildup of such an association, recombination breaks it down in the next generation. The situation is even worse in a third scenario in which different alleles at the mating locus code for different colour preferences, with one set of alleles coding for 'mate with a red individual' and another set coding for 'mate with a blue individual'.

Dieckmann and Doebeli (1999) show that the process can work in a quantitative genetic model corresponding to the second scenario described above (Fig. 8.2). Their model incorporates one ecological trait (a multilocus analogue of the 'A trait'), one marker trait (e.g. blue phenotypes at one extreme, red phenotypes at the other), and one mating trait (a high value corresponds to 'mate with an individual of the same colour as yourself', a low value to 'mate with an individual of the opposite colour as yourself', and a middle value with 'mate randomly with respect to colour'). The ecological trait evolves to a mean value corresponding to the peak in carrying capacity at which intermediate phenotypes have a fitness disadvantage. Inevitably, a weak linkage disequilibrium arises by chance between the marker trait and the ecological trait (e.g. large phenotypes are blue more often than red, and small phenotypes are more often red than blue). An increase in the mating locus is then favoured which strengthens the association between marker and ecological traits, eventually leading to two reproductively isolated populations. Other types of ecological interactions, including shared predation and shared mutualism, may also lead to selection against intermediate phenotypes (Brown and Vincent, 1992; Matsuda et al., 1996; Doebeli, 1996; Doebeli and Dieckmann, 2000;

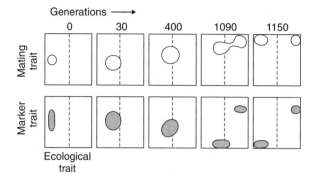

Fig. 8.2 A simulation of competitive speciation. The distributions of three quantitative traits are depicted through time. The ecological trait determines the use of resources, which are modelled with a carrying capacity curve like that in Fig. 6.2. The location of the peak in carrying capacity is indicated by the vertical dashed line. The marker is a neutral trait that becomes used as the basis of assortative mating. The mating trait indicates strength and sign of assortative mating: a high value indicates a strong tendency for individuals to mate with other individuals having the same value of the marker trait (assortative mating); a low value indicates a strong tendency for unlike individuals to mate; and a middle value indicates random mating, the mean at the start. Little happens before the ecological trait arrives at a point corresponding to the peak in carrying capacity (generation 400), at which time the trait experiences disruptive selection. Stochastic fluctuations in genotype frequencies eventually create a small degree of linkage disequilibrium between the marker trait and the ecological trait (lower panel of generation 400) which causes marker-based assortative mating to increase, eventually precipitating a split in the population (generation 1150) yielding two reproductively isolated species. After Dieckmann and Doebeli (1999), with permission of *Nature* and the authors.

Abrams, 2000), and in theory also favour reinforcement or full-blown speciation in sympatry.

There are no experimental models of this process in sexual species. Asexual species maintained in simple environments in the laboratory often undergo splitting that leads to stable polymorphisms of ecologically divergent strains (Helling *et al.*, 1987; Rosenzweig *et al.*, 1994; Rainey and Travisano, 1998). These systems therefore illustrate that the ecological dynamics required for interactive speciation are plausible, but they lack the most difficult step in sexual species: the evolution of reproductive isolation.

8.2.3 *Ecology and sexual selection*

A great deal of interest has recently focussed on how sexual selection, differential mating success of phenotypes within populations, facilitates the evolution of pre- and postmating isolation between populations and hence speciation. Much attention has centred on the role of mate preference and the evolution of premating isolation.

Diverging preferences generate disparities in the probability of mating that may disadvantage intermediate phenotypes already suffering under natural selection (Liou and Price, 1994). Also, biases in mate preferences tend to be self-reinforcing and, once established, may 'run away' to arbitrary extremes (e.g. Lande, 1981, 1982). The result is an amplification of levels of divergence in mate preferences initiated by other processes. Sexual selection arising from intersexual conflict over mating and fertilization may also give rise to pre- and/or postmating isolation (e.g. Rice, 1998).

Theoretical models of speciation involving sexual selection can be classified according to the role played by divergent natural selection between environments (Table 8.1). Consider the scenario in which males display, females choose males according to the size of male traits displayed, and speciation accompanies divergence in female preferences. (The amount of reproductive isolation resulting per increment of divergence depends on assumptions about the shape of the female preference function.) Under this scenario only two models of sexual selections yield divergence without divergent natural selection. Under the classical 'Fisherian' model, females gain no advantage or disadvantage by choosing, and male traits and female preferences coevolve to arbitrary equilibrium values at which natural and sexual selections on the male traits are balanced. An infinite number of such equilibria exist and populations diverge between alternate equilibria by genetic drift (Fig. 4.6). (Under some conditions the Fisher process may lead to cycling preferences instead; Pomiankowski and Iwasa, 1998.)

The 'chase-away' model of sexual selection also does not require divergent natural selection stemming from environment (e.g. Holland and Rice, 1998). In this model, females evolve resistance to traits in males that increase male fitness at the expense of female fitness, rather like predator and prey. For example, seminal fluid proteins of male *Drosophila* that mediate sperm competition among males, or that induce female reluctance to remate, are hazardous to females. In an experiment in which female evolution was arrested, evolution in the males experiencing male–male competition for fertilizations led to reduced female fitness (Rice, 1996). In contrast, elimination of male–male competition produced the reverse effect (Holland and Rice, 1999). One can imagine that over the long term, mutations yielding benefits to males and counter-adaptations in females would accumulate, and that the precise sequence of appearance of these mutations is unlikely to be duplicated in different allopatric populations of the same species, even those inhabiting similar external environments. The inevitable result may be pre- or postmating isolation between them, independently of environment.

Divergent natural selection between environments plays a critical role at some stage in all the other models of speciation by sexual selection (Table 8.1). Mate preferences may diverge if there is spatial variation in selection on the male trait (Lande, 1982), as an incidental effect of divergent natural selection on sensory systems (Ryan and Rand, 1993), or as a result of changes between environments in the most efficient signals (Endler, 1992; Endler and Basolo, 1998; Schluter and Price, 1993). Because the displays that are the target of female choice are essentially arbitrary, however, even

Table 8.1 Models of speciation incorporating sexual selection. The models are grouped according to whether divergent natural selection between environments is the ultimate cause of divergence in mate preference (a variant of ecological speciation) and those in which divergence is initiated by other processes (nonecological speciation). Reinforcement, which may involve sexual selection (Liou and Price, 1994), is not listed but it may be a part of either ecological or nonecological speciation depending on the initial causes of reduced hybrid fitness

Model	Mechanism of initial divergence of preferences	Role of sexual selection in divergence of preferences	Role of divergent natural selection	Source
Ecological speciation				
Fisherian with spatial variation in selection	Natural selection	Amplify differences	Divergent optima in male trait	Lande (1982)
Pre-existing bias	Correlated response to natural selection	None	Divergence in sensory traits	Ryan and Rand (1993)
Sensory drive with contrasting environments	Natural selection	Various	Modify preferences	Endler (1992); Schluter and Price (1993)
Competitive speciation	Natural selection	Amplify differences; reduce hybrid fitness	Reduce fitness of intermediates	Van Doorn et al. (1998); Kondrashov and Kondrashov (1999)
Nonecological speciation				
Fisherian runaway	Drift, mutation	Amplify differences	None	Lande (1981); Kirkpatrick (1982); Payne and Krakauer (1997); Pomiankowski and Iwasa (1998); Higashi et al. (1999); Turner and Burrows (1995)
Chase-away	Mutation	Amplify differences	None	Rice (1998); Holland and Rice (1998)

those models that invoke divergent natural selection often also allow for the possibility of divergence via genetic drift, for example if different displays yield equivalent benefits to females. Reinforcement may involve sexual selection (Liou and Price, 1994; Kirkpatrick and Servedio, 1999), and it may be regarded as a part of ecological or nonecological speciation depending on the initial causes of reduced hybrid fitness.

The great variety of models in which it plays a role suggests that sexual selection is not a unitary process driving speciation. Rather it may play a role in most models of speciation, whether ecological or nonecological, sympatric or allopatric, and whether fully a by-product of divergent natural selection or additionally involving reinforcement. A demonstration that sexual selection is involved in the speciation process therefore does not by itself much restrict the range of speciation models under investigation. The mechanisms driving the evolution of mate preference (or resistance) must still be identified.

8.2.4 Remarks

The above is not an exhaustive list of speciation models and I have overlooked finer distinctions between the types of models listed. Nevertheless, it highlights some mechanisms by which divergent natural selection between environments may lead to reproductive isolation, and how sexual selection may assist the process. For example, one speciation scenario not discussed is the evolution of reproductive isolation arising from genetic changes accompanying a coevolutionary 'arms race' between a species and its enemies or mutualists (the 'Red Queen' process; Van Valen, 1973). The process is like that of antagonistic coevolution between the sexes, described above. It should perhaps be regarded as nonecological speciation if the primary cause of differentiation between populations is the order of appearance of different advantageous mutations, but ecological speciation if the primary cause of divergence is change in the constellation of natural enemies.

The various hypotheses of speciation by divergent natural selection elaborate the idea of ecological speciation presented by Dobzhansky, Mayr, and others. Several laboratory experiments have verified the plausibility of the general idea. This does not mean the idea explains most speciation events in adaptive radiation because other processes may also give rise to new species. Moreover, there is no reason to think that the mechanism of speciation will be the same in different radiations or even at different stages within the same radiation. Given enough time, genetic drift, accumulation of different advantageous mutations, and other nonecological processes will inevitably lead to reproductive isolation between separate populations even if they inhabit similar environments. However, divergent natural selection can cause reproductive isolation to evolve rapidly, probably more quickly than by most other processes. The question is: does it commonly do so?

How should we proceed to test the ecological speciation hypothesis in nature? The bewildering diversity of possible speciation mechanisms presents a sizeable challenge to the researcher interested in distinguishing them. One profitable approach may be to devise general tests that do not rely too heavily on model particulars. For example, it is possible to test a role for divergent natural selection whatever the details, and this would be a significant first step. The subsequent step is then to determine the mechanism by which divergent selection exerted its effects. As the next section will show, more progress has been made in meeting the first of these two challenges.

8.3 Tests of ecological speciation

Under the ecological theory of adaptive radiation, reproductive isolation evolves ultimately from divergent natural selection. Speciation is rapid because reproductive isolation evolves quickly when populations are in contrasting environments and divergent natural selection is strong. This hypothesis makes a variety of predictions, some of which have been addressed by data. Here I review tests bearing on the evolution of pre- and postmating isolation.

8.3.1 *Niche shift and rate of evolution of reproductive isolation*

A straightforward test of ecological speciation compares the rate at which reproductive isolation evolves between regions differing in the frequency and strength of divergent natural selection. For example, are speciation rates on islands higher than on mainlands (Fig. 7.7) because reproductive isolation evolves more quickly there? Extremely limited data are available from laboratory measures of reproductive isolation in *Drosophila* (Fig. 8.3). The average strength of pre- and postmating isolation is similar between species or subspecies of Hawaiian picture-winged flies (*planitibia* subgroup) than between continental *Drosophila* of similar age, where age is measured by genetic distance. The sample size of Hawaiian observations is small and the observations are not phylogenetically independent. Moreover, both measures of reproductive isolation were made in the laboratory and so do not include isolation requiring a specific ecological context. Additional caution is warranted in this case given that so little information exists on the frequency and strength of divergent natural selection on ecologically relevant traits in Hawaiian and continental *Drosophila*. Nevertheless, the results show the kind of data needed to compare rates of evolution of reproductive isolation.

A second test compares the strength of pre- and postmating isolation among equal-age pairs of sister populations varying in the degree of similarity of their environments. If ecological speciation is the norm, then we would predict that reproductive isolation should evolve most quickly between sister populations undergoing extensive ecological differentiation, assuming that the latter reflects strong divergent natural selection. Unfortunately, forces other than divergent natural selection can also produce a correlation between niche shift and reproductive isolation. Chance dispersal may create a habitat shift in one of two sister populations and instantly produce premating isolation through an alteration of breeding time or location. Tests of the second prediction

198 • The ecological basis of speciation

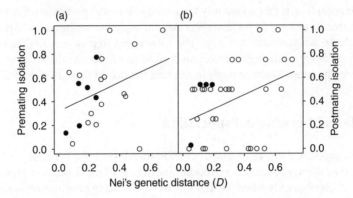

Fig. 8.3 Levels of (a) pre- and (b) postmating isolation (0 = none, 1 = full) between allopatric populations and species of *Drosophila* in Hawaii (●) and on continents (○). Only species having Nei's (1972) genetic distance less than 0.75 are included. Postmating isolation is postzygotic isolation (hybrid sterility and inviability). Regression lines are based on continental points only and are not corrected for phylogeny. The data are from Coyne and Orr (1989, 1997), and were kindly supplied by H.A. Orr and J.A. Coyne.

should therefore consider only cases in which selection is the probable cause of niche shift. Alternatively, if the niche shift itself was caused by other agents, then the test should compare the portion of reproductive isolation that accumulated after the shift.

This second prediction has not been systematically tested with species in nature. Indirect evidence in support is the observation that many of the youngest species on earth are strongly differentiated ecologically, such as Galápagos finches (Grant, 1986) and postglacial fishes (Schluter, 1996b; Table 3.2), but the pattern lacks a control and there are exceptions. For example, young, closely related species of cichlid fishes in Lakes Malawi and Victoria seem to show only subtle differences in feeding behaviour and morphology whereas they are often strongly differentiated in colour (Turner, 1994; Seehausen and Bouton, 1997; Seehausen *et al.*, 1998, 1999; Albertson *et al.*, 1999).

8.3.2 'Background' rates of nonecological speciation

Polyploid speciation is a relatively common and recognizable process of speciation in plants (Niklas, 1997; V. Grant, 1981; Ramsey and Schemske, 1998). The mechanism is unambiguously nonecological because reproductive isolation between diploid parent and tetraploid descendant populations, for example, does not build by divergent natural selection (though the success and subsequent evolution of the tetraploid species may depend greatly on ecological circumstances; Rieseberg, 1997; Ramsey and Schemske, 1998). Rather, reproductive isolation arises automatically via the low fertility of triploid hybrids. For these reasons, instances of polyploid speciation may be useful as a barometer of nonecological speciation.

If speciation is rapid during adaptive radiation because of strong selection accompanying shifts to new environments (which leads to more rapid evolution of reproductive isolation), then the background per capita rate of polyploid speciation should remain unaltered even as the total speciation rate rises. However, if speciation rates are up because of higher population persistence in novel environments, then rates of polyploid speciation should rise with the overall rate of speciation. Distinguishing these alternatives would provide a novel test of changing speciation mechanisms in adaptive radiation.

Polyploid speciation has not been described for any speciation event within the best-known plant radiations in Hawaii. The Hawaiian silversword alliance is itself tetraploid, but the genome doubling event predates the ancestry of all extant Hawaiian species. The shortage of polyploid speciation events within the most spectacular radiations might indicate that polyploidy is unimportant relative to ecological mechanisms in producing new species in adaptive radiations. However, this conclusion would be premature since measures of polyploid speciation rates are lacking from control taxa on mainlands.

8.3.3 Ecological basis of postmating isolation

More direct evidence for ecological speciation would be provided by a demonstration that ecological mechanisms directly reduce the fitness of hybrids between sympatric species that occasionally interbreed yet lack strong genetic mechanisms of postmating isolation. By 'ecological' mechanisms I mean postmating isolation stemming from traits in the hybrid that are disadvantageous in natural environments (Price and Waser, 1979; Shields, 1982; Waser, 1993; Schluter, 1998). Ecological postmating isolation may arise because an intermediate phenotype renders the hybrid less efficient at capturing prey in the wild, or because intermediate defences leave the hybrid susceptible to predation and parasitism. In contrast, 'genetic' mechanisms of postmating isolation (Dobzhansky's 'discordant gene patterns') result from the breakup of favourable gene combinations in the parent species (epistasis) and from interactions between parental alleles leading to underdominance (Lynch, 1991; Waser, 1993). The chief practical difference is that ecological mechanisms of postmating isolation are detectable in the wild but usually vanish in the laboratory (i.e. if food is uniform and predators are absent), whereas genetic mechanisms are largely independent of environment and should be detectable in most laboratory environments as well as in the wild.

Genetic mechanisms of postmating isolation between the youngest species of an adaptive radiation are frequently weak or lacking, as indicated by high viability and fertility of hybrids in laboratory settings. Examples include Hawaiian and many other *Drosophila* (Templeton, 1989; Coyne and Orr, 1989), Hawaiian silverswords (Carr and Kyhos, 1981) (and indeed many perennial flowering plants; V. Grant, 1981; Gill, 1989; Rieseberg and Wendel, 1993; Mcnair and Gardner, 1998), some East African cichlid fishes (Fryer and Iles, 1972; Seehausen *et al.*, 1997), postglacial fishes (McPhail, 1984; Wood and Foote, 1990; Hatfield and Schluter, 1999), and

Heliconius butterflies (McMillan *et al.*, 1997). Most studies have examined only the F1 hybrids, whereas genetic mechanisms of postmating isolation are typically most pronounced in F2 and backcross hybrids (e.g. Lynch 1991). However, in some of these examples fitness is also high in backcrosses and/or the F2 generation, such as *Heliconius* butterflies (McMillan *et al.*, 1997), three-spine sticklebacks (Hatfield and Schluter, 1999), and Lake Victoria cichlids (Seehausen *et al.*, 1997). The persistence of such species implies that some form of selection against intermediates takes place in the wild, possibly natural selection stemming from environment.

Evidence of speciation in sympatry likewise implies some form of divergent natural selection against intermediates, possibly ecological (but not necessarily: it may arise instead from divergent sexual selection; e.g. Higashi *et al.*, 1999). The most compelling examples of sympatric speciation are the cichlids from Barombi Mbo and other crater lakes in Cameroon, West Africa (Fig. 8.4). Phylogenetic studies using mtDNA indicate that the species within a crater basin constitute a monophyletic group: all are more closely related to species within the crater basin than any is to candidate ancestral cichlid species outside the basin. The only alternative to sympatric speciation is that the species in the crater formed from a series of separate invasions to the crater by an ancestral form, but that all trace of multiple ancestry was erased by mtDNA gene flow among species within the crater. mtDNA gene flow is common between young sympatric fish species, and has more than once led researchers astray (Taylor *et al.*, 1997). Nevertheless, the possibility seems far-fetched in the crater lake cichlids given that no candidate immediate ancestors for the newest species inside the

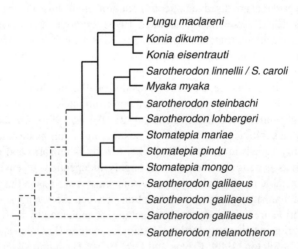

Fig. 8.4 Phylogeny of cichlids of Barombi Mbo crater lake, based on mtDNA sequences (Schliewen *et al.*, 1994). Bold lines indicate species within the crater lake. Dashed lines indicate closely related species and subspecies from nearby drainages. Modified from Schliewen *et al.* (1994).

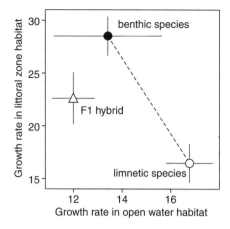

Fig. 8.5 Mean growth rates (mg/day) of two three-spine stickleback species (limnetic and benthic) from Paxton Lake, British Columbia, and their F1 hybrids. The dotted line connects the means of the two parent species. Modified from Hatfield and Schluter (1999).

crater can be found outside of it (Schliewen *et al.*, 1994). The mechanism of selection driving sympatric speciation in cichlids has not been discovered.

Direct measures of ecological selection against hybrids are few. F1 hybrids between the limnetic and benthic species of three-spine stickleback have a high fitness in the laboratory but an intermediate phenotype that compromises their ability to acquire food from the two main habitats of native lakes. Their ability to seize and retain small, evasive zooplankton in open water is inferior to that of the limnetic species. Their rate of intake in the littoral zone is lower than that of the benthic species, mainly because the F1 hybrids do not take the larger prey (Schluter, 1993). These feeding differences translate into slower growth of F1 hybrids relative to either of its parent species when in the habitat of that parent (Fig. 8.5).

Ecological selection also implies that the strength of postmating isolation should change as environments change, perhaps to the point of increasing hybrid fitness and causing the collapse of the parental species pair. Something like this happened in a natural experiment on a Galápagos island. Grant and Grant (1992, 1993) recorded the fate of offspring of crosses between *Geospiza fuliginosa* (small ground finch) and *G. fortis* (medium ground finch) over 20 years on the island of Daphne Major. *G. fuliginosa* is an uncommon but regular immigrant to the island, and many adults hybridize with *G. fortis*. The offspring are intermediate in beak size between the parent species and consume mainly small, soft seeds also eaten by *G. fuliginosa* (Grant and Grant, 1996). Small seeds are typically much less abundant than the large, hard seeds eaten by *G. fortis*, and hybrid survival is correspondingly poor (Fig. 8.6). However, food conditions were dramatically changed in the years after record rains associated with an El Niño event, and this elevated hybrid survival to a level not less than pure

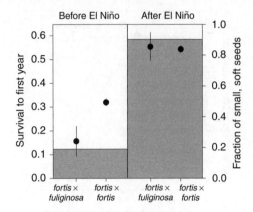

Fig. 8.6 Survival of F1 hybrids between two Galápagos ground finches (*Geospiza fortis* and *G. fuliginosa*) before and after the major El Niño event of 1982–83. Survival is the fraction of fledglings still alive after one year. Hybrid survival is compared with that of pure *G. fortis* (pure offspring of *G. fuliginosa* were rare). Vertical bars denote standard errors. Shaded columns (right-hand scale) indicate the abundance (mg/m^2) of small, soft seeds as a proportion of all available seeds. Modified from Schluter (1998), with permission of Oxford University Press, using the data of Grant and Grant (1992, 1993).

G. fortis. Hybrids suffered no reduction in fertility through this period (they mated mainly to *G. fortis*). The pair of species essentially ceased to be biological species on Daphne. *G. fuliginosa* persists only because of recurrent immigration from another island.

A similar dynamic may be under way in the hybrid zone between the plants *Ipomopsis aggregata* and *I. tenuituba* (Campbell *et al.*, 1997). The two species differ in corolla width, and intermediate corolla widths are at a disadvantage when both the major pollinators (hummingbirds and hawkmoths) are present. However, hawkmoths are rare in the zone and *I. aggregata* appears to be advancing at the expense of *I. tenuituba*.

Several other studies have explored the ecological basis of hybrid fitness. Life histories of host races of the apple maggot *Rhagoletis pomonella*, particularly the timing of diapause, appear to be adaptations to the divergent phenologies of their different host plants, and hybrid offspring of individuals that switch hosts are heavily disadvantaged as a consequence (Feder, 1998; Filchak *et al.*, 1999). Craig *et al.* (1997) showed that F1 and F2 hybrids between two host races of the fly, *Eurosta solidaginis*, survived poorly on the host plants of their parents, although the pattern of fitnesses was complex. An extended season of leaf production by hybrids between *Populus* species may explain the far higher levels of insect herbivory they experience (Floate *et al.*, 1993). Reciprocal transplants along an elevational gradient of two subspecies of sagebrush, *Artemisia tridentata*, and their hybrids indicated that each of the three populations has highest fitness in its own environment, with the hybrids most fit at intermediate elevation (Wang *et al.*, 1997). Hybrids in the last two studies were not F1 or F2 hybrids, but rather were individuals from populations of hybrid origin.

A tendency for hybrids to do worse than the parents in the habitat of those parents is not universal. For example, F1 and F2 hybrids between the irises *Iris fulva* and *I. hexagona* grow as well or better than the parent species in the native habitats (Emms and Arnold, 1997). However, the species are characterized by strong genetic incompatibilities (hybrid *Iris* have low pollen fertility and poor germination). Clearly, ecological factors may play a role even when genetic incompatibilities are present, but their role in the origin of these incompatibilities is unclear. In contrast, the ecological component of hybrid fitness may yield more direct evidence of the role of divergent natural selection between environments in the origin of reproductive isolation when genetic incompatibilities have not yet built up.

These examples suggest that when genetic mechanisms of postmating isolation are weak, hybrids in nature may often be selected against because they fall between the niches of their parents. They are best adapted to intermediate environments that do not exist. The significance of such mechanisms to speciation is twofold. First, any environmental agent that preferentially removes hybrids helps forestall the collapse of sympatric species by hybridization. Second, such mechanisms favour further divergence between species and may have contributed to their origin.

8.3.4 *Phenotypic divergence and degree of reproductive isolation*

Evidence that speciation resulted from divergent natural selection is gained when traits conferring adaptation to alternative environments are the basis of reproductive isolation, or are genetically correlated with traits that are the basis of reproductive isolation. For example in the monkey flower (*Mimulus guttatus*), alleles conferring tolerance to soils contaminated with copper are lethal when combined in the offspring of crosses with plants from uncontaminated soils (Macnair and Christie, 1983). Reproductive isolation between two other monkey flowers, *M. cardinalis* and *M. lewisii*, is also associated with differences in floral traits that attract different pollinators. *M. lewisii* has broad, flat, pink petals with yellow nectar guides, small nectar volume, and is primarily pollinated by bumblebees, whereas *M. cardinalis* has a narrow tubular corolla and large nectar rewards and is pollinated by hummingbirds. These divergent adaptations to different pollinators appear to be the basis of premating isolation, as pollinators used artificial F2 hybrids according to the proportion of genes from the preferred parent (Schemske and Bradshaw, 1999).

Body size is strongly divergent between sympatric stickleback species, and several lines of evidence suggest that this difference is the result of contrasting natural selection between the habitats they mainly exploit (Schluter and McPhail, 1992; Schluter, 1998; Nagel and Schluter, 1998). In no-choice laboratory mating trials, body size was found to strongly affect the probability of interspecific hybridization. Interspecific spawnings occurred only between the largest individuals of the smallest species (limnetics) and the smallest individuals of the largest species (benthics) (Nagel and Schluter, 1998). Similar results have been found in other instances of size differentiation between closely related species. Body size is also an important component of assortative mating between sockeye and kokanee salmon (Foote and Larkin, 1988). In Galápagos finches, beak and body size and shape, which are strongly selected for efficient food exploitation, are also used as cues in interspecific mate discrimination (Ratcliffe and Grant, 1983).

Reproductive isolation may evolve as a correlated response to life history divergence in many insects (Miyatake and Shimizu, 1999). For example, development time in populations of *Drosophila mojavensis* has diverged between populations on different cactus hosts, and this trait is genetically correlated with behavioural traits causing reproductive isolation (Etges, 1998). Postmating isolation between the apple and hawthorn races of the apple maggot is linked to changes in the timing and duration of diapause (Feder, 1998). Orr's (1996) study of grasshoppers along an elevational gradient indicated that levels of life history divergence between populations at different elevations were correlated with the strength of postmating isolation between them.

8.3.5 Parallel speciation

One of the most compelling forms of evidence suggesting an ecological basis to speciation is the parallel evolution of traits determining premating isolation in different populations experiencing similar selection environments ('parallel speciation'; Schluter and Nagel, 1995). Repetitive shifts in traits causing reproductive isolation support the hypothesis of ecological speciation when they occur under similar environmental conditions, because nonecological mechanisms of speciation (e.g. genetic drift) are unlikely to produce a consistent association with environment. Parallel evolution has long been the basis of comparative methods for testing natural selection on morphology, physiology and behaviour (Harvey and Pagel, 1991) and the same logic can be used to test a role for selection in speciation.

Consider an ancestral species that independently gives rise to multiple populations in two new types of environments at the periphery of its range. Parallel speciation by natural selection occurs when reproductive isolation evolves between descendant populations adapting to different environments, but not between descendant populations adapting to the same type of environment. As an example, sympatric limnetic and benthic species of three-spine sticklebacks have arisen independently as many as four times in separate lakes (Schluter and McPhail, 1992; Taylor *et al.*, 1997; Taylor and McPhail, manuscript). Laboratory experiments indicate that reproductive isolation is strong between populations differing in phenotype whether they are from the same lake (e.g. Paxton Lake limnetic vs Paxton Lake benthic) or from different lakes (e.g. Paxton Lake limnetics vs Enos Lake benthics) (Fig. 8.7). However, premating isolation is absent between populations from different lakes having similar phenotypes (e.g. Paxton Lake limnetic vs Enos Lake limnetic). This pattern strongly implicates

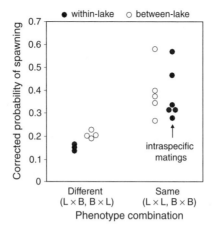

Fig. 8.7 Parallel evolution of premating isolation in benthic (B) and limnetic (L) three-spine sticklebacks. Each point represents a different pair of populations measured in no-choice laboratory trials. Filled symbols on the right are mating probabilities within populations. Open symbols next to them are from trials involving independent populations inhabiting the same type of environment in different lakes (e.g. the benthic species from Paxton Lake vs the benthic species from Priest Lake). Points on the left are from trials in which a limnetic and a benthic, either from the same lake or from different lakes, were paired. Independent populations inhabiting the same type of environment have a common mating preference, whereas those from different environments mate at a lower rate. Modified from Rundle et al. (2000).

natural selection in the origin of stickleback species, even though the details of the process that brought it about remain to be determined.

Parallel speciation is not confined to sticklebacks that are only a few thousand years old, but is also seen among populations separated for one or more millions of years. Around the northern hemisphere, hundreds of coastal stream-resident stickleback populations have formed and diverged independently from nearby marine populations (Orti et al., 1994). Different stream-resident populations share a variety of phenotypic characteristics such as small size, deep bodies, reduced armour, and greenish body colouration (Bell and Foster, 1994). They also appear to share the characteristics that determine mating compatibility. Mating trials using one stream and one marine population from each side of the Pacific Ocean (British Columbia and Japan) revealed that spawning frequency between individuals from the two regions is higher when they are from the same type of environment than when they are from different environments (McKinnon et al., manuscript).

Freshwater amphipods, *Hyalella azteca*, occur in two types of environments. Habitats with predatory sunfish contain a small-bodied ecotype whereas habitats lacking fish predators contain a large-bodied ecotype. Size differences between populations are genetically based (Wellborn, 1994) and are unrelated to electrophoretic distance (i.e. to phylogeny) (McPeek and Wellborn, 1998). Mating trials show that independent populations of the same ecotype readily interbreed whereas populations of different ecotype do not (McPeek and Wellborn, 1998). Again, environment not history predicts reproductive isolation.

Funk (1998) examined levels of premating isolation between populations of the leaf beetle, *Neochlamisus bebbianae*, which exploits different host plants in different parts of its range. Mating trials were carried out between individuals from two populations adapted to maple (*Acer*), one adapted to birch (*Betula*), and one from willow (*Salix*). Leaves from the host plants were present in half the trials and absent from the other half. Reproductive isolation was strong between maple and birch populations, and between maple and willow populations (isolation between the willow and birch populations was not tested), but not between the two maple populations. This is despite the fact that mtDNA divergence was greater between the two maple populations tested than between one of the maple populations and the willow population. Invariably, the populations on different hosts were strongly isolated whether or not they were closely related. Environment was the only consistent predictor of mating compatibilities.

8.3.6 A hypothetical selection experiment

The following experiment, which has never been carried out, would constitute another strong test of ecological speciation between two native species. The idea is to determine whether natural selection would rebuild premating isolation between species whose genomes have been blended by hybridization (Schluter, 1998; Coyne and Orr, 1998). The design would have two steps. The first creates a hybrid population by crossing two ecologically divergent species. This hybrid population is then taken through a series of generations intended to eradicate linkage disequilibrium between genes inherited from the same parent species. The hybrids should show some postmating isolation from both parent species, but its severity should be less than that between the parent species themselves.

The second step places lines established from this hybrid blend into environments similar to one or the other of those of their wild parents. The hypothesis of ecological speciation predicts that divergent natural selection will cause reproductive isolation to rebuild between hybrid lines placed in different environments. It also predicts that reproductive isolation should decay simultaneously between the parent species and the hybrid lines raised in its environment. The experiment is reminiscent of selection experiments on *Drosophila* by Dodd (1989; Fig. 8.1) and Kilias *et al.* (1980), but its goal is to test for selection's role in the formation of two real species in nature. The experiment requires that genetic mechanisms of postmating isolation are not too strong so that hybrids can be formed and bred with little loss of parental alleles.

8.3.7 Ecotones and hybrid zones

A further test examines geographical gradients in adaptive characters differing between species that form narrow hybrid zones where their ranges abut (Hewitt, 1989).

If postmating isolation is a by-product of interactions between alleles at the loci responsible for adaptive trait differences between species, then the location at which trait means shift from predominantly one extreme to predominantly the other extreme should coincide with the location of the zone in which hybrid breakdown occurs (e.g. Orr, 1996). Alternatively, if the genes responsible for hybrid breakdown are not those responsible for adaptive trait differences between the species, then the geographical gradients in ecologically important traits should wander away from the locations of hybrid zones.

The majority of hybrid zones are broadly associated with environmental transitions (Hewitt, 1989), and the locations of transitions in the means of ecologically important traits often coincide with the location of these zones. This supports the hypothesis that hybrid breakdown is the result of fixation of genes responsible for differences between the species in adaptive characters. The problem with this evidence, however, is that nonecological hypotheses predict the same pattern if there has not been sufficient time for geographical gradients in ecologically significant traits to become uncoupled from zones of hybrid breakdown.

8.3.8 *Remarks*

The trouble with speciation is that it may occur by a variety of mechanisms. Indeed, given enough time speciation is inevitable between allopatric populations even if external environments are the same and divergent natural selection is absent. Testing the ecological theory requires that we determine whether, in adaptive radiation, reproductive isolation typically evolves sooner by divergent natural selection. Another problem is that differences between the species that contribute to reproductive isolation continue to accumulate after speciation is already completed, perhaps at an accelerating rate (the 'snowballing effect'; Orr, 1995). Therefore testing the ecological theory requires that we investigate the earliest stages of the process.

The present section has listed a variety of approaches one might take to test the hypothesis of ecological speciation. Most of these have been attempted only once or twice, and a few of them have not yet been tried at all. This underscores the preliminary state of the field, and means that we are unable to generalize from the observations that have accumulated thus far. Nevertheless, results from a number of studies are promising and indicate that ecological speciation is a viable hypothesis of speciation in adaptive radiation, but it is too soon to say whether it is the dominant mechanism.

8.4 Divergent sexual selection

Theoretical models suggest ways in which sexual selection may contribute to speciation (Table 8.1). In most roles, speciation involving sexual selection is an important extension of the theory that reproductive isolation evolves ultimately from divergent natural selection. In other roles, however, sexual selection enhances the rate of nonecological speciation. Distinguishing these processes for speciation in adaptive radiation is the major challenge.

The main observation fuelling the idea that sexual selection drives nonecological speciation in adaptive radiation is that closely related species in some of the most spectacular adaptive radiations, such as the Hawaiian *Drosophila* and the cichlid fishes of the East African Great Lakes, differ more in colour and other secondary sexual characteristics than in morphological traits linked to resource exploitation. This has led to the suggestion that divergent sexual selection may generate new, ecologically similar species at a high rate in the absence of differences in resources or environment. Ecological and morphological divergence in adaptive radiation result later when competition between similar species leads to ecological character displacement (Galis and Metz, 1998). On the other hand, the observation may be incorrect: resource differences driving divergence may simply be subtle. Or, divergence in mating signals may be driven by differences between environments in factors other than food, such as predators or interference competitors. These considerations indicate that distinguishing ecological from nonecological mechanisms of speciation involving sexual selection may not be easy.

My goal here is to briefly assess progress in testing whether sexual selection is involved in speciation, and whether it is an agent of divergent natural selection stemming from difference in environment or instead operates independently of environment.

8.4.1 Comparative tests of sexual selection in speciation

Some evidence for sexual selection in speciation is derived from a small number of comparative studies linking species richness of clades with indirect measures of the strength of sexual selection acting within populations (Table 8.2). In most cases the studies support a link between sexual selection and speciation (the nonavian

Table 8.2 Comparative tests of a link between sexual selection and speciation, mainly using sister-group comparison. Results in support indicate that taxa presumably subject to stronger sexual selection (more dimorphic, more ornamented, more polygamous) contain more species than control taxa. 'No support' indicates that the taxon with greater sexual dimorphism (dioecious flowers) has fewer species than monoecious sister taxa

Taxon	Trait	Results	Source
Passerine birds	Sexual dichromatism	Support	Barraclough *et al.* (1995)
Birds	Feather ornamentation	Support	Møller and Cuervo (1998)
Birds	Mating system	Support	Møller and Cuervo (1998)
Birds	Mating system	Support	Mitra *et al.* (1996)
Angiosperm plants	Monoecy/dioecy	No support	Heilbuth (2000)

example is the sole exception: sexual dimorphism in flowers is associated with significantly lower species richness). The traits being compared between lineages are not direct measures of the rate of preference evolution but presumably they are indicators. The reasons why preferences evolve at a higher rate in some lineages than others are not known. Divergent natural selection is not entirely ruled out, but it is not easy to see why strength of divergent natural selection should vary with mating system.

A second line of evidence for sexual selection in speciation comes from the observation that the shape of male genitalia is among the most rapidly evolving traits in insects (Eberhard, 1985). Two hypotheses involving sexual selection are consistent with this finding. In the 'lock and key' model, population differentiation in male genitalia are the outcome of reinforcement or reproductive character displacement, enhanced premating isolation driven by selection to minimize the risk of producing inferior hybrid offspring. The alternative hypothesis is that the trend results from a 'chase-away' arms race between males and females of a species (see Section 8.2.3). The outcome in both cases is greater reproductive isolation. In agreement with the second hypothesis, Arnqvist (1998) showed that the rate of male genital divergence between species is correlated with mating system: divergence in clades of species in which females mate multiple times is double that in clades in which females mate only once. The chase-away hypothesis is also more consistent with observations that sperm and other male reproductive proteins are among the most rapidly evolving proteins in *Drosophila* (Aguade *et al.*, 1992; Thomas and Singh, 1992; Nurminsky *et al.*, 1998). This appears to be part of the reason why male offspring of interpopulation or interspecies crosses are more often sterile than female offspring (Wu and Davis, 1993; Wu *et al.*, 1996; Orr, 1997). If the 'chase-away' interpretation is correct, then differences between environments would seem to matter little in the evolution of this particular form of reproductive isolation. The only uncertainty is over whether the chase-away process frequently yields reproductive isolation before other mechanisms accomplish it first.

Comparative evidence is beginning to emerge suggesting the importance of sexual selection in speciation of cichlid fishes of the East African Great Lakes (Seehausen *et al.*, 1997). The cichlid species of Lakes Malawi and Victoria are very young, and genetic mechanisms of postmating isolation are weak or nonexistent. Males are usually brightly coloured and the colours of sympatric sister species usually differ. One is usually red or yellow and the other is blue. These colours correspond to the location of absorbance peaks of the retinal pigments. Red and blue also stand out against background light in clear water. By manipulating light spectra, Seehausen *et al.* (1997) showed that colour is the basis of assortative mating between the two Lake Victoria sister species, *Pundamilia nyererei* and *P. pundamilia*. Differences in mating patterns correspond to differences between the species in the relative heights of absorbance peaks of the retinal pigments. The red species is more sensitive to red light whereas the blue species is more sensitive to blue light. The two species mated indiscriminantly under monochromatic light and the species pair essentially collapsed there. In rock-dwelling cichlids of Lake Victoria, colours evolve to more vivid extremes in clear water and species diversity is higher there than in murky water, suggesting that species persistence requires a broad light spectrum.

Divergent sexual selection has not been measured in these cichlids (only assortative mating has) but it is implicated by the rapid turnover of male colouration between closely related species, and also by the finding of genetic variation in colour-based preference of individual females in a polymorphic species, *Neochormis omnicaerulus*, that Seehausen *et al.* (1999) regard as possibly a transient stage in speciation by sexual selection. The mechanism driving divergence in female preference for colour is unknown, and the findings do not distinguish between ecological and nonecological models of speciation by sexual selection (Table 8.1). The correlation between water transparency and cichlid diversity does not yet rule out the possibility that colour is involved in the coexistence of these territorial cichlids for reasons other than sexual selection by male and female choice.

8.4.2 Measurements of divergent sexual selection

If divergent sexual selection is involved in speciation, then we should be able to measure it. Even better, it may sometimes be possible to correlate the direction of sexual selection with agents of environment if these are driving divergence in mate preference. Only modest progress has been made here.

One of the clearest examples of divergent sexual selection is from the pied flycatcher, *Ficedula hypoleuca* (Sætre *et al.*, 1997). In populations in which males have the typical black-and-white plumage, females prefer black-and-white males over brown males. In contrast, in populations where brown males predominate, females prefer to mate with brown males rather than black-and-white males (Fig. 8.8). This difference in the direction of female preference is correlated with the presence and absence of the collared flycatcher *Ficedula albicollis*, the sister species to the pied flycatcher and whose males have a striking black-and-white plumage. Two selective forces may explain the preference for brown males in sympatry. First, the collared flycatcher is dominant over the pied flycatcher in competition for nesting sites, and black-and-white pied flycatcher males suffer more interspecific interference than brown males (Alatalo *et al.*, 1994). Female pied flycatchers mating with brown males may thus experience less interference around the nest site than those mating with black-and-white males. Second, a preference for brown males in the zone of sympatry leads to a lower probability of hybridization (Alatalo *et al.*, 1994; Sætre *et al.*, 1997). Therefore, divergent natural selection is probably driving divergence of mate preference between sympatric and allopatric populations of the pied flycatcher, and also between sympatric pied and collared flycatchers. Whether this has produced any degree of reproductive isolation between sympatric and allopatric pied flycatchers as a by-product is unknown.

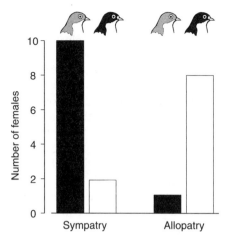

Fig. 8.8 Divergent sexual selection on male colour pattern in the pied flycatcher, *Ficedula hypoleuca*. The mating success of males of two morphs, brown and black-and-white, is compared between areas where the collared flycatcher *Ficedula albicollis* is present (sympatry) and absent (allopatry). Divergent selection on preferences in the zone of sympatry may stem from aggression by collared flycatchers (Alatalo *et al.*, 1994) and/or reduced fitness of hybrids (reinforcement; Sætre *et al.*, 1997). Modified from Sætre *et al.* (1997), with permission of *Nature* and the authors.

Reinforcement, the enhanced levels of premating isolation driven by the reduced fitness of hybrid offspring, should generally create divergent sexual selection between young sympatric species. Furthermore, the process may be common (e.g. Howard, 1993). However, reinforcement usually takes place in the latter stages of the speciation process, and whether it is an ingredient of ecological speciation depends on whether divergent natural selection created the initial differences that later cause low hybrid fitness and/or assortative mating. This is unknown in most cases. Sticklebacks may represent a case of reinforcement driven by ecologically based selection against hybrids (Rundle and Schluter, 1998).

Another source of evidence of divergent sexual selection comes from studies of mating success of hybrids between young sympatric species. Frequently the hybrids are found to suffer a mating disadvantage (Stratton and Uetz, 1986; Krebs, 1990; Price and Boake, 1995; Rolán-Alvarez *et al.*, 1995; Davies *et al.*, 1997; McMillan *et al.*, 1997; Noor, 1997; True *et al.*, 1997; Vamosi and Schluter, 1999). The magnitude of this disadvantage is sometimes larger than other known components of postmating isolation between species. Such a finding is consistent with the idea that divergent sexual selection played a role in the origin of species, although it does not identify the factors driving mate preference evolution. A problem with this form of evidence, however, is that reduced hybrid mating success may be an incidental consequence of genomic incompatibilities between species that arose for other reasons.

8.4.3 Remarks

Strong sexual selection need not lead to speciation, as the guppy *Poecilia reticulata* can attest (Magurran, 1998). Despite strong sexual selection in many guppy populations, and high variation between populations in degree of female preference for male colouration, strong reproductive isolation between populations seems to be absent (Houde, 1997). Part of the reason may be that the colour preference in females is open-ended (Price, 1998). But it must be the case that mate preferences diverge at a high rate in some taxa and not others. The key is to understand the forces that cause mating preferences to diverge in the first place, and here little progress has been made. It is possible that natural selection plays a major role in determining the rate of preference evolution, but there is as yet little evidence for this.

In theory, sexual selection may generate new species at a high rate for reasons having nothing to do with divergent natural selection between environments (West-Eberhard, 1983; Dominey, 1984). The Fisher runaway process is one way in which this might happen (but Price (1998) regards this as unlikely, at least in birds). The 'chase-away' mechanism is more plausible, and evidence consistent with this hypothesis is mounting. While the evidence is still at an early stage, the possibility of speciation without divergent natural selection between environments would represent a substantial departure from classical ecological views on the causes of speciation in adaptive radiation.

8.5 Discussion

The idea was prominent by about the middle of the past century that divergent natural selection between populations in distinct ecological environments was a principal cause of speciation in adaptive radiation ('ecological speciation' hypothesis; Mayr, 1942; Dobzhansky, 1951; Simpson, 1953). At the time no compelling examples were presented and no evidence from nature was forwarded in support of the hypothesis. The idea was popular because it was plausible and it fit well with the observation that speciation and morphological differentiation often went hand-in-hand. It explained rapid speciation under conditions of high ecological opportunity.

Today, speciation remains the least understood part of adaptive radiation. The main stumbling block has been that other mechanisms leading to the evolution of reproductive isolation are so difficult to rule out, except perhaps speciation by polyploidy. As a result, over the past 50 years only a few compelling candidate cases of ecological speciation have emerged, a glaring reminder of the distance still to go. Nevertheless, progress has been made in a number of taxa. Research is nearing the stage where we will be able to point to at least two species in nature and confidently state that they formed by divergent natural selection. In this sense the original theorists of adaptive radiation were at least partly right, but the generality of the mechanism remains to be determined.

The evidence in favour of ecological speciation includes measurements of ecological selection against hybrids, the finding that traits under divergent natural selection are the basis of reproductive isolation (or are genetically correlated with traits that form the basis of reproductive isolation), and that mating preferences sometimes evolve in parallel between populations evolving independently in similar environments. Observations that stable sympatry between ecologically differentiated species is achieved despite gene flow between them, and evidence for sympatric speciation of ecologically differentiated species, also lend support to the hypothesis. The latter observations are incomplete, however, until the mechanisms are identified (e.g. sexual selection against hybrids may also promote persistence in the face of gene flow).

Ecological speciation remains a viable explanation for the high rate of speciation in adaptive radiation: species originate more rapidly when divergent natural selection is strong. Evidence for this, however, is mixed. Divergent natural selection can clearly cause reproductive isolation to evolve much sooner than most nonecological mechanisms, and it may allow stable sympatry after a period of time too brief for genetic drift and other processes to effect much change. However, speciation rates may also rise during adaptive radiation if ecological opportunity improves the viability of populations and reduces extinction. More evidence is needed to distinguish the contribution of these alternative mechanisms.

The addition of sexual selection to theories of speciation represents perhaps the single greatest recent theoretical advance in the area of speciation in adaptive radiation. Evidence is emerging that sexual selection is indeed involved in many speciation events, and that rates of speciation are correlated with indirect indicators of sexual selection. However, sexual selection is probably a part of most modes of speciation, and finding evidence for it does not much reduce the diversity of possible ways in which new species may have formed. In particular, empirical research has only begun to distinguish two major views of how sexual selection may contribute to speciation. One is that divergent sexual selection is a consequence of divergent natural selection on mate preferences stemming from external environment. Cases of divergent mate preferences in sympatry (generated via reinforcement or direct interactions between populations) exemplify this first hypothesis but may only explain the later stages of the speciation process. The second view is that sexual selection generates new species by a mechanism largely independent of environment. Of the mechanisms that have been suggested, evidence for the 'chase-away' process looks most promising, in which reproductive isolation arises as a by-product of antagonistic coevolution between the sexes. In most examples of adaptive radiation in which high ecological diversity is accompanied by many species, the causes of divergence in mate preferences are unknown.

The implications of rapid, nonecological speciation by sexual selection in adaptive radiation are profound. For example, it has been suggested that sexual selection can generate many new, ecologically identical species that then undergo ecological and phenotypic divergence via ecological character displacement only later (Galis and Metz, 1998). Under this hypothesis, speciation creates the conditions that generate divergent natural selection (i.e. many competing species). This would represent a complete reversal of the causal chain encoded in the ecological theory, whereby divergent natural selection is a precondition for speciation. There is presently no evidence that high species production generates divergent natural selection that leads to phenotypic divergence. Speciation without environment-based divergent natural selection may merely produce many largely allopatric, ecological equivalents that remain as such and contribute little to ecological and morphological diversification (Price, 1998).

9

Divergence along genetic lines of least resistance

> Almost any change is likely to have others carried along with it, more or less incidentally, by genetic correlation, by effects of an established growth pattern, or by secondary adaptation to the change, itself.
>
> —Simpson (1953)

9.1 Introduction

The idea that genetic variation in a population is limited, that traits covary genetically, and that the results of natural selection are consequently biased, is older than the ecological theory of adaptive radiation (Maynard Smith *et al.*, 1985). Simpson (1953) mentioned the idea (see the above quote) but made no attempts to determine whether in any adaptive radiation 'the realized path of evolution [was] more likely, in strictly genetic terms, than the paths not taken' (Futuyma *et al.*, 1995). A conceptual framework for addressing the question had not yet been developed and the ecological theory did not incorporate one. Yet, incorporating such a framework would benefit the study of adaptive radiation for at least one very practical reason: if genetics favours certain paths of evolution over others, then the direction and rate of phenotypic divergence is partly predictable. In this chapter I evaluate the success of recent attempts to predict divergence in adaptive radiation from genetic measurements of populations.

The genetic theory of natural selection on suites of continuously varying traits really began with the seminal paper by Lande (1979). Under assumptions to be discussed below, the path of short-term evolution is predictable from measures of directional selection and genetic covariance. This theory might be useful over the longer term, and this possibility forms the basis of the present chapter. My goal is not to review the genetic basis of phenotypic divergence and speciation (an area presently making tremendous gains; e.g. see Jones, 1998; Bradshaw *et al.*, 1998; Schemske and Bradshaw, 1999). Rather, I consider whether some knowledge of genetic variation in continuously varying traits permits us to predict better the direction and scope of adaptive radiations.

I begin with a brief summary of the genetic theory and its assumptions. Then I review existing evidence of genetic bias and its duration. My examples dwell mainly on morphological change. In principle, speciation is subject to the same rules. I discuss only briefly predictions that might be made concerning the evolution of reproductive isolation.

9.2 Quantitative genetic framework

Quantitative changes are the most regular feature of adaptive radiation at low taxonomic levels. Closely related populations and species usually differ in phenotypic traits by degree rather than by kind. These same traits tend to vary continuously within populations, and additive genetic variation at multiple loci is responsible for much of it (Falconer and Mackay, 1996; Roff, 1997). Evolution of such traits is therefore best handled using the framework of quantitative genetics.

Quantitative genetics ignores the effects of each locus and focusses on the net effects of all the genes that influence traits. The crucial genetic measure of an individual is its 'breeding value', the mean phenotype of all its potential offspring, assuming random mating, expressed as twice the deviation from the population mean. Falconer and Mackay (1996), Lynch and Walsh (1998), and Roff (1997) provide comprehensive reviews of the field. Below I give a sketch of the most relevant aspects for predicting biases in divergence.

9.2.1 The infinitesimal model

The theory of divergence I will apply is the standard one based on the 'infinitesimal model' in which breeding values for a trait are determined by the summed effects of many loci, each of which contributes a small part. Breeding values in populations are assumed to be normally distributed on some scale (e.g. logarithmic scale for most external dimensions). A mutation is just as likely to increase the value of the trait as decrease it, and mutations are not intrinsically advantageous or deleterious.

Recent research has suggested that many quantitative traits are actually influenced by relatively few genes, some of which have large effects (although the methods are biased in favour of detecting large effects; Lynch and Walsh, 1998), and whose effects on the phenotype may be strongly nonadditive (Falconer and MacKay, 1996; Roff, 1997). The consequences of these violations of assumptions are not serious over the short term, but prediction of long-term evolution are compromised because changes in gene frequencies will inevitably change the genetic parameters on which prediction depends (Turelli and Barton, 1994). This is the first of several important reasons why we might expect the predictive power of the theory to erode with time since divergence.

Other traits, such as survival, may violate the assumption of symmetric mutation and conform better to an alternative 'deleterious mutation' model in which the majority of mutations are unconditionally harmful and reduce trait value (Charlesworth, 1990; Kondrashov and Turelli, 1992; Houle, 1991; Rowe and Houle, 1996). Selection gradually removes deleterious mutations but the supply is continually replenished especially if the number of loci affecting the trait is large (Lynch *et al.*, 1999). Fitness components, and even fitness itself, may be best described by this model, as well as certain other traits for which 'more is always better' (other things being equal) such as general physiological efficiency or physical condition.

In this chapter I focus on morphological traits which seem to satisfy the assumptions of the infinitesimal model reasonably well. Presumably evolutionary shifts in traits such as beak size, in response to changing seed availabilities, would not typically involve increases in the frequencies of unconditionally deleterious alleles. Unfortunately, this does not completely solve the problem because deleterious mutations affecting fitness may have pleiotropic effects on morphology, and if such alleles are abundant they will tend to upset expectations from the simple theory. In the face of this uncertainty, it is best to adopt an empirical attitude and ask: does the theory predict reasonably well when applied to nature? The results shown below suggest that for morphological traits (and sometimes even for life history traits; Mitchell-Olds, 1996) the answer is often yes, at least coarsely.

9.2.2 V_A matters more than h^2

The fundamental aspects of adaptive evolution according to the infinitesimal theory are as follows. Evolutionary change in the mean $\bar{z}$ of a single trait across one generation in response to a bout of natural selection is proportional to the additive genetic variance V_A:

$$\Delta \bar{z} = V_A \beta. \tag{9.1}$$

The quantity β is the intensity of directional selection, and $\Delta \bar{z}$ is the change in $\bar{z}$ between generations, i.e. the response to selection. As discussed in Chapter 5, β is the steepness of the adaptive landscape at the original population mean $\bar{z}$. Additive genetic variance V_A is the component of phenotypic variance that determines average resemblance between parents and their offspring (i.e. excluding nongenetic variance and effects produced by dominance and epistasis). Guides to estimating V_A are found in Falconer and Mackay (1996) and Lynch and Walsh (1998).

An alternate equation is more often seen in textbooks:

$$\Delta \bar{z} = h^2 s, \tag{9.2}$$

where h^2 is the heritability of z (the additive genetic proportion of total phenotypic variance V_P) and s is the selection differential (the difference in the mean values before and after selection in the parent generation). This equation is mathematically equivalent to the first, since $h^2 = V_A/V_P$ and $s = V_P \beta$. But the latter equation leads to confusion when applied to natural selection because s is easily misinterpreted as selection intensity and h^2 is then regarded as an index of evolutionary potential. For example, one might conclude from eqn (9.2) that traits of low heritability, such as life history traits (Mousseau and Roff, 1987), should be less responsive to natural selection than traits of high heritability, such as morphological traits. However, low heritability does not necessarily mean that genetic variance is low. Additive genetic variance in life history traits (if measured as the squared coefficient of additive genetic variation) is no less than that in morphological traits, and therefore by this measure they are no less evolvable (Houle, 1992). Life history traits simply have a higher environmental component of variance (Price and Schluter, 1991; Houle, 1992) which has no immediate influence on the expected response to a given selection intensity β (eqn (9.1)).

9.2.3 Selection and genetic covariance

The multivariate analogue of eqn (9.1) is

$$\Delta \bar{z} = G\beta \qquad (9.3)$$

(Lande, 1979). G is now the genetic covariance matrix, a matrix whose diagonal elements $V_{A1}, V_{A2}, \ldots, V_{Am}$ are the genetic variances of the m traits. Its off-diagonal elements Cov_{Aij} are genetic covariances between pairs of traits i and j. $\Delta \bar{z}$ is the vector of evolutionary changes in the means of the m individual traits $\Delta \bar{z}_1, \Delta \bar{z}_2, \ldots, \Delta \bar{z}_m$. The selection gradient β is a vector whose elements $\beta_1, \beta_2, \ldots, \beta_m$ are the intensities of directional selection on individual traits. β points in the direction of the steepest rise in mean fitness (Chapter 5).

The crucial aspect of eqn (9.3) is that evolution in each trait is the result not only of selection directly on that trait but also of selection on other traits with which it covaries genetically. The predominant cause of genetic covariance is pleiotropy, in which allelic variation at a locus influences more than one trait. Linkage disequilibrium (nonrandom assortment of alleles at different loci) is a secondary cause of genetic covariance except in unusual cases.

Figure 9.1 shows what a genetic covariance matrix looks like (the upper triangular portion, not shown, is the mirror image of the lower triangular portion). The illustration reveals that leaf and flower dimensions in wild radish vary widely in the pattern of genetic association. Covariance is illustrated using ellipses. Each ellipse surrounds 95% of the estimated breeding values. Genetic covariance Cov_{Aij} between two traits i and j is a product of the genetic correlation between them (r_{Aij}) and the genetic standard deviations:

$$\text{Cov}_{Aij} = r_{Aij}\sqrt{V_{Ai}}\sqrt{V_{Aj}}. \qquad (9.4)$$

For a given genetic variance, a strong genetic correlation leads to a high genetic covariance. For example, the covariance between leaf width and leaf length is greater than that between leaf width and petal width (Fig. 9.1). Selection on leaf width is therefore expected to produce a stronger indirect response in leaf length than in petal width. Covariance also increases with genetic variance. For example, genetic

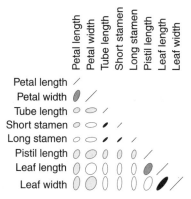

Fig. 9.1 Visualization of a genetic covariance matrix G for floral and leaf traits in wild radish, *Raphanus raphanistrum*. Ellipses surround 95% of breeding values for each pair of traits assuming a bivariate normal distribution. Lines along the diagonal represent additive genetic variance. Strength of genetic correlation is indicated by shading. All traits are drawn to the same scale so that ellipse sizes are directly comparable. G was computed using squared coefficients of additive genetic variance and covariance from data in Conner and Via (1993). Drawn using the ELLIPSE program (Murdoch and Chow, 1994).

covariance between leaf width and petal width is greater than that between short and long stamen filament lengths even though the genetic correlation between the pair of leaf traits is weak. The indirect response of petal width to selection on leaf length is therefore expected to be greater than the indirect response of short stamen length to an equal selection pressure applied to long stamen length.

Genetic variance and covariance are the genetic parameters governing phenotypic evolution by natural selection. Their effect on a population is to bias the path of evolution away from the direction of greatest fitness increase (Fig. 9.2; Lande, 1979; Via and Lande, 1985). The path to higher mean fitness is instead a curved one that in its initial stages follows the direction of greatest genetic variance, g_{max} (the long axis of the ellipse of breeding values). I refer to this direction as the genetic 'line of least resistance'. If genetic constraints are not severe, the bias decays with time and an optimum is eventually attained. However, if the genetic covariance between traits is sufficiently strong that little or no genetic variation exists in directions other than g_{max}, the population will evolve uphill in the direction of g_{max} and then stall on the adaptive hillside well below the optimum (Kirkpatrick and Lofsvold, 1992). As I show below, severe constraints do exist in natural populations in at least some directions even when all individual traits possess substantial levels of genetic variance.

When more than one optimum is present, genetic covariance may steer the population away from a nearby peak and toward a more distant one closer in direction to that of g_{max} (Fig. 9.2(b)). Even though selection initially favours an increase in the second trait z_2, and even favours a slight decrease in z_1, z_1 increases through time because it covaries genetically with z_2. The result is a substantially altered

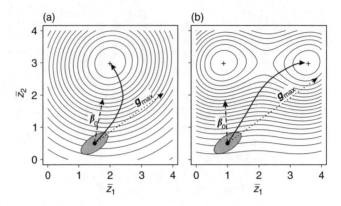

Fig. 9.2 Genetic covariance biases the direction of evolution in two traits, z_1 and z_2. Panels depict the initial mean of a population (filled circle) and the path of this mean over subsequent generations of selection (solid curved line) when there are one (a) and two adaptive peaks (b). Ellipses surround 95% of the breeding values in each population. The dotted line is g_{max}, the direction of maximum genetic variance within the population. Contours are increments of mean fitness $\overline{W}$ on the adaptive landscape. Adaptive peaks are indicated by '+'. β_0 is the selection gradient at the start, the direction of steepest ascent in mean fitness. Note that the path of evolution does not follow the gradient, but is initially biased in the direction of g_{max}.

course that may move populations between the domains of different adaptive peaks (Price *et al.*, 1993).

9.2.4 Genetic degrees of freedom

A large number of studies have now been conducted on a great many phenotypic traits. Basically, this work reveals that most traits are heritable in wild populations (Roff and Mousseau, 1987; Houle, 1992). Therefore most traits are capable of at least short-term change under natural selection, a result borne out by the large number of studies of artificial selection on single characters (Charlesworth *et al.*, 1982; Maynard Smith *et al.*, 1985; Bell, 1997; Roff, 1997) or pairs of traits (Weber, 1992). Yet the capacity of individual traits to change freely of others is reduced by genetic covariance between traits, which is also ubiquitous (Roff, 1997).

The concept of genetic 'degrees of freedom' is useful for summarizing bias and restriction potentially imposed by genetic covariance (Kirkpatrick and Lofsvold, 1992). This concept is visualized in Fig. 9.3, which plots the amount of genetic variance present in each of m fully independent directions. Each direction is a composite trait, a vector whose coefficients describe a linear combination of the original m traits. The first independent direction (g_{max}) by definition accounts for more genetic variance than any other. The second direction is perpendicular to the first and accounts for most of the remaining variance, and so on through m total directions.

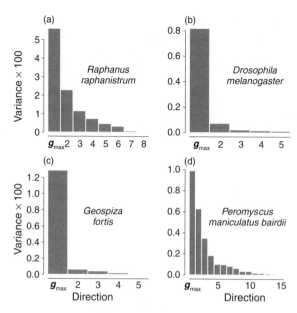

Fig. 9.3 Additive genetic variance in fully independent directions (combinations of trait values) of four G matrices. g_{max} is the most variable direction, '2' is the most variable direction orthogonal to the first, and so on. (a) Floral and leaf measurements in wild radish (Conner and Via, 1993; see Fig. 9.1); (b) body and wing dimensions in population of fruit flies (Shaw et al., 1995); (c) beak and body measurements of the Galápagos medium ground finch (Boag, 1983); (d) skull measurements in deer mice (Lofsvold, 1986). All calculations are based on additive genetic covariance of ln-transformed traits (b, c) or squared coefficients of additive genetic covariance (a, d). Variance of each direction is the corresponding eigenvalue of G. If necessary, G was 'bent' to ensure that all eigenvalues were nonnegative (Hayes and Hill, 1981).

A haphazard sample of G matrices from the published literature shows that genetic variance is usually tied up in a few directions, while the remaining directions may have little (Fig. 9.3). This unevenness may be described using the Levins' diversity index

$$L = 1 \bigg/ \sum p_i^2, \qquad (9.5)$$

where p_i is the proportion of total additive variance accounted for by direction i. $L = 1$ if all variance is in the first direction, and $L = m$, the number of traits, if variance is equitably distributed. L is therefore useful as a measure of effective genetic degrees of freedom. For the data sets analysed in Fig. 9.3, L is computed as (a) 2.9, (b) 1.3, (c) 1.2, and (d) 4.4. These numbers are remarkably low, on the order of 1/4 to 1/3 the number of traits in total. Therefore, additive genetic variance of individual traits is an inadequate measure of evolvability because genetic degrees of freedom are often far fewer than the number of heritable traits. The fewer the genetic degrees of freedom, the greater the potential bias in evolutionary response to selection.

Why genetic variance is greater in some directions than others is not fully understood, but ultimately a combination of mutation and selection must be responsible. A discussion of progress in understanding the maintenance of genetic variance is beyond the scope of this chapter. I ask instead what the consequences of existing patterns are for the direction and rate of divergence. Possibly, a better understanding of the forces that shape genetic variance will eventually improve our ability to make such predictions.

9.2.5 Short-term prediction in a Galápagos finch

The genetic theory outlined above should correctly predict evolutionary change across one generation of selection provided that genetic parameters are known for all the traits under selection. But a great difficulty stands in the way of applying the equations to nature: the true selection gradient β is unknown, indeed it is unknowable. This is because estimated selection intensities are biased if important traits under directional selection are left out of the analysis, traits phenotypically correlated with the subset of traits measured (Mitchell-Olds and Shaw, 1987). The predicted evolutionary response to selection will also then be biased if the missing traits covary genetically with the measured traits.

This problem of missing variables is potentially severe. Several instances are known in which single-variable predictions (i.e. based on eqn (9.1) or (9.2)) were badly off because selection on correlated traits was not accounted for. The worst cases involve heritable traits apparently under directional selection (e.g. clutch size, body mass, breeding date, and tarsus length in birds) that nevertheless consistently fail to show an evolutionary response (Price *et al.*, 1988; van Noordwijk *et al.*, 1988; Alatalo *et al.*, 1990; Price and Liou, 1989). It turned out that the bulk of selection was not on the measured trait at all but on an unmeasured, nonheritable 'body condition' that was phenotypically correlated with the measured trait, giving the appearance of selection on the latter. These worst-case examples involve traits having very different levels of genetic determination. Things are not expected to go as badly when the heritabilities of traits under selection are more equal. The discrepancy between observed and expected responses in the above example was resolved when a multivariate framework was adopted. In general, analysis of a manageable subset of traits, though imperfect, should usually be an improvement over analyses of one trait at a time (Lande and Arnold, 1983).

The ultimate test of the theory's utility is whether it works. Only one quantitative test has been carried out on a wild population, by Grant and Grant (1995) in a study of the medium Galápagos ground finch, *Geospiza fortis*. The outcome was wildly successful (Fig. 9.4). The authors studied two episodes of selection on an island population, both associated with periods of high mortality during food shortage. Increases in the means of all six traits were accurately predicted in the first episode. Reductions in these same

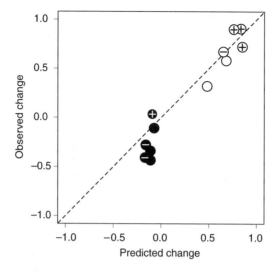

Fig. 9.4 Predicted and observed evolutionary changes after two major episodes of selection on a population of Galápagos ground finch, *Geospiza fortis*. Open circles correspond to changes in means for six traits over the 1976/77 drought (mass, wing length, tarsus length, beak length, beak width, and beak depth). Filled circles indicate change in the same six traits over a wet year, 1984/85. The '+' ('−') symbol indicates that selection favoured an increase (decrease) in the trait, judged from estimates of the selection gradients, β. Unsigned traits showed little or no selection. All changes are in standard deviation units. The diagonal line indicates $Y = X$. Modified from Grant and Grant (1995).

six traits were successfully predicted in the second episode, although the accuracy of these predictions was more variable than in the first. The accuracies of predictions based on the multivariate selection theory were slightly better than predictions of traits analysed one at a time assuming no genetic covariance (Grant and Grant, 1995). These results suggest that problems created by missing variables were not acute.

This Galápagos finch study is significant because it confirms the utility of measures of genetic covariance for short-term prediction. For example, selection apparently favoured a decrease in mean beak width (indicated by a minus sign in one of the open symbols in Fig. 9.4; Price *et al.*, 1984a; Grant and Grant, 1995), yet the trait was predicted to increase because of its strong and positive genetic covariance with other traits for which an increase in size was favoured.

9.2.6 *Remarks*

Prediction from one generation to the next is impressive, but the quantitative genetic approach would be more useful for the study of adaptive radiation if predictions held over longer time frames. The theory tells us that long-term prediction should succeed only if G remains constant through time, which is unfortunately highly unlikely.

Changes in gene frequencies and new patterns of selection and mutation will ultimately change G in ways that cannot be predicted *a priori* (Turelli and Barton, 1994; Shaw *et al.*, 1995). But as I show in the next section, a few crude predictions hold for millions of years, the time frame of many adaptive radiations in the narrow sense.

A potential weakness of the quantitative genetics approach is that genetic parameters are statistical and thereby ignore the genetic details. Eventually we would like to know more about the genetic basis of the most useful quantities, such as differences between species and genetic covariance. Studies of quantitative trait loci (QTLs) and candidate genes are slowly opening the black box of statistical genetics, but the work is still in a descriptive phase and no general principles have yet emerged that improve our abilities to predict directions of divergence. Orr (1998*b*) has identified general principles regarding the distribution of effect sizes of individual genes on the phenotype and on mean fitness as an evolving population approaches a new optimum. These principles apply only to new mutations whose effects are not necessarily deducible from measurements of standing variation within populations.

While simplistic, the quantitative genetic approach may be the only theoretical framework for predicting evolutionary response in quantitative traits under natural selection. It therefore makes sense to incorporate the approach, despite its simplistic assumptions, into an ecological theory of adaptive radiation if the data warrant.

9.3 Divergence along genetic lines of least resistance

This section addresses whether there is evidence of genetic bias, and therefore predictability, over the time periods required for adaptive radiation. Testing this idea is perilous because we do not have measurements of the historical selection pressures. As well, theory alone provides no clear answer to the question of whether long-term prediction is possible even when we know β. One reason is that we are not confident that V_A and G are sufficiently stable over time to allow prediction over millions or even thousands of generations (Lande, 1980; Turelli, 1988; Shaw *et al.*, 1995; but see Arnold, 1992). Genetic parameters are themselves evolvable and respond to selection and mutation (Lande, 1980; Cheverud, 1984, 1996; Wilkinson *et al.*, 1990; Arnold, 1992; Stanton and Young, 1994; Jernigan *et al.*, 1994; Shaw *et al.*, 1995) and this too may limit the insights that can be obtained by measuring standing genetic variance and covariance in contemporary populations.

A second reason is that unless genetic constraints are severe, a population in the neighbourhood of one adaptive peak will eventually climb it anyway. Genetic constraints affect the outcome only temporarily (Fig. 9.2(a); Lande, 1979; Via and Lande, 1985; Zeng, 1988). On the other hand, the effects of genetic constraints may endure if there are multiple adaptive peaks (Fig. 9.2(b)). And genetic constraints may often indeed be severe, as indicated by a shortage of genetic variance in one or more morphological dimensions (Fig. 9.3; Kirkpatrick and Lofsvold, 1992; Gomulkiewicz and Kirkpatrick, 1992). This issue of long-term bias cannot therefore be decided by theory, so empirical tests are needed. This section reviews several tests.

9.3.1 Phenotypic divergence along g_{max}

Lande's (1979) multivariate selection theory makes at least three qualitative predictions about the long-term direction of phenotypic divergence under natural selection. The easiest way to visualize them is to focus on divergence in relation to 'g_{max}', the direction of greatest additive variance within populations. First, if genetic covariance biases evolution then the direction of divergence between populations and species that split only recently from a common ancestor should be close to g_{max} (Fig. 9.2). The size of the bias, measured by the angle θ between g_{max} and the direction of divergence, cannot be determined without knowledge of the location of the optimum. But if the direction of selection is random with respect to the orientation of g_{max}, then initial divergence between populations and species should still show a tendency to follow g_{max}. Later I discuss how the interpretations are affected if the location of the optimum is not random with respect to g_{max}.

The second prediction is that the bias to the direction of evolution should be temporary and diminish with time (Fig. 9.5). An increase in the value of θ with time is expected even if genetic covariances remain constant, but it may be hastened if the covariances change as well. The third prediction is that progress should be relatively slow if selection favours divergence in a direction markedly different from that of g_{max}, at least in the initial stages. This is because relatively little genetic variance exists in directions other than g_{max}, which reduces the size of the evolutionary response expected from a given intensity of directional selection.

To test these predictions, Schluter (1996b) gathered estimates of G for morphological traits in several vertebrate species: a limnetic species of three-spine stickleback; a population of Galápagos medium ground finch *Geospiza fortis*; the song sparrow *Melospiza melodia*; the collared flycatcher *Ficedula albicollis*; and two species of *Peromyscus* mice (references and details in Schluter, 1996b). The genetically most variable direction, g_{max}, was estimated in each of these 'focal' species. This direction was then compared with the direction of the differences in the same traits between the focal species and one or more of its congeneric species. Data were available for at least five traits from each species. Subsets of five traits were used if more than five were available. The comparison was restricted to vertebrates because the ecological relevance of the morphological traits was known or suspected in each case. Traits were ln-transformed or converted to proportions of the mean so that the scales of measurement of the different traits were comparable. All statistical tests were corrected for phylogeny, as detailed in Schluter (1996b).

The analysis showed that for these traits and taxa, evolution was usually biased in the direction of g_{max} (Fig. 9.6), confirming the first prediction. Each θ in the graph is the angle between g_{max} of the focal species and the direction of the difference between that species and a close relative. θ is plotted against the time separating the focal species from its close relative. Time was estimated using allozyme differences. The mean θ was 0.61 rad (18°), which is well below the random expectation of 1.18 rad

226 • Divergence along genetic lines of least resistance

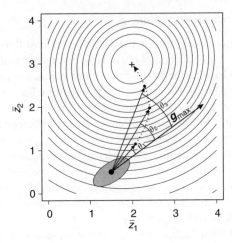

Fig. 9.5 A bias toward g_{max} in the direction of divergence is expected to decay with time. Contours are increments of mean fitness on an adaptive landscape (maximum at +). The ellipse at lower left represents the distribution of breeding values in a fixed, ancestral population. The dotted line indicates the path of evolution in a second population derived from the first. The bias toward g_{max} is measured as the angle θ at three temporary stages, indicated by filled circles along the path of divergence. θ is the angle between g_{max} and the line connecting the mean of the derived population from the mean of the ancestral population. The decay in the bias is indicated by an increase in θ with time. Modified from Schluter (1996b).

(68°). Individual values of θ are nevertheless variable, and several approach or exceed the random expectation (Fig. 9.6).

The smallest values of θ tended to occur between the most recently diverged species, suggesting a decay in the bias with time. However, this trend was not significant. If real, the trend indicates that bias in the direction of divergence endures until species are at least 0.3 units of Nei's distance apart, roughly four million years in birds (Zink, 1991).

The rate of divergence between species was inversely related to θ (Fig. 9.7), in accord with the third prediction. The greater the departure between the direction of evolution and the direction of g_{max}, the slower the rate of evolution. This trend showed no tendency to weaken with time, implying that this constraint endures for considerably longer than four million years.

Mitchell-Olds (1996) carried out a similar analysis on two life history traits, age and size of reproduction, using ten wild populations of birdsrape mustard, *Brassica rapa*. The two traits genetically covaried positively within populations as well as between populations. Divergence between populations was close to the direction of maximum genetic variance within populations, g_{max}.

Interpretation of these results as evidence for long-term genetic constraints hinges on the assumption that the direction of divergent natural selection is random with

Divergence along genetic lines of least resistance • 227

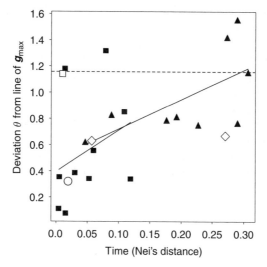

Fig. 9.6 The angle θ between the direction of evolution and g_{max} in relation to time (X-axis) and compared with the random expectation (dashed line). Each observation contrasts the focal species of a given clade with one of its relatives. Time is in units of Nei's (1978) allozyme distance. Symbols refer to different taxa: sticklebacks (○), Galápagos finches (■), flycatchers (□), sparrows (▲), and mice (◇). Solid lines are least-squares regressions within the two largest clades (Galápagos finches and sparrows). Modified from Schluter (1996*b*), with permission of the Society for the Study of Evolution.

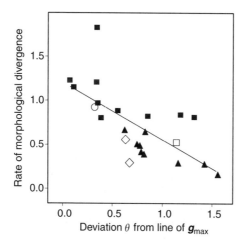

Fig. 9.7 The rate of evolution between species in relation to the deviation from g_{max}. Each observation contrasts the focal species of a given clade with one of its relatives. Symbols refer to different taxa: sticklebacks (○), Galápagos finches (■), flycatchers (□), sparrows (▲), and mice (◇). The line is an unweighted linear regression. Rate of divergence was calculated as adjusted residuals from a regression through the intercept of morphological distance against time. From data in Schluter (1996*b* and references therein).

respect to the direction of greatest genetic variance, g_{max}. If not, then an alternative interpretation is possible: that selection rather than genetic covariance is responsible for the bias in the direction of divergence. Under this alternative explanation, the direction of divergence is close to g_{max} because both are driven by selection.

Natural selection almost surely modifies genetic variance and covariance in populations (Cheverud, 1984, 1996; Wilkinson *et al.*, 1990; Arnold, 1992; Stanton and Young, 1994; Jernigan *et al.*, 1994; Shaw *et al.*, 1995); indeed, in theory, persistent selection is partly responsible for the maintenance of **G** (Lande, 1980) and also g_{max}. The fact that selection moulds genetic parameters is therefore not inconsistent with the idea that these parameters bias subsequent evolution of the mean. The observation that greater amounts of time lead to greater deviation from g_{max} is one indication that selection alone is not the basis of the pattern. Nevertheless, teasing apart the precise roles of selection and genes at various stages of divergence requires more information on selection than is currently available.

9.3.2 Genetic variance and the directions of host shift in leaf beetles

Futuyma *et al.* (1995) asked a related question: whether genetic variance in host use by *Ophraella* leaf beetles biased realized host shifts within the genus. Each of the 12 species of *Ophraella* utilizes a different host plant (usually only a single plant species) within the Asteraceae. Both larvae and adults consume the leaves of the same plants. The phylogenies of *Ophraella* species and their host plants indicate that beetles have switched hosts multiple times through their history. Most transitions have occurred between related plants having similar secondary compounds in the leaves (Futuyma *et al.*, 1995).

The expectation tested by Futuyma *et al.* was that the history of host transitions in *Ophraella* should be reflected in the amounts of expressed genetic variance in feeding traits immediately after a new host is colonized. If genetic variance in feeding performance is absent when a new population is established on a particular new host, then a successful transition to that host is unlikely (perhaps because inability to adapt to a new host increases risk of extinction). To test this hypothesis, the authors transplanted larvae and adults between host plants and measured leaf consumption. The prediction was that beetles transplanted to the host of a closely related *Ophraella* species (representing a host shift that actually occurred in prehistory) would more often exhibit genetic variance in leaf consumption rate than beetles transplanted to the host of a distantly related species (representing a host shift that never occurred in prehistory).

Results were consistent with these expectations (Table 9.1), albeit not significantly. Transplanted beetles survived and consumed the novel food in 29 of 39 experiments. Considering these survivors only, genetic variance was detected in 12 of 14 transplants to hosts used by a near relative (86%), whereas genetic variance was detected in a lower percentage of transplants to hosts of distant relatives (9 of 15, or 60%) (one-tailed Fisher exact test, $P = 0.11$, assuming that each transplant is independent).

Table 9.1 Additive genetic variance in leaf consumption by *Ophraella* leaf beetles experimentally transplanted to nonnative hosts. Rows indicate whether the novel host plant is presently consumed by a close relative or by a distant relative. Counts in each row indicate the number of experiments in which genetic variance was (+) or was not (−) detected in the survivors, and those experiments in which the beetles died or consumed only trace amounts of leaf, leaving genetic variance undefined. The results are based on four beetle species each transplanted to five novel host plants. Transplants were conducted on both adults and juveniles in separate experiments, for a total of 39 experiments. Data are from Futuyma et al. (1995)

	Genetic variance in surviving beetles		Nonsurvivors
	+	−	
Close relative	12	2	2
Distant relative	9	6	8

Genetics aside, the history of host shifts may also have been influenced by selection. That is, selection may have favoured some shifts and not others. A crude index of the role of selection might be gained by contrasting experiments in which the transplanted beetles survived (first two columns of Table 9.1) with those in which the beetles died or failed to consume more than trace amounts of leaf (third column of Table 9.1). Survival was indeed higher in transplants to hosts of close relatives (14 of 16 experiments, 88%) than in transplants to hosts of more distantly related beetles (15 of 23, or 65%), albeit not significantly (one-tailed Fisher exact test, $P = 0.08$).

The above tests consider selection and genetic constraint separately, whereas both may have contributed to the history of host shifts on *Ophraella*. In support, genetic variance was detected in 12 of 16 experiments involving transplants to close relatives (75%), whereas none was detected, or the beetles died, in the remaining 4. In contrast, genetic variance was detected in only 9 of 23 transplants to hosts of more distant relatives (39%), whereas none was detected, or the beetles died, in the remaining 14 (one-tailed Fisher exact test, $P = 0.02$, assuming independence).

9.3.3 Nongenetic evidence of genetic bias

If direct measures of genetic covariance are lacking, then phenotypic covariance might sometimes suffice to test for genetic bias in divergence. In the morphological data set from vertebrates, the direction of maximum phenotypic variance, p_{max}, was not the same as g_{max} but it was nearly as successful at predicting the direction of divergence between species (Schluter, 1996b). Similarity of G and P (and of g_{max} and p_{max}) is not assured, however, and in many instances they may be very different (Willis et al., 1991). In the sticklebacks, for example, gill raker number covaries genetically with most other traits, especially gape width and gill raker length, but the corresponding phenotypic covariances are weak and the trait does not contribute to p_{max}. G and P for life history traits are often expected to be different (Charlesworth, 1990; Houle, 1991).

Kluge and Kerfoot (1973) showed that phenotypically the most variable traits within species are also the most variable among species (but see Rohlf et al., 1983). This too is consistent with genetic bias if phenotypic variance reflects underlying genetic variance. In *Cerion* land snail populations, phenotypic variation is dominated by the negative covariance between the size and number of whorls in the adult shell. Interpopulation and interspecific differences in the genus point in the same direction (Gould, 1989b); thus p_{max} and divergence are correlated. In this case, a negative covariance is the result of the geometry of shell growth and mode of coiling in *Cerion* (Gould, 1989b) and is likely to be genetically based.

Somewhat paradoxically, environmental covariance has also been used to test for genetic bias in divergence. Cheverud (1982) has observed that genetic and environmental correlations between traits are often similar, which he attributes to a tendency for effects of both genotype and environment to be mediated by the same developmental pathways. Alberch and Gale (1985) treated developing limb buds of an *Ambystoma* salamander with an inhibitor of mitosis, reducing the number of cell divisions, limb size, and the final number of toes. Environmental variation so induced paralleled variation seen among extant salamander genera, particularly the identity of missing digits. This result is evidence for genetic bias on divergence if we are confident that genetic perturbations would have had a similar effect on variation.

In the vertebrate data set analysed by Schluter (1996b), e_{max}, the direction of maximum environmental variance within populations (computed from the environmental covariance matrix, E, where $E = P - G$), is correlated with g_{max}. However e_{max} was a poorer predictor of species divergence than g_{max}. This may be partly because E has more sampling error than G and P, since it is computed as their difference and therefore compounds the sampling error of both.

9.3.4 *Remarks*

Differentiation among populations is expected to mirror patterns of genetic covariance within populations when genetic drift is the cause of divergence. The results reviewed here suggest that measurements of genetic parameters within populations also roughly predict the principal directions of divergence when natural selection is the cause. The effects of genes are detected as biases in direction. These predictions concerning bias are not as powerful as the quantitative predictions permitted from selection and covariance one generation at a time (Fig. 9.4). Nevertheless, they provide some insights into the direction and rate of phenotypic divergence in adaptive radiation.

Moreover, the genetic bias is reasonably long-lasting. In the vertebrates the influence on rate seemed not to decay with time, whereas the direction of divergence was random with respect to g_{max} within about four million years. A greater sample size is necessary to confirm these trends. If they hold, then decay in the predictability of the direction of divergence is probably not fully explained by changes in G with time, since most such changes would affect predictability of the rate of divergence as well as the direction. The decay may result instead because local optima are eventually attained (Fig. 9.2(a)). On the other hand, it is not clear that climbing a single local optimum would require four million years. Perhaps the bias is long-lived because there are multiple local optima (Fig. 9.2(b)).

Further testing is needed to distinguish this genetic explanation for the bias from the alternative hypothesis that direction and rate of divergence are predictable from g_{max} because both are moulded by the same selection pressures. Neither explanation can be ruled out with present data, although a decay in the magnitude of the bias seems more consistent with a genetic influence. A study of multiple taxa, all of which have diverged across the same environmental gradient, but which have different patterns of genetic covariance, might allow the hypotheses to be distinguished.

9.4 Divergent natural selection in retrospect

Quantitative genetic theory offers the promise of another remarkable (but more controversial) insight into phenotypic divergence: the backcalculation of net selection gradients accumulated over the time since species were separated from a common ancestor. The calculations, as I show below, are straightforward, but most quantitative geneticists would consider the results dubious at best. This is because many of the assumptions required are unlikely to hold over the time spans involved, and the estimates are not expected to be robust to alterations (Shaw *et al.*, 1995).

While these concerns are serious, even fatal if accuracy is an objective, the approach still has some merit as an exploratory tool. Its main utility lies in demonstrating how present-day differences between species might not closely mirror the history of selection pressures that led to those differences. In this spirit I demonstrate some of its uses below.

9.4.1 Retrospective selection estimation

The calculations are obtained from a simple rearrangement of eqn (9.3). Consider two populations or species, a and b, that diverge through time after splitting from a common ancestor. If we summed the divergent natural selection pressures they experienced, generation after generation, we would obtain their current differences in mean phenotype:

$$\bar{z}_b - \bar{z}_a = G \sum_1^t (\boldsymbol{\beta}_{bi} - \boldsymbol{\beta}_{ai}). \qquad (9.6)$$

The left side of this equation is the present difference between species a and b in their trait means, t generations after splitting from a common ancestor. $\sum_1^t (\boldsymbol{\beta}_{bi} - \boldsymbol{\beta}_{ai})$, which I will abbreviate as $\Delta\boldsymbol{\beta}$, is the net selection gradient over this time period. It is the sum of differences in selection experienced each generation. This quantity can be estimated by rearranging:

$$\Delta\boldsymbol{\beta} = \boldsymbol{G}^{-1}\bar{z}_b - \boldsymbol{G}^{-1}\bar{z}_a \qquad (9.7)$$

(Lande, 1979). Examples where this approach has been used may be found in Price et al. (1984b), Schluter (1984), and Dudley (1996b).

This approach has important limitations. Applying eqns (9.6) and (9.7) to real species requires the assumption that G has remained constant since the species last shared a common ancestor. This assumption is testable by comparing genetic covariances among the descendant species and also by testing whether the results are robust to any variations detected. This will not take account of transient changes in G that occurred during divergence (Shaw et al., 1995). The method also assumes that all genetically covarying traits under selection are included in the analysis. This is the standard assumption of multiple regression (of which eqn (9.7) is a special case). Satisfying it is simply impractical. Most analyses of selection are based on a few traits chosen because they are easy to measure and thought to be important *a priori*. Failure to include all traits will bias the estimates of $\Delta\boldsymbol{\beta}$ (Mitchell-Olds and Shaw, 1987). The sensitivity of results to violations of these assumptions (especially the latter) is the main reason any results should be treated with caution. The method also assumes that differences between population means are wholly genetic, which can be tested in a common garden experiment.

9.4.2 Retrospective selection on finch beaks

Use of the method may nevertheless help intuition about how patterns of divergence in suites of traits may have departed from historical patterns of divergent natural selection. For example, consider the beaks of the Galápagos ground finches, *Geospiza* (Fig. 1.1). Most of the variation between the species is in overall beak size (Fig. 1.1(a)), separating species with longer and deeper beaks from species with smaller and more slender beaks. The greatest differences in beak occur between the robust *G. magnirostris* and its close relative, the sparrow-sized *G. fuligonosa*, with the mid-sized *G. fortis*, closely related to the other two, falling in between. *Geospiza* beaks also differ in shape but this variation is less in absolute terms (short arrow in Fig. 9.8(a)) than differences in size. Without consideration of genetics, one

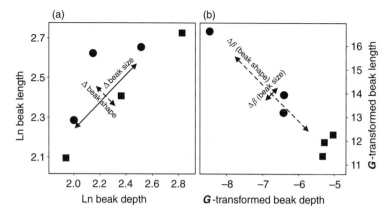

Fig. 9.8 Mean beak length and depth of Galápagos ground finch species (a) compared with the same means after a transformation that takes account of genetic variances and covariance between the two traits (b). The difference between any two points in (b) represents the net amount of selection on beak length and depth separating species means. Arrows in (a) show that the greatest differences between species are in overall beak size, separating long, deep beaks from short, shallow beaks. Nevertheless, most selection has been on beak shape rather than size: the greatest net selection differentials separate beaks that are long relative to their depth (upper left in (b)) from beaks that are short relative to their depth (lower right in (b)). Axes in (b) were obtained by multiplying the vector of trait means for each species by G^{-1} using eqn (9.7). Filled symbols distinguish the clade with the three most closely related species, *Geospiza fuliginosa*, *G. fortis*, and *G. magnirostris* (■), from the rest of the *Geospiza* (●). Genetic variances and covariance were from a population of *G. fortis* (Boag, 1983).

might surmise from this pattern that selection on size has been the dominant feature in the history of the group, and that selection on beak shape was weaker or less frequent.

However, beak length and depth covary positively genetically (Boag, 1983) and this has probably been the case throughout the divergence of *Geospiza*, if phenotypic covariance of present-day species is a guide. Positive phenotypic covariance between the two traits characterizes every known population (e.g. Schluter, 1984; Grant and Grant, 1994). Positive genetic covariance also characterizes the two other populations of *Geospiza* in which G has been estimated (Price et al., 1984b; Grant and Grant, 1989). Consequently it is 'easier', genetically speaking, to change both beak length and beak depth in the same direction (increase both together or decrease both together) than to increase one of the two traits and decrease the other.

Retrospective selection estimates help to quantify the difference between patterns of absolute variation and the net amounts of selection needed to generate them. To illustrate the effect, I multiplied the vector of trait means (mean beak depth and mean beak length) for each species by the inverse of G, the genetic covariance matrix. The resulting transformed vectors are plotted in Fig. 9.8(b). In contrast to Fig. 9.8(a), the difference between any two species in Fig. 9.8(b) represents the net amount of divergent natural selection needed to generate their difference in beak depth and length.

The picture from Fig. 9.8(b) departs from that in Fig. 9.8(a) in that from the perspective of selection, the greatest differences between species are in beak shape, not beak size. Estimates of net selection separating the three youngest species, which are the most different from one another overall in absolute terms (Fig. 9.8(a)), are trivial by comparison. These estimates of net selection on beak length and depth are obviously rough. Other traits have been left out, for simplicity, and the possibility of historical changes in G has not been incorporated. Nevertheless, one general conclusion from this analysis would probably hold: whereas the most conspicuous differences among the *Geospiza* are in beak size, most selection has been exerted on beak shape. Adding the third beak dimension, beak width, to the analysis does not alter this conclusion.

The blunt, conical beak shape of the youngest trio of species, *G. magnirostris*, *G. fuliginosa*, and *G. fortis*, is the most derived condition relative to the probable state of the last common ancestor of the present-day Galápagos finches (Fig. 1.1). If the molecular phylogeny is correct, then the warbler-like beak of the three basal lineages was the starting point for the tree finches and ground finches that came later. The positive genetic covariance between length and depth of beak may be one reason why the blunt, conical beak came relatively late in the diversification of the *Geospiza*.

9.4.3 Remarks

If phenotypic divergence is really biased toward g_{max} because of patterns of genetic variance and covariance in populations, and if selection is indeed slower in directions other than g_{max}, then G must contain *some* useful information about selection that ought to be explored. For example, a systematic bias toward g_{max} implies that the direction of divergence in adaptive radiation regularly departs from the direction favoured by selection (Fig. 9.2). This fact is readily appreciated by a depiction of retrospective selection, however crude. This departure is probably the most robust conclusion to be drawn from the analysis of retrospective selection on finch beaks (Fig. 9.8). A more specific interpretation is that the ranking of importance of beak size and shape in the diversification of the Galápagos finches, as judged from absolute differences, is deceptive. Whereas beak size is the most variable trait combination, to generate the variety of beaks seen in *Geospiza* apparently required much more selection on beak shape than size.

Retrospective analyses force us to contemplate interpretations of the selective forces that depart from the differences between species actually observed. Lack of variation in a trait does not mean that it was free of persistent directional selection. Conversely, change in a given trait does not imply that directional selection favoured it. Some general features of the history of selection may often be extracted, and a richer understanding of adaptive radiations may be the outcome.

9.5 Discussion

The quantitative genetic theory for phenotypic evolution extrapolates from variation within populations to variation between means of populations and species. The above results suggest that such extrapolation is often justified and may contribute to our understanding of the kinds of changes taking place in suites of phenotypic traits during adaptive radiation. The direction of phenotypic divergence is biased toward the genetic 'line of least resistance', at least in the early stages. The more different the path of evolution from this line of least resistance, the slower its progress. These results, along with backcalculations of the net selection pressures separating species, suggest that the path of divergence in adaptive radiation is not typically in the direction most favoured by natural selection, at least in the early stages. Rather, the path is deflected in the direction of least resistance. Later evolution seems to eliminate this bias, perhaps because local optima are eventually climbed.

These conclusions partly fulfil the initial promises of quantitative genetics. Although precise prediction of long-term phenotypic evolution is out of the question, a genetic 'footprint' is recognizable and incorporating it improves prediction of the course of divergence. This is not to say that we understand the origin and maintenance of the genetic patterns on which the predictions depend. The underlying genetic basis of g_{max} is obscure. Eventually it may be possible to uncover the most important genes responsible for variation in this trait and general rules governing their action. Such principles would also be worth including in a further revised ecological theory of adaptive radiation especially if the result is better prediction.

The methods were applied here to phenotypic traits influencing exploitation of environment, but they may be just as informative when the trait is a component of reproductive isolation. Under one version of the by-product model of speciation (Chapter 8), reproductive isolation is fully an unselected consequence of divergent natural selection on genetically correlated traits (Rice and Hostert, 1993). In this case, speciation itself is a 'realized path of evolution more likely, in strictly genetic terms, than the paths not taken' (Futuyma *et al.*, 1995). What is not clear, however, is whether the process can be predicted from measurements of genetic variance within populations, as was done for phenotypic traits. I am unaware of studies of reproductive isolation equivalent to those reviewed above. Nevertheless, instances of speciation are often explained in terms of genetic covariance, for example between the probability of mating and changes in the timing of various life history events (e.g. Etges, 1998; Miyatake and Shimizu, 1999). The approach is less likely to be successful when reproductive isolation evolves via other mechanisms such as the accumulation of incompatible alleles in isolated populations. This is because alleles that generate hybrid breakdown between populations are unlikely to coexist within populations (e.g. Orr, 1995). Genetic variance that persists within populations cannot then anticipate how populations will diverge. The same may be true of premating isolation that evolves by reinforcement. The study of the genetic and phenotypic basis of reproductive isolation is in its early stages, and we do not yet have much information on how much genetic variation in reproductive isolation exists in natural populations.

10

The ecology of adaptive radiation

> *In adaptive radiation ...all the modes and all the factors of evolution are inextricably woven. The total process cannot be made simple, but it can be analysed in part. It is not understood in all its appalling intricacy, but some understanding is in our grasp, and we may trust our own powers to obtain more.*
> —Simpson (1953)

10.1 Finale

What are the ecological causes of adaptive radiation? The last major synthesis of ideas led to the 'ecological theory' that phenotypic divergence and rapid speciation in adaptive radiation are ultimately the outcome of divergent natural selection arising from differences between environments and competition for resources. In this final chapter, I summarize our state of knowledge of the ecology of adaptive radiation, and judge the theory in light of the evidence presented in the previous chapters. I also review some of the major questions that remain and thus the work that lies ahead.

10.2 General features of adaptive radiation

Some of the greatest recent advances in our knowledge of events taking place during a typical adaptive radiation derive from better estimation of changes in phenotypes and resource use through time. The sample of well-described cases is still small. A tentative picture may be drawn of the usual events in adaptive radiation by combining findings from studies of the best-known cases with broader ecological studies of recently diverged species.

Tests of correlated divergence between phenotype, use of environment, and performance in those environments have been more widely and rigorously applied. Quantitative tests of rapid speciation have likewise increased. It is clear that species diversity in many lineages has accumulated in episodic bursts accompanied by ecological and phenotypic differentiation, and it makes sense to focus on these episodes. Some uncertainty nevertheless persists over what 'rapid speciation' really represents and how it should be recognized. Speciation rates rise in adaptive radiation either because reproductive isolation evolves more rapidly or because more populations avoid extinction long enough to attain the status of species. The relative importance of these two alternatives is unknown. Distinguishing changing speciation rates from changing species extinction rates is also problematic.

Continuous expansion to new environments is the most general overriding feature of all adaptive radiations. The pattern of this divergence between populations and species is partly predictable from principles of quantitative genetic covariation. The initial stages of divergence between populations and species tend to occur along directions of maximum genetic variance within populations, a tendency that seems to decay with time. Genetic constraint probably underlies this bias, but tests are needed to distinguish this hypothesis from the alternative possibility that similarity of paths of divergence and directions of genetic variance results because both are shaped by the same selection pressures.

In the midst of this expansion to new environments, a great deal of repetition is seen among lineages undergoing divergence in similar settings, particularly when ancestral forms are closely related. Parallel evolution is evidently common in adaptive radiation. The sequence of appearance of new phenotypes on one island or in one lake may echo those observed in others, and in some cases replicate them closely. The order in which divergence occurs along different ecological dimensions (prey size, habitat, pollinators) may also be partly repeatable from lineage to lineage within broad taxonomic categories. Such findings suggest that one adaptive radiation is very often like others in broad outline. Adaptive radiations of distantly related lineages show few other broad similarities in patterns of niche evolution. The hypothesis that the founders of adaptive radiation are typically generalists receives little widespread support. Nor is there compelling evidence that in general later descendants are increasingly specialized, at least in adaptive radiations at low taxonomic level.

Initial increments of divergence between populations and species are virtually instantaneous, a rate that is not sustained throughout expansion. Nevertheless, moderate rates of phenotypic differentiation are usually maintained for several million years thereafter, and these yield most of the phenotypic variation characterizing a radiation. Phenotypic diversity may continue to rise even as species diversity starts to fall and adaptive radiation nears its end. It is not possible to tell whether new niches are still being colonized at the same rate at this stage, or whether species are instead accumulating new traits beneficial in niches acquired earlier. The events taking place at the end of adaptive radiation are less well described than at earlier stages.

Many of these and other features of adaptive radiation are still sketchy and require further study. It is possible that addition of more lineages will reduce the generality of some of the trends proposed herein and reveal others that I have discounted. Trends in still other features, such as geographic range size, relative rates of speciation and extinction, and the precise timing of increments of phenotypic divergence and speciation have scarcely been addressed. Still other features of adaptive radiations are more idiosyncratic (granivorous finches re-evolving from a warbler finch itself descended from a previous granivorous finch). These represent the 'appalling intricacies' of adaptive radiation (Simpson, 1953) that we may also hope someday to explain.

10.3 Fate of the ecological theory

Inquiry into the validity of the ecological theory is not complete but much progress has been made. The theory has been subjected to a large array of tests, as the previous chapters indicate, and except for several important extensions mentioned below the theory so far has survived largely intact. Several parts of the theory can be regarded as essentially confirmed, while evidence for other parts and/or their alternatives is still wanting.

10.3.1 *Divergent natural selection and environment*

The best documented part of the theory is probably the idea that features of environment and resources generate alternative fitness peaks that bring about divergence in mean phenotype. Direct measurements of selection and at least one estimate of adaptive landscapes estimated from measurements of resource availability favour the idea that phenotypically divergent species that use different environments have indeed been driven apart by selection. Much additional evidence for the existence of fitness peaks is indirect, coming from comparisons of population differences against the neutral expectation (mainly the Q_{ST} method) and from reciprocal transplant experiments. Increasingly, information is also available on the environmental factors responsible for adaptive peaks and valleys. Most of this evidence is still obtained by correlative methods rather than by predictive approaches or experiments. The adaptive landscape, a multidimensional surface having many hills and valleys determined by features of environment, remains a dramatic and effective depiction of the forces shaping phenotypic evolution in adaptive radiation.

Most of the studies suggesting the presence of a fitness valley, even those directly measuring divergent natural selection and evolution, provide a limited perspective of the shape of adaptive landscapes as a whole. Yet, understanding the full *distribution* of mean phenotypes that are the outcome of an adaptive radiation requires a broader view of these landscapes. To this end, a combination of approaches to estimate landscapes may be most revealing. First a landscape is estimated from resources, and then some of its features are tested using direct measurements of natural selection in the vicinity of populations occupying single points. A further benefit is that the combination incidentally yields a strong test of whether the environmental features driving divergence have been correctly identified.

Divergence purely via mutation and drift can be entirely ruled out. A more useful alternative to divergent natural selection, drift along adaptive ridges, is also not well supported in general, but many individual studies have not ruled it out. This is mainly because much of the evidence has been obtained for the purposes of testing selection and adaptation in a broad sense but not divergent selection specifically. It is hoped that appreciation of the distinction between these concepts will lead to more frequent and direct tests of divergent natural selection and measurement of adaptive landscapes.

How populations move from the domain of one adaptive peak to that of another in the adaptive landscape of mean phenotypes is still largely speculative. A small amount of evidence suggests that the topography of adaptive landscapes varies in space and time because of variations in resource supply. This provides a simple mechanism for rapid adaptive peak shift purely by changed natural selection pressures. The contribution of genetic drift in mean phenotype to shifts between adaptive peaks is completely unknown, but theory indicates that its role is likely to be small.

10.3.2 *Competition, ecological opportunity, and divergence*

Further evidence of divergent natural selection comes from a growing list of cases of ecological character displacement between closely related species. Here the agent of divergence is interaction between species rather than contrasting external environments. Our understanding of character displacement is vastly greater now than 50 years ago. Not only has the number of alleged cases multiplied, but the strength of evidence from individual studies is also increasing. There can be little doubt that competition for resources is an important process in many adaptive radiations.

This evidence for character displacement nevertheless raises many further challenges, in part because the data do not fully answer the question of the role of interactions between species in adaptive radiation. One reason is that the evidence for character displacement is comprised of only positive cases. The size of the list, and its rate of growth, permits us to say that competition for resources commonly favours divergence but not whether it is universal or even that it characterizes the majority of cases. The list itself is not representative of taxa in nature, being heavily biased toward carnivores. We should not presume thereby that phenotypic divergence of species of herbivores, detritivores, or plants are less often affected by competition for resources than carnivores, but the bias requires us to entertain this as a possibility. Perhaps competition does not favour divergence if species compete only weakly or diffusely, or if the conditions under which competition is occurring do not favour divergence even if competition is strong. On the other hand, perhaps character displacement is simply more difficult to detect at lower trophic levels.

We do not know whether other interactions between close relatives, such as direct interference, interspecific predation, and apparent competition, are any less important than competition in promoting divergence. The theory of phenotypic divergence by other ecological interactions is underdeveloped, and little effort has been expended in the search for possible cases in nature. It is possible that a number of the cases of character displacement recorded from the literature are actually the outcome of one or more of these alternative interactions, since resource competition has not often been confirmed as the underlying mechanism.

Ecological opportunity, meaning lots of unexploited or underexploited resources, is a vital precondition for adaptive radiation according to the ecological theory. Ecological opportunities might be encountered by a lineage colonizing a remote archipelago, surviving a mass extinction, or acquiring an evolutionary novelty ('key innovation') that enables use of resources otherwise little exploited. I have concentrated on the first and third of these possibilities.

Tests of the dependence of phenotypic differentiation and speciation on ecological opportunity are largely consistent with the theory. Phenotypic diversity is often higher in Hawaiian and Galápagos lineages than in their mainland relatives, although this pattern has been quantified in very few cases. Cases of character displacement between species in different taxa is also consistent with the idea that divergence in one lineage is constrained by the presence in the same environment of species in other lineages. Recent speciation rates also tend to be higher in taxa in depauperate environments including remote archipelagoes and newly formed lakes. The pattern is not universal, however, as speciation rates are sometimes much higher in species-rich mainland faunas. The reasons for this remain obscure and difficult to test. Unfortunately, because ecological opportunity is difficult to recognize and quantify *a priori*, one's tests rely upon imperfect surrogates (e.g. depauperate environments). It is not yet known whether rapid speciation under conditions of high ecological opportunity involves faster evolution of reproductive isolation.

The discovery of candidate key innovations that appear to promote speciation and/or reduce extinction represents a major advance. Presently the best examples are the hypocone in mammals, flower nectar spurs in angiosperm plants, resin canals in plants, and unidentified traits accompanying the transition to phytophagy in insects. The significance of these finds is appreciated all the more when one realizes that a substantial number of independent acquisitions by different lineages of a similar novelty is required even to test whether or not an association with elevated speciation rate is statistically sound. Such replication is not available for most of the key innovations that have been proposed over the past 50 years. Unfortunately, tests of key innovations have measured only speciation rates whereas the ecological theory predicts a link between key innovations and adaptive radiation, of which speciation is but a part. Quantifying effects on phenotypic differentiation is therefore an important next step.

Stronger evidence is required that greater ecological opportunity, and not some other mechanism, is the reason behind the higher speciation rates accompanying acquisition of a novel trait. Field and experimental studies of contemporary species hold considerable promise in providing answers. Ecological opportunity itself needs to be better understood. The ecological theory emphasized freedom from competition but at least one of the candidate key innovations (resin canals) implicates reduced susceptibility to natural enemies instead. If real, then the concept of ecological opportunity needs to be expanded to incorporate other interactions. The contribution of predation and other interactions to ecological opportunity is uncertain and largely untested.

10.3.3 *Ecological speciation*

Over 140 years after Darwin's (1859) book, *On the origin of species by means of natural selection*, evidence is at last accumulating that some species in nature have originated by means of divergent natural selection. Under the hypothesis of ecological speciation, the agents of divergent natural selection leading to rapid speciation in adaptive radiation are the same as those generating phenotypic differences.

Knowledge of the role of divergent natural selection in natural speciation is advancing steadily. The strongest tests are those demonstrating that environmental agents are responsible for low hybrid fitness, that degree of assortative mating is correlated with degree of differentiation in phenotype and environment, and that premating isolation has evolved in parallel independently in similar environments. Less is known about the role of different environments in producing unconditional postzygotic isolation (e.g. hybrid sterility and inviability) but perhaps this evolves late in the speciation cycle anyway, at least when speciation is rapid. Reproductive isolation may evolve solely as an incidental by-product of divergent natural selection, but evidence suggests that the process may be completed in sympatry by reinforcement or by more direct interactions. One good case of sympatric speciation now exists, and ecological mechanisms are implied (if not yet proved). Hints of similar processes are evident in other cases.

It is too early to tell whether most species originating in adaptive radiation have arisen by ecological speciation or in some other way. A number of nonecological mechanisms of speciation are feasible, including polyploidy and the fixation of different advantageous mutations in isolated populations inhabiting similar environments. There is little indication that polyploid speciation is common in adaptive radiations even in plants. Some evidence for reproductive isolation arising from fixation of divergent advantageous mutations comes from indirect evidence of one special case, the 'chase-away' arms race between the sexes. It is not yet clear whether reproductive isolation builds quickly enough by nonecological mechanisms to play a role in rapid speciation in adaptive radiation.

Sexual selection has been observed in many natural populations and its consequences for speciation in adaptive radiation are potentially enormous. Indirect comparisons have correlated speciation rate with the evolution of secondary sexual characteristics, so a link to adaptive radiation may follow. Field studies have detected reduced mating success in hybrids, constituting direct evidence for divergent sexual selection. From an ecological perspective, the most important issues concern the mechanisms driving divergence in mate preference (i.e. whether they stem from environment or from other factors instead), and here little progress has been made. Some evidence is accumulating that premating reproductive isolation may evolve as a by-product of a chase-away arms race between the sexes. The rate of this process has not been quantified, and whether it leads to reproductive isolation faster than ecological mechanisms is not known.

Mechanisms of speciation also have implications for phenotypic diversification in adaptive radiation. Two extreme perspectives on speciation and phenotypic evolution help illuminate the issues at stake. In the first, phenotypic diversification in adaptive radiation is limited principally by ecological opportunity. When these opportunities abound, divergent natural selection causes differentiation in phenotypic traits influencing use of environments, and speciation ultimately follows. The opposite perspective is that speciation is driven by mechanisms having little to do with differences between environments or ecological opportunity. Ecological opportunity merely allows many species generated by nonecological speciation to persist. Phenotypic divergence may occur alongside this process, or it may occur later following secondary contact in response to interspecific competition or other interactions. This second perspective is about as radical a departure from the ecological theory of adaptive radiation as can be imagined.

Unfortunately, the data from wild populations is not yet able to distinguish these scenarios. Growing evidence for ecological speciation in some lineages favours the first scenario, whereas preliminary hints of nonecological speciation leaves open the possibility of the second. It is entirely possible that adaptive radiation occurs both ways.

10.3.4 The next 50 years

Although the theory needs to be expanded in parts, none of its most significant claims have yet been overthrown. On the whole, it should be regarded as one of the most highly successful theories of evolution ever advanced. Over the past 50 years, we have learned the most about divergent natural selection, competitive interactions, and ecological speciation. We have learned much less about the factors driving adaptive peak shift, the mechanisms of key innovations, and the role of speciation. The most significant extensions to the theory of the past 50 years have been the claim that interactions other than competition also drive phenotypic divergence; the idea that sexual selection is a vital part of the process of speciation; and the incorporation of principles of quantitative genetics, especially to predict biases in the course of divergence. The summary above reveals many of the questions for research that remain. In addition to testing the most significant extensions to the ecological theory, we require much more knowledge of the topography of adaptive landscapes, and the mechanisms of divergent natural selection, key innovations and speciation.

The next 50 years will see more refinements to, and applications of, concepts of rapid speciation and correlated divergence of phenotype and use of environment. The result will be more complete descriptions of adaptive radiations in nature. As well, the role and frequency of adaptive radiation in evolution as a whole will be more firmly established. We now have a clearer concept of the features of adaptive radiation and some of their causes, but attempts will be made to assess the portion of earth's diversity this process accounts for. Adaptive radiation is surely common, but has all taxonomic and phenotypic diversity on earth originated in this way?

The next 50 years will also see great strides in uncovering the genetic basis of adaptation and reproductive isolation. This work will have huge implications for our understanding of the genetics of phenotypic divergence and speciation in adaptive radiation. Questions that will soon be addressed include: Are species differences the result of changes at many loci of small effect or a few loci of large effect? How much of adaptive radiation represents sorting of standing genetic variation and how much the input of new mutations? Does parallel evolution (including parallel speciation) in different parts of a clade result from changes at identical loci? What is the genetic basis of g_{max}, the direction of greatest genetic variance within populations? Answers to these questions will inevitably lead to a fuller integration of genetics into the ecological theory.

The ecological theory has been a fruitful guide to the study of adaptive radiation. Its accuracy in many respects is all the more staggering when we realize how little evidence was available, at the time it was formulated, for any of the processes it incorporates. There is little doubt that the theory will continue to serve as a starting point for new investigations into the dramatic episodes of speciation and differentiation that have generated much of the ecological diversity on earth.

REFERENCES

Abbott, I., Abbott, L. K., and Grant, P. R. 1977. Comparative ecology of Galápagos ground finches (*Geospiza* Gould): evaluation of the importance of floristic diversity and interspecific competition. *Ecological Monographs*, **47**, 151–84.
Abouheif, E. 1999. A method for testing the assumption of phylogenetic independence in comparative data. *Evolutionary Ecology Research*, **1**, 895–1020.
Abrams, P. A. 1986. Character displacement and niche shift analysed using consumer-resource models of competition. *Theoretical Population Biology*, **29**, 107–60.
Abrams, P. A. 1987. Alternative models of character displacement: I. Displacement when there is competition for nutritionally essential resources. *American Naturalist*, **130**, 271–82.
Abrams, P. A. 1989. The importance of intraspecific frequency-dependent selection in modelling competitive coevolution. *Evolutionary Ecology*, **3**, 215–20.
Abrams, P. A. 1990. Mixed responses to resource densities and their implications for character displacement. *Evolutionary Ecology*, **4**, 93–102.
Abrams, P. A. 1996. Evolution and the consequences of species introductions and deletions. *Ecology*, **77**, 1321–8.
Abrams, P. A. 2000. Character shifts of prey species that share predators. *American Naturalist*, **156** *(Supplement)*, in press.
Abrams, P. A. and Matsuda, H. 1996. Positive indirect effects between prey species that share predators. *Ecology*, **77**, 610–16.
Abramsky, Z., Rosenzweig, M. L., Pishow, B., Brown, J. S., Kotler, B., and Mitchell, W. A. 1990. Habitat selection: an experimental field test with two gerbil species. *Ecology*, **71**, 2358–69.
Adams, D. C. and Rohlf, F. J. 2000. Ecological character displacement in *Plethodon*: biomechanical differences found from a geometric morphometric analysis. *Proceedings of the National Academy of Sciences, USA*, **97**, 4106–11.
Aguade, M., Miyashita, N., and Langley, C. H. 1992. Polymorphism and divergence of the mst 355 male accessory gland gene region. *Genetics*, **132**, 755–70.
Alatalo, R. V., Gustafsson, L., and Lundberg, A. 1986. Interspecific competition and niche changes in tits (*Parus* spp.); evaluation of nonexperimental data. *American Naturalist*, **127**, 819–34.
Alatalo, R. V., Gustafsson, L. 1988. Genetic component of morphological differentiation in coal tits under competitive release. *Evolution*, **42**, 200–3.
Alatalo, R. V., Gustafsson, L., and Lundberg, A. 1990. Phenotypic selection on heritable size traits: environmental variance and genetic response. *American Naturalist*, **135**, 464–71.
Alatalo, R. V., Gustafsson, L., and Lundberg, A. 1994. Male coloration and species recognition in sympatric flycatchers. *Proceedings of the Royal Society of London B, Biological Sciences*, **256**, 113–18.

Alberch, P. and Gale, E. A. 1985. A developmental analysis of an evolutionary trend: digital reduction in amphibians. *Evolution*, **39**, 8–23.

Albertson, R. C., Markert, J. A., Danley, P. D., and Kocher, T. D. 1999. Phylogeny of a rapidly evolving clade: The cichlid fishes of Lake Malawi, East Africa. *Proceedings of the National Academy of Sciences, USA*, **96**, 5107–10.

Allmon, W. D. 1992. A causal analysis of stages in allopatric speciation. *Oxford Surveys in Evolutionary Biology*, **8**, 219–57.

Amadon, D. 1950. The Hawaiian honeycreepers (Aves, Drepaniidae). *Bulletin of the American Museum of Natural History*, **95**, 157–268.

Angerbjörn, A. 1986. Gigantism in island populations of wood mice (*Apodemus*) in Europe. *Oikos*, **47**, 47–56.

Antonovics, J. and Primack, R. B. 1982. Experimental ecological genetics in *Plantago*. VI. The demography of seedling transplants of *P. lanceolata*. *Journal of Ecology*, **70**, 55–75.

Armbruster, W. S. 1988. Multilevel comparative analysis of the morphology, function, and evolution of *Dalechampia* blossoms. *Ecology*, **69**, 1746–61.

Armbruster, W. S. 1990. Estimating and testing the shapes of adaptive surfaces: the morphology and pollination of *Dalechampia* blossoms. *American Naturalist*, **135**, 14–31.

Armbruster, W. S. 1993. Evolution of plant pollination systems: hypotheses and tests with the neotropical vine *Dalechampia*. *Evolution*, **47**, 1480–505.

Armbruster, W. S., Edwards, M. E., and Debevec, E. M. 1994. Floral character displacement generates assemblage structure of Western Australian triggerplants (*Stylidium*). *Ecology*, **75**, 315–29.

Armbruster, W. S. and Baldwin, B. G. 1998. Switch from specialized to generalized pollination. *Nature (London)*, **394**, 632.

Arnold, S. J. 1983. Morphology, performance and fitness. *American Zoologist*, **23**, 347–61.

Arnold, S. J. 1992. Constraints on phenotypic evolution. *American Naturalist*, **140** *(Supplement)*, S85–S107.

Arnqvist, G. 1998. Comparative evidence for the evolution of genitalia by sexual selection. *Nature (London)*, **393**, 784–8.

Arthur, W. 1982. The evolutionary consequences of interspecific competition. *Advances in Ecological Research*, **12**, 127–87.

Askew, R. R. 1961. On the biology of the inhabitants of oak galls of Cynipidae (Hymenoptera) in Britain. *Transactions of the Society for British Entomology*, **14**, 237–68.

Asquith, A. 1995. Evolution of *Sarona* (Heteroptera, Miridae). In *Hawaiian biogeography* (ed. W. W. Wagner and V. A. Funk), pp. 90–120. Smithsonian Institution Press, Washington D.C.

Avise, J. C. and Walker, D. 1998. Pleistocene phylogeographic effects on avian populations and the speciation process. *Proceedings of the Royal Society of London B, Biological Sciences*, **265**, 457–63.

Ayala, F. 1975. Genetic differentiation during the speciation process. *Evolutionary Biology*, **8**, 1–75.

Ayala, F., Tracey, M., Hedgecock, D., and Richmond, R. C. 1974. Genetic differentiation during the speciation process in *Drosophila*. *Evolution*, **28**, 576–92.

Baker, R. and DeSalle, R. 1997. Multiple sources of character information and the phylogeny of Hawaiian drosophilids. *Systematic Biology*, **46**, 654–73.

Baldwin, B. C. 1997. Adaptive radiation of the Hawaiian silversword alliance: congruence and conflict of phylogenetic evidence from molecular and non-molecular investigations. In

Molecular evolution and adaptive radiation (ed. T. J. Givnish and K. J. Sytsma), pp. 104–28. Cambridge University Press, Cambridge.

Baldwin, B. G. and Robichaux, R. H. 1995. Historical biogeography and ecology of the Hawaian Silversword Alliance (Asteraceae). In *Hawaiian biogeography* (ed. W. W. Wagner and V. A. Funk), pp. 259–87. Smithsonian Institution Press, Washington D.C.

Baldwin, B. G. and Sanderson, M. J. 1998. Age and rate of diversification of the Hawaiian silversword alliance (Compositae). *Proceedings of the National Academy of Sciences, USA*, **95**, 9402–6.

Bambach, R. K. 1977. Species richness in marine benthic habitats through the Phanerozoic. *Paleobiology*, **3**, 152–67.

Barker, A. M. and Maczka, C. J. M. 1996. The relationship between host selection and subsequent larval performance in three free-living graminivorous sawflies. *Ecological Entomology*, **21**, 317–27.

Barraclough, T. G., Harvey P. H., and Nee, S. 1995. Sexual selection and taxonomic diversity in passerine birds. *Proceedings of the Royal Society of London B, Biological Sciences*, **259**, 211–15.

Barrett, S. C. H. and Graham, S. W. 1997. Adaptive radiation in the aquatic plant family Pontederiaceae: insights from phylogenetic analysis. In *Molecular evolution and adaptive radiation* (ed. T. J. Givnish and K. J. Sytsma), pp. 225–58. Cambridge University Press, Cambridge.

Barton, N. H. 1989. Founder effect speciation. In *Speciation and its consequences* (ed. D. Otte and J. A. Endler), pp. 229–56. Sinauer, Sunderland, Mass.

Barton, N. H. 1998. Natural selection and random genetic drift as causes of evolution on islands. In *Evolution on islands* (ed. P. R. Grant), pp. 102–23. Oxford University Press, Oxford.

Barton, N. H. and Rouhani, S. 1987. The frequency of shifts between alternative equilibria. *Journal of Theoretical Biology*, **125**, 397–418.

Baum, D. A. and Larson, A. 1991. Adaptation reviewed—a phylogenetic methodology for studying character macroevolution. *Systematic Zoology*, **40**, 1–18.

Behnke, R. J. 1972. The systematics of salmonid fishes of recently glaciated areas. *Journal of the Fisheries Research Board of Canada*, **29**, 639–71.

Bell, G. 1997. *Selection: the mechanism of evolution*. Chapman & Hall, New York.

Bell, M. A. and Foster, S. A. 1994. Introduction to the evolutionary biology of the threespine stickleback. In *Evolutionary biology of the threespine stickleback* (ed. M. A. Bell and S. A. Foster), pp. 1–27. Oxford University Press, Oxford.

Benkman, C. W. 1989. On the evolution and ecology of island populations of red crossbills. *Evolution*, **43**, 1324–30.

Benkman, C. W. 1991. Predation, seed size partitioning and the evolution of body size in seed-eating finches. *Evolutionary Ecology*, **5**, 118–27.

Benkman, C. W. 1993. Adaptation to single resources and the evolution of crossbill (*Loxia*) diversity. *Ecological Monographs*, **63**, 305–25.

Benkman, C. W. 1999. The selection mosaic and diversifying coevolution between crossbills and lodgepole pine. *American Naturalist*, **153** *(Supplement)*, S75–S91.

Benkman, C. W. and Lindholm, A. K. 1991. The advantages and evolution of a morphological novelty. *Nature (London)*, **349**, 519–20.

Bennington, C. C. and Mcgraw, J. B. 1995. Natural selection and ecotypic differentiation in *Impatiens pallida*. *Ecological Monographs*, **65**, 303–23.

Benton, M. J. 1983. Large-scale replacements in the history of life. *Nature (London)*, **302**, 16–17.
Benton, M. J. 1987. Progress and competition in macroevolution. *Biological Reviews*, **62**, 305–38.
Benton, M. J. 1996a. Testing the roles of competition and expansion in tetrapod evolution. *Proceedings of the Royal Society of London B, Biological Sciences*, **263**, 641–6.
Benton, M. J. 1996b. On the nonprevalence of competitive replacement in the evolution of tetrapods. In *Evolutionary paleobiology* (ed. D. Jablonski, D. H. Erwin and J. H. Lipps), pp. 185–210. Chicago University Press, Chicago.
Benzing, D. A. 1987. Major patterns and processes in orchid evolution: a critical synthesis. In *Orchid biology: reviews and perspectives, IV* (ed. J. Arditti), pp. 33–78. Cornell University Press, Ithaca, New York.
Bernatchez, L. and Dodson, J. J. 1990. Allopatric origin of sympatric populations of lake whitefish (*Coregonus clupeaformis*) as revealed by mitochondrial-DNA restriction analysis. *Evolution*, **44**, 1263–71.
Bernatchez, L., Vuorinen, J. A., Bodaly, R. A., and Dodson, J. J. 1996. Genetic evidence for reproductive isolation and multiple origins of sympatric trophic ecotypes of whitefish (*Coregonus*). *Evolution*, **50**, 624–35.
Bernatchez, L., Chouinard, A., and Lu, G. 1999. Integrating molecular genetics and ecology in studies of adaptive radiation: whitefish, *Coregonus* sp., as a case study. *Biological Journal of the Linnean Society*, **68**, 173–94.
Bernatchez, L. and Wilson, C. C. 1998. Comparative phylogeneography of Nearctic and Palearctic fishes. *Molecular Ecology*, **7**, 431–52.
Björklund, M. 1991. Patterns of morphological variation among cardueline finches (Fringillidae: Carduelinae). *Biological Journal of the Linnean Society*, **43**, 239–48.
Bjorkman, O. and Holmgren, P. 1963. Adaptability of the photosynthetic apparatus to light intensity in ecotypes from exposed and shaded habitats. *Physiologia Plantarum*, **16**, 889–914.
Boag, P. T. 1983. The heritability of external morphology in the Darwin's finches (Geospizinae) of Daphne Major Island, Galápagos. *Evolution*, **37**, 877–94.
Boag, P. T. and Grant, P. R. 1981. Intense natural selection in a population of Darwin's finches (Geospizinae) in the Galápagos. *Science (Washington, D.C.)*, **214**, 82–5.
Boag, P. T. and Grant, P. R. 1984. The classic case of character release: Darwin's finches (*Geospiza*) on Isla Daphne Major, Galápagos. *Biological Journal of the Linnean Society*, **22**, 243–87.
Bodaly, R. A. 1979. Morphological and ecological divergence within the lake whitefish *Coregonus clupeaformis* species complex in Yukon territory. *Journal of the Fisheries Research Board of Canada*, **36**, 1214–22.
Boulding, E. G. and Van Alstyne, K. L. 1993. Mechanisms of differential survival and growth of two species of *Littorina* on wave-exposed and on protected shores. *Journal of Experimental Marine Biology and Ecology*, **169**, 139–66.
Bouton, N., Witte, F., van Alpen, J. J. M., Schenk, A. and Seehausen, O. 1999. Local adaptations in populations of rock-dwelling haplochromines (Pisces: Cichlidae) from southern Lake Victoria. *Proceedings of the Royal Society of London B, Biological Sciences*, **266**, 355–60.
Bowman, R. I. 1961. Morphological differentiation and adaptation in the Galápagos finches. *University of California Publications in Zoology*, **58**, 1–302.

Bradshaw, A. D. 1972. Some of the evolutionary consequences of being a plant. *Evolutionary Biology*, **5**, 25–47.

Bradshaw, H. D., Jr., Otto, K. G., Frewen, B. E., McKay, J. G., and Schemske, D. W. 1998. Quantitative trait loci affecting differences in floral morphology between two species of monkeyflower (*Mimulus*). *Genetics*, **149**, 367–82.

Brodie, E. D. III. 1992. Correlational selection for color pattern and antipredator behavior in the garter snake *Thamnophis ordinoides*. *Evolution*, **46**, 1284–98.

Brodie, E. D. III, Moore, A. J., and Janzen, F. J. 1995. Visualizing and quantifying natural selection. *Trends in Ecology & Evolution*, **10**, 313–8.

Bronstein, J. L. 1994. Our current understanding of mutualism. *Quarterly Review of Biology*, **69**, 31–51.

Brooks, D. R., O'Grady, R. T., and Glen, D. R. 1985. Phylogenetic analysis of the Digenea (Platyhelminthes: Cercomeria) with comments on their adaptive radiation. *Canadian Journal of Zoology*, **63**, 411–43.

Brown, J. H. and Munger, J. C. 1985. Experimental manipulation of a desert rodent community: food addition and species removal. *Ecology*, **66**, 1545–63.

Brown, J. S. and Vincent, T. L. 1987. Coevolution as an evolutionary game. *Evolution*, **41**, 66–79.

Brown, J. S. and Vincent, T. L. 1992. Organization of predator-prey communities as an evolutionary game. *Evolution*, **46**, 1269–83.

Brown, W. L., Jr. and Wilson, E. O. 1956. Character displacement. *Systematic Zoology*, **5**, 49–64.

Burnell, K. L. and Hedges, S. B. 1990. Relationships of West Indian *Anolis* (Sauria: Iguanidae): an approach using slow-evolving protein loci. *Caribbean Journal of Science*, **26**, 7–30.

Butlin, R. 1989. Reinforcement of premating isolation. In *Speciation and its consequences* (ed. D. Otte and J. A. Endler), pp. 85–110. Sinauer, Sunderland, Mass.

Cabrera, V., Gonzales, A. M., Larruga, J. M., and Gullon, A. 1983. Genetic distance and evolutionary relationships in the *Drosophila obscura* group. *Evolution*, **37**, 675–89.

Campbell, D. R., Waser, N. M., and Ackerman, M. E. J. 1997. Analysing pollinator-mediated selection in a plant hybrid zone: hummingbird visitation patterns on three spatial scales. *American Naturalist*, **149**, 295–315.

Carleton, M. D. and Eshelman, R. E. 1979. A synopsis of fossil grasshopper mice, genus *Onychomys*, and their relationship to recent species. *Papers on Paleontology*, **21**, 1–63.

Carlquist, S. 1974. *Island biology*. Columbia University Press, New York.

Carlquist, S. 1980. *Hawaii, a natural history*. 2nd ed. Pacific Tropical Botanical Garden, Lawai, Hawaii.

Carr, G. D. and Kyhos, D. W. 1981. Adaptive radiation in the Hawaiian silversword alliance (Compositae: Madiinae). I. Cytogenetics of spontaneous hybrids. *Evolution*, **35**, 543–56.

Carr, G. D., Robichaux, R. H., Witter, M. S., and Kyhos, D. W. 1989. Adaptive radiation of the Hawaiian silversword alliance (Compositae–Madiinae): a comparison with Hawaiian picture-winged *Drosophila*. In *Genetics, speciation and the founder principle* (ed. L. V. Giddings, K. Y. Kaneshiro, and W. W. Anderson), pp. 79–97. Oxford University Press, Oxford.

Carroll, S. P., Klassen, S. P., and Dingle, H. 1998. Rapidly evolving adaptations to host ecology and nutrition in the soapberry bug. *Evolutionary Ecology*, **12**, 955–68.

Carson, H. L. and Kaneshiro, K. Y. 1976. *Drosophila* of Hawaii: systematics and ecological genetics. *Annual Review of Ecology and Systematics*, **7**, 311–45.
Carson, H. L., Ashburner, M., and Thompson, J. N. 1981. The Genetics of *Drosophila*, Vol. 3a. Academic Press, New York.
Carson, H. L. and Templeton, A. R. 1984. Genetic revolutions in relation to speciation phenomena: the founding of new populations. *Annual Review of Ecology and Systematics*, **15**, 97–131.
Case, T. J. 1979. Character displacement and coevolution in some *Cnemidophorus* lizards. *Fortschritte der Zoologie*, **25**, 235–82.
Case, T. J. and Sidell, R. 1983. Pattern and chance in the structure of model and natural communities. *Evolution*, **37**, 832–49.
Chambers, J. M., Cleveland, W. S., Kliener, B., and Tukey, P. A. 1983. *Graphical methods for data analysis*. Wadsworth, Pacific Grove, California.
Charlesworth, B. 1990. Optimization models, quantitative genetics, and mutation. *Evolution*, **44**, 520–38.
Charlesworth, B., Lande, R., and Slatkin, M. 1982. A neo-Darwinian commentary on macroevolution. *Evolution*, **36**, 474–98.
Charlesworth, B., Schemske, D. W., and Sork, V. L. 1987. The evolution of plant reproductive characters; sexual versus natural selection. In *The evolution of sex and its consequences* (ed. S. C. Stearns), pp. 317–35. Birkhäuser Verlag, Basel.
Chase, M. W. and Hills, H. G. 1992. Orchid phylogeny, flower sexuality and fragrance-seeking. *BioScience*, **42**, 43–9.
Chase, M. W. and Palmer, J. D. 1997. Leapfrog radiation in floral and vegetative traits among twig epiphytes in the orchid subtribe Oncidiinae. In *Molecular evolution and adaptive radiation* (ed. T. J. Givnish and K. J. Sytsma), pp. 331–52. Cambridge University Press, Cambridge.
Chase, V. C. and Raven, P. H. 1975. Evolutionary and ecological relationships between *Aquilegia formosa* and *A. pubescens* (Ranunculaceae), two perennial plants. *Evolution*, **29**, 474–86.
Cheetham, A. H. and Jackson, J. B. C. 1995. Process from pattern: tests for selection versus random change in punctuated bryozoan spseciation. In *New approaches to speciation in the fossil record* (ed. D. H. Erwin and R. L. Anstey), pp. 185–207. Columbia University Press, New York.
Cheverud, J. M. 1982. Phenotypic, genetic, and environmental morphological integration in the cranium. *Evolution*, **36**, 499–516.
Cheverud, J. M. 1984. Quantitative genetics and developmental constraints on evolution by selection. *Journal of Theoretical Biology*, **110**, 155–72.
Cheverud, J. M. 1996. Developmental integration and the evolution of pleiotropy. *American Zoologist*, **36**, 44–50.
Chiba, S. 1996. Ecological and morphological diversification within single species and character displacement in *Mandarina*, endemic land snails of the Bonin Islands. *Journal of Evolutionary Biology*, **9**, 277–91.
Chiba, S. 1999*a*. Accelerated evolution of land snails *Mandarina* in the oceanic Bonin Islands: evidence from mitochondrial DNA sequences. *Evolution*, **53**, 460–71.
Chiba, S. 1999*b*. Character displacement, frequency-dependent selection, and divergence of shell colour in land snails *Mandarina* (Pulmonata). *Biological Journal of the Linnean Society*, **66**, 465–79.

Cho, S. 1997. *Molecular phylogenetics of Heliothinae (Lepidoptera: Noctuidae) using elongation factor 1-alpha and dopa decarboxylase.* Ph.D. Dissertation, University of Maryland.

Christidis, L., Schodde, R., and Baverstock, P. R. 1988. Genetic and morphological differentiation and phylogeny in the Australo-Papuan scrubwrens (*Sericornis,* Acanthizidae). *Auk,* **105,** 616–29.

Clarke, B. 1962. *Balanced polymorphism and the diversity of sympatric species.* Systematics Association Publications, London.

Clarke, B., Johnson, M. S., and Murray, J. 1996. Clines in the genetic distance between two species of island land snails: how 'molecular leakage' can mislead us about speciation. *Philosophical Transactions of the Royal Society of London B, Biological Sciences,* **351,** 773–84.

Clarke, B., Johnson, M. S., and Murray, J. 1998. How 'molecular leakage' can mislead us about island speciation. In *Evolution on islands* (ed. P. R. Grant), pp. 181–95. Oxford University Press, Oxford.

Coddington, J. A. 1988. Cladistic tests of adaptational hypotheses. *Cladistics,* **4,** 3–22.

Cody, M. L. 1973. Character convergence. *Annual Review of Ecology and Systematics,* **4,** 189–211.

Cohan, F. M. and Hoffman, A. A. 1986. Genetic divergence under uniform selection. II Different responses to selection for knockdown resistance to ethanol among *Drosophila melanogaster* populations and their replicate lines. *Genetics,* **114,** 145–63.

Cohan, F. M., Hoffman, A. A., and Gayley, T. W. 1989. A test of the role of epistasis in divergence under uniform selection. *Evolution,* **43,** 766–74.

Cole, B. J. 1981. Overlap, regularity, and flowering phenologies. *American Naturalist,* **117,** 993–7.

Collins, T. M., Frazer, K., Palmer, A. R., Vermeij, G. J., and Brown, W. M. 1996. Evolutionary history of northern hemisphere *Nucella* (Gastropoda, Muricidae): molecular, morphological, ecological, and paleontological evidence. *Evolution,* **50,** 2287–304.

Colbourne, J. K., Hebert, P. D. N., and Taylor, D. J. 1997. Evolutionary origins of phenotypic diversity in *Daphnia.* In *Molecular evolution and adaptive radiation* (ed. T. J. Givnish and K. J. Sytsma), pp. 163–88. Cambridge University Press, Cambridge.

Colwell, R. K. 1986. Community biology and sexual selection: lessons from hummingbird flower mites. In *Community ecology* (ed. J. Diamond and T. J. Case), pp. 406–24. Harper and Rowe, New York.

Colwell, R. K. and Winkler, D. W. 1984. A null model for null models in biogeography. In *Ecological communities: conceptual issues and the evidence* (ed. D. R. Strong, D. S. Simberloff, L. G. Abele, and A. B. Thistle), pp. 344–59. Princeton University Press, Princeton, N.J.

Connell, J. H. 1980. Diversity and the coevolution of competitors, or the ghost of competition past. *Oikos,* **35,** 131–8.

Connell, J. H. 1983. On the prevalence and relative importance of interspecific competition: evidence from field experiments. *American Naturalist,* **122,** 661–96.

Conner, J. and Via, S. 1993. Patterns of phenotypic and genetic correlations among morphological and life-history traits in wild radish, *Raphanus raphanistrum. Evolution,* **47,** 704–11.

Conway Morris, S. 1998. *The crucible of creation: the Burgess Shale and the rise of animals.* Oxford University Press, Oxford.

Coyne, J. A., Barton, N. H., and Turelli, M. 1997. Perspective: a critique of Sewall Wright's shifting balance theory of evolution. *Evolution*, **51**, 643–71.
Coyne, J. A. and Orr, H. A. 1989. Patterns of speciation in *Drosophila*. *Evolution*, **43**, 362–81.
Coyne, J. A. and Orr, H. A. 1998. The evolutionary genetics of speciation. *Philosophical Transactions of the Royal Society of London B, Biological Sciences*, **353**, 287–305.
Coyne, J. A. and Orr, H. A. 1997. "Patterns of speciation in *Drosophila*" revisited. *Evolution*, **51**, 295–303.
Coyne, J. A., Barton, N. H., and Turelli, M. 2000. Is Wright's shifting balance process important in evolution? *Evolution*, **54**, 306–17.
Coyne, J. A. and Price, T. Manuscript. No evidence for sympatric speciation in birds.
Coyne, J. A. and Orr, H. A. Manuscript. *Speciation*. Sinauer, Sunderland, Mass.
Cracraft, J. 1985. Biological diversification and its causes. *Annals of the Missouri Botanical Garden*, **72**, 794–822.
Craig, T. P., Horner, J. D., and Itami, J. K. 1997. Hybridization studies on the host races of *Eurosta solidaginis*: implications for sympatric speciation. *Evolution*, **51**, 552–60.
Crepet, W. L. 1984. Advanced (constant) insect pollination mechanisms: pattern of evolution and implications vis-a-vis angiosperm diversity. *Annals of the Missouri Botanical Garden*, **71**, 607–30.
Crespi, B. J. and Sandoval, C. P. 2000. Phylogenetic evidence for the evolution of ecological specialization in *Timema* walking-sticks. *Journal of Evolutionary Biology*, **13**, 249–62.
Crowder, L. B. 1984. Character displacement and habitat shift in a native cisco in southeastern Lake Michigan: evidence for competition? *Copeia*, **1984**, 878–83.
Crowder, L. B. 1986. Ecological and morphological shifts in Lake Michigan fishes: glimpses of the ghost of competition past. *Environmental Biology of Fishes*, **16**, 147–56.
Cullings, K. W., Szaro, T. M., and Bruns, T. D. 1996. Evolution of extreme specialization within a lineage of ectomycorrhizal epiparasites. *Nature (London)*, **379**, 63–5.
Culver, D. C., Kane, T. C., and Fong, D. W. 1995. *Adaptation and natural selection in caves*. Harvard University Press, Cambridge, Mass.
Darwin, C. R. 1842. *Journal of researches into the geology and natural history of the various countries visited during the voyage of H.M.S. 'Beagle', under the command of Captain Fitzroy, R. N. from 1832 to 1836*. Henry Colborn, London.
Darwin, C. R. 1859. *On the origin of species by means of natural selection*. John Murray, London.
Davies, M. S. and Snaydon, R. W. 1976. Rapid population differentiation in a mosiac environment. *Heredity*, **36**, 59–66.
Davies, N., Aiello, A., Mallet, J., Pomiankowski, A., and Silberglied, R. E. 1997. Speciation in two neotropical butterflies: extending Haldane's rule. *Proceedings of the Royal Society of London B, Biological Sciences*, **264**, 845–51.
Day, T. and Taylor, P. D. 1996. Evolutionary stable versus fitness maximizing life histories under frequency-dependent selection. *Proceedings of the Royal Society of London B, Biological Sciences*, **263**, 333–8.
Dayan, T., Simberloff, D., Tchernov, E., and Yom-Tov, Y. 1989*a*. Interspecific and intraspecific character displacement in mustelids. *Ecology*, **70**, 1526–39.
Dayan, T., Tchernov, E., Yom-Tov, Y., and Simberloff, D. 1989*b*. Ecological character displacement in Saharo-Arabian *Vulpes*: outfoxing Bergmann's rule. *Oikos*, **55**, 263–72.

Dayan, T., Simberloff, D., Tchhernov, E., and Yom-Tov, Y. 1990. Feline canines: community wide character displacement among the small cats of Israel. *American Naturalist*, **136**, 39–60.

Dayan, T., Simberloff, D., Tchhernov, E., and Yom-Tov, Y. 1992. Canine carnassials: character displacement in the wolves, jackals and foxes of Israel. *Biological Journal of the Linnean Society*, **45**, 315–31.

Dayan, T. and Simberloff, D. 1994*a*. Character displacement, sexual dimorphism, and morphological variation among british and Irish mustelids. *Ecology*, **75**, 1063–73.

Dayan, T. and Simberloff, D. 1994*b*. Morphological relationships among coexisting heteromyids: An incisive dental character. *American Naturalist*, **143**, 462–77.

DeAngelis, D. L., Persson, L., and Rosemond, A. D. 1996. Interaction of productivity and consumption. In *Food webs* (ed. G. A. Polis and K. O. Winemiller), pp. 109–12. Chapman & Hall, New York.

DeSalle, R. 1995. Molecular approaches to biogeographic analysis of Hawaiian Drosophilidae. In *Hawaiian biogeography* (ed. W. W. Wagner and V. A. Funk), pp. 1272–89. Smithsonian Institution Press, Washington D.C.

Diamond, J. 1986. Evolution of ecological segregation in the New Guinea montane avifauna. In *Community ecology* (ed. J. Diamond and T. J. Case), pp. 98–125. Harper and Rowe, New York.

Diamond, J., Pimm, S. L., Gilpin, M. E., and LeCroy, M. 1989. Rapid evolution of character displacement in myzomelid honeyeaters. *American Naturalist*, **134**, 75–708.

Dieckmann, U. and Doebeli, M. 1999. On the origin of species by sympatric speciation. *Nature (London)*, **400**, 354–7.

Dobler, S., Mardulyn, P., Pasteels, J. M., and Rowell-Rahier, M. 1996. Host-plant switches and the evolution of chemical defense and life history in the leaf beetle genus *Oreina*. *Evolution*, **50**, 2373–86.

Dobzhansky, T. 1937. *Genetics and the origin of species*. Columbia University Press, New York.

Dobzhansky, T. 1951. *Genetics and the origin of species. 3rd ed*. Columbia University Press, New York.

Dodd, D. M. B. 1989. Reproductive isolation as a consequence of adaptive divergence in *Drosophila pseudoobscura*. *Evolution*, **43**, 1308–11.

Dodd, M. E., Silvertown, J., and Chase, M. 1999. Phylogenetic analysis of trait evolution and species diversity variation among angiosperm families. *Evolution*, **53**, 732–44.

Doebeli, M. 1996. An explicit genetic model for ecological character displacement. *Ecology*, **77**, 510–20.

Doebeli, M. and Dieckmann, U. 2000. Evolutionary branching and sympatric speciation caused by different types of ecological interactions. *American Naturalist*, **156** *(Supplement)*, in press.

Dominey, W. 1984. Effects of sexual selection and life history on speciation: species flocks in African cichlids and Hawaiian *Drosophila*. In *Evolution of fish species flocks* (ed. A. A. Echelle and I. Kornfield), pp. 231–49. University of Maine Press, Orono, Maine.

Drossel, B. and McKane, A. 1999. Ecological character displacement in quantitative genetic models. *Journal of Theoretical Biology*, **3**, 363–76.

Dudley, S. A. 1996*a*. Differing selection on plant physiological traits in response to environmental water availability: a test of adaptive hypotheses. *Evolution*, **50**, 92–102.

Dudley, S. A. 1996*b*. The response to differing selection on plant physiological traits: evidence for local adaptation. *Evolution*, **50**, 103–10.

Dunham, A. E., Smith, G. R., and Taylor, J. N. 1979. Evidence for ecological character displacement in western American catostomid fishes. *Evolution*, **33**, 877–96.

Dynes, J., Magnan, P., Bernatchez, L., Rodriguez, M.A. 1999. Genetic and morphological variation between two forms of lacustrine brook charr. *Journal of Fish Biology*, **54**, 955–72.

Eadie, J. M., Broekhoven, L., and Colgen, P. 1987. Size ratios and artifacts: Hutchinson's rule revisited. *American Naturalist*, **129**, 1–17.

Eberhard, W. G. 1985. *Sexual selection and animal genitalia*. Harvard University Press, Cambridge, Mass.

Ehleringer, J. R. and Clark, C. 1988. Evolution and adaptation in *Encelia* (Asteraceae). In *Plant evolutionary biology* (ed. L. D. Gottlieb and S. K. Jain), pp. 221–48. Chapman & Hall, New York.

Ehrlich, P. R. and Raven, H. 1964. Butterflies and plants: a study in coevolution. *Evolution*, **18**, 586–608.

Eisses, K., Van Dijk, H., and Van Delden, W. 1979. Genetic differentiation within the *melanogaster* species subgroup of the genus *Drosophila*. *Evolution*, **33**, 1063–8.

Eldridge, J. L. and Johnson, D. H. 1988. Size differences in migrant sandpiper flocks: ghosts in the ephemeral guilds. *Oecologia*, **77**, 433–44.

Emerson, S. B. and Arnold, S. J. 1989. Intra- and interspecific relationships between morphology, performance, and fitness. In *Complex organismal functions: integration and evolution in vertebrates* (ed. D. B. Wake and G. Roth), pp. 295–314. Wiley, Chichester.

Emmons, L. H. 1980. Ecology and resource partitioning among nine species of African rain forest squirrels. *Ecological Monographs*, **50**, 31–54.

Emms, S. K. and Arnold, M. L. 1997. The effect of habitat on parental and hybrid fitness: transplant experiments with Louisiana Irises. *Evolution*, **51**, 1112–19.

Endler, J. A. 1977. *Geographic variation, speciation, and clines*. Princeton University Press, Princeton, N.J.

Endler, J. A. 1986. *Natural selection in the wild*. Princeton University Press, Princeton, N.J.

Endler, J. A. 1992. Signals, signal conditions, and the direction of evolution. *American Naturalist*, **139** *(Supplement)*, S125–S53.

Endler, J. A. and Basolo, A.L. 1998. Sensory ecology, receiver biases and sexual selection. *Trends in Ecology & Evolution*, **13**, 415–20.

Eriksson, O. and Bremer, B. 1992. Pollination systems, dispersal modes, life forms, and diversification rates in Angiosperm families. *Evolution*, **46**, 258–66.

Etges, W. J. 1998. Premating isolation is determined by larval rearing substrates in cactophilic *Drosophila mojavensis*. IV. Correlated responses in behavioral isolation to artificial selection on a life-history trait. *American Naturalist*, **152**, 129–44.

Evans, D. R., Hill, J., Williams, T. A., and Rhodes, I. 1985. Effects of coexistence in the performances of white clover-perennial ryegrass mixtures. *Oecologia*, **66**, 536–9.

Evans, D. R., Hill, J., Williams, T. A., and Rhodes, I. 1989. Coexistence and the productivity of white clover-perennial ryegrass mixtures. *Theoretical and Applied Genetics*, **77**, 65–70.

Falconer, D. S. 1981. *Introduction to quantitative genetics*. 2nd ed. Longman, New York.

Falconer, D. S. and Mackay, T. F. C. 1996. *Introduction to quantitative genetics*. Longman, New York.

Farrell, B. D. 1998. "Inordinate fondness" explained: why are there so many beetles? *Science (Washington, D.C.)*, **281**, 555–9.

Farrell, B. D., Dussourd, D. E., and Mitter, C. 1991. Escalation of plant defense: do latex and resin canals spur plant diversification? *American Naturalist*, **138**, 881–900.

Fear, K. K. and Price, T. 1998. The adaptive surface in ecology. *Oikos*, **82**, 440–8.

Feder, J. L. 1998. The apple maggot fly, *Rhagoletis pomonella*: flies in the face of conventional wisdom about speciation? In *Endless forms: species and speciation* (ed. D. Howard and S. Berlocher), pp. 130–44. Oxford University Press, Oxford.

Felsenstein, J. 1981. Skepticism towards Santa Rosalia, or why are there so few kinds of animals? *Evolution*, **35**, 124–38.

Felsenstein, J. 1982. Numerical methods for inferring evolutionary trees. *Quarterly Review of Biology*, **57**, 379–404.

Felsenstein, J. 1985. Phylogenies and the comparative method. *American Naturalist*, **125**, 1–15.

Fenchel, T. 1975. Character displacement and co-existence in mud snails (Hydrobiidae). *Oecologia*, **20**, 19–32.

Fenchel, T. and Kofoed, L. H. 1976. Evidence for exploitative interspecific competition in mud snails (Hydrobiidae). *Oikos*, **27**, 367–76.

Fenderson, O. C. 1964. Evidence of subpopulations of lake whitefish, *Coregonus clupeaformis*, involving a dwarfed form. *Transactions of the American Fisheries Society*, **93**, 77–94.

Ferguson, A. and Taggart, J. B. 1991. Genetic differentiation among the sympatric brown trout (*Salmo trutta*) populations of Lough Melvin, Ireland. *Biological Journal of the Linnean Society*, **43**, 221–37.

Ficken, R. W., Ficken, M. S., and Morse, D. 1968. Competition and character displacement in two sympatric pine-dwelling warblers (*Dendroica*, Parulidae). *Evolution*, **22**, 307–14.

Filchak, K. E., Feder, J. L., Roethele, J. B., and Stolz, U. 1999. A field test for host-plant dependent selection on larvae of the apple maggot fly, *Rhagoletis pomonella*. *Evolution*, **53**, 187–200.

Fisher, R. A. 1930. *The genetical theory of natural selection*. Oxford University Press, Oxford.

Fisher, R. A. 1936. The measurement of selective intensity. *Proceedings of the Royal Society of London B, Biological Sciences*, **121**, 109–22.

Fjeldså, J. 1983. Ecological character displacement and character release in grebes Podicipedidae. *Ibis*, **125**, 463–81.

Floate, K. D., Kearsley, M. J. C., and Whitham, T. G. 1993. Elevated herbivory in plant hybrid zones: *Chrysomela confluens*, *Populus* and phenological sinks. *Ecology*, **74**, 2056–65.

Foote, C. J. and Larkin, P. A. 1988. The role of male choice in the assortative mating of anadromous and non-anadromous sockeye salmon (*Oncorhynchus nerka*). *Behaviour*, **106**, 43–62.

Foote, M. 1992. Paleozoic record of morphological diversity in blastozoan echinoderms. *Proceedings of the National Academy of Sciences, USA*, **89**, 7325–9.

Foote, M. 1993. Discordance and concordance between morphological and taxonomic diversity. *Paleobiology*, **19**, 185–204.

Fox, L. R. and Morrow, P. A. 1981. Specialization: species property or local phenomenon. *Science (Washington, D.C.)*, **188**, 887–92.

Francisco-Ortega, J., Jansen, R. K., and Santo-Guerra, A. 1996. Chloroplast DNA evidence of colonization, adaptive radiation, and hybridization in the evolution of the Macaronesian flora. *Proceedings of the National Academy of Sciences, USA*, **93**, 4085–90.

Francisco-Ortega, J., Crawford, D. J., Santo-Guerra, A., and Jansen, R. K. 1997. Origin and evolution of *Argyranthemum* (Asteraceae: Anthemideae) in Macaronesia. In *Molecular evolution and adaptive radiation* (ed. T. J. Givnish and K. J. Sytsma), pp. 406–31. Cambridge University Press, Cambridge.

Fry, J. D. 1993. The 'general vigor' problem: can antagonistic pleiotropy be detected when genetic covariances are positive? *Evolution*, **47**, 329–33.

Fry, J. D. 1996. The evolution of host specialization: are trade-offs overrated? *American Naturalist*, **148** *(Supplement)*, S84–S107.

Fryer, G. and Iles, T. D. 1972. *The cichlid fishes of the Great Lakes of Africa.* Oliver and Boyd, Edinburgh.

Fulton, M. and Hodges, S. 1999. Floral isolation between *Aquilegia formosa* and *A. pubescens*. *Proceedings of the Royal Society of London B, Biological Sciences*, **266**, 2247–52.

Funk, D. J. 1998. Isolating a role for natural selection in speciation: host adaptation and sexual isolation in *Neochlamisus bebbianae* leaf beetles. *Evolution*, **52**, 1744–59.

Futuyma, D. J. 1986. *Evolutionary biology*. 2nd ed. Sinauer, Sunderland, Mass.

Futuyma, D. J. and Moreno, G. 1988. The evolution of ecological specialization. *Annual Review of Ecology and Systematics*, **19**, 207–33.

Futuyma, D. J., Keese, M. C., and Funk, D. J. 1995. Genetic constraints on macroevolution: The evolution of host affiliation in the leaf beetle genus *Ophraella*. *Evolution*, **49**, 797–809.

Futuyma, D. J. and Mitter, C. 1996. Insect-plant interactions: the evolution of component communities. *Philosophical Transactions of the Royal Society of London B, Biological Sciences*, **351**, 1361–6.

Galen, C., Shore, J. L. and Deyoe, H. 1991. Ecotypic divergence in apline *Polemonium viscosum*: genetic structure, quantitative variation, and local adaptation. *Evolution*, **45**, 1218–28.

Galis, F. and Druckner, E. G. 1996. Pharyngeal biting mechanisms in centrarchid and cichlid fishes: insights into a key evolutionary innovation. *Journal of Evolutionary Biology*, **9**, 641–70.

Galis, F. and Metz, J. A. J. 1998. Why are there so many cichlid species? *Trends in Ecology & Evolution*, **13**, 1–2.

Ganders, F. R. 1989. Adaptive radiation in Hawaiian *Bidens*. In *Genetics, speciation and the founder principle* (ed. L. V. Giddings, K. Y. Kaneshiro, and W. W. Anderson), pp. 99–112. Oxford University Press, Oxford.

Garland, T. R., Jr. and Losos, J. B. 1994. Ecological morphology of locomotor performance in squamate reptiles. In *Ecological morphology* (ed. P. C. Wainwright and S. M. Reilly), pp. 240–302, Chicago University Press, Chicago.

Gautier-Hion, A., Duplantier, J. M., Quris, R., Feer, F., Sourd, C., Decoux, J. P., Dubost, G., Emmons, L., Erard, C., Hecketsweiler, P., Moungazi, A., Roussilhon, C., and Thiollay, J. M. 1985. Fruit characters as a basis of fruit choice and seed dispersal in a tropical forest vertebrate community. *Oecologia*, **65**, 324–37.

Geary, D. H. 1990. Patterns of evolutionary tempo and mode in the radiation of *Melanopsis* (Gastropoda; Melanopsidae). *Paleobiology*, **16**, 492–511.

Gibbs, H. L. and Grant P. R. 1987a. Oscillating selection on a population of Darwin's finch. *Nature (London)*, **327**, 511–13.

Gibbs, H. L. and Grant P. R. 1987b. Adult survivorship in Darwin's ground finch (*Geospiza*) populations in a variable environment. *Journal of Animal Ecology*, **56**, 797–813.

Gilbert, L. E. 1975. Ecological consequences of a coevolved mutualism between butterflies and plants. In *Coevolution of animals and plants* (ed. L. E. Gilbert and P. H. Raven), pp. 210–40. University of Texas Press, Austin.

Gilbert, L. E. 1983. Coevolution and mimicry. In *Coevolution* (ed. D. J. Futuyma and M. Slatkin), pp. 263–81. Sinauer, Sunderland, Mass.

Gill, D. E. 1989. Fruiting failure, pollinator inefficiency, and speciation in orchids. In *Speciation and its consequences* (ed. D. Otte and J. A. Endler), pp. 458–81. Sinauer, Sunderland, Mass.

Gingerich, P. D. 1983. Rates of evolution: effects of time and temporal scaling. *Science (Washington, D.C.)*, **222**, 159–61.

Gingerich, P. D. 1985. Species in the fossil record: concepts, trends, and transitions. *Paleobiology*, **11**, 27–41.

Gittenberger, E. 1991. What about non-adaptive radiation? *Biological Journal of the Linnean Society*, **43**, 263–72.

Gittleman, J. L. and Kot, M. 1990. Adaptation statistics and a null model for estimating phylogenetic effects. *Systematic Zoology*, **39**, 227–41.

Givnish, T. J. 1997. Adaptive radiation and molecular systematics: issues and approaches. In *Molecular evolution and adaptive radiation* (ed. T. J. Givnish and K. J. Sytsma), pp. 1–54. Cambridge University Press, Cambridge.

Givnish, T. J. 1998. Adaptive plant evolution on islands: classical patterns, molecular data, new insights. In *Evolution on islands* (ed. P. R. Grant), pp. 281–304. Oxford University Press, Oxford.

Givnish, T. J., Sytsma, K. J., Hahn, W. J., and Smith, J. F. 1995. Molecular evolution, adaptive radiation, and geographic speciation in *Cyanea* (Campanulaceae, Lobelioideae). In *Hawaiian biogeography* (ed. W. W. Wagner and V. A. Funk), pp. 299–337. Smithsonian Institution Press, Washington D.C.

Givnish, T. J., Systsma, K. J., Smith, J. F., Hahn, W. J., Benzing, D. H., and Burkhardt, E. M. 1997. Molecular evolution and adaptive radiation in *Brocchinia* atop tepuis of the Guyana Shield. In *Molecular evolution and adaptive radiation* (ed. T. J. Givnish and K. J. Sytsma), pp. 259–311. Cambridge University Press, Cambridge.

Givnish, T. J. and Sytsma, K. J. (editors). 1997. *Molecular evolution and adaptive radiation*. Cambridge University Press, Cambridge.

Gomulkiewicz, R. and Kirkpatrick, M. 1992. Quantitative genetics and the evolution of reaction norms. *Evolution*, **46**, 390–411.

Gorbushin, A. M. 1996. The enigma of mud snail shell growth: asymmetrical competition or character displacement? *Oikos*, **77**, 85–92.

Gotelli, N. J. and Bossert, W. H. 1991. Ecological character displacement in a variable environment. *Theoretical Population Biology*, **39**, 49–62.

Gotelli, N. J. and Graves, G. R. 1996. *Null models in ecology*. Smithsonian Institution Press, Washington D.C.

Goudet, J. 1999. An improved procedure for testing the effects of key innovations on rate of speciation. *American Naturalist*, **153**, 549–55.

Gould, S. J. 1983. The hardening of the modern synthesis. In *Dimensions of Darwinism* (ed. M. Greene), pp. 71–93. Cambridge University Press, Cambridge.

Gould, S. J. 1989*a*. *Wonderful life; the Burgess shale and the nature of history.* W. W. Norton, New York.

Gould, S. J. 1989*b*. A developmental constraint in *Cerion*, with comments on the definition and interpretation of constraint in evolution. *Evolution*, **43**, 516–39.

Gould, S. J. and Vrba, E. 1982. Exaptation—a missing term in the science of form. *Paleobiology*, **8**, 4–15.

Grahame, J. and Mill, P. J. 1989. Shell shape variation in *Littorina saxatilis* and *L. arcana*: a case of character displacement? *Journal of the Marine Biological Association of the United Kingdom*, **69**, 837–55.

Grant, B. R. and Grant, P. R. 1989. *Evolutionary dynamics of a natural population.* Chicago University Press, Chicago.

Grant, B. R. and Grant, P. R. 1993. Evolution of Darwin's finches caused by a rare climatic event. *Proceedings of the Royal Society of London B, Biological Sciences*, **251**, 111–17.

Grant, P. R. 1972. Convergent and divergent character displacement. *Biological Journal of the Linnean Society*, **4**, 39–68.

Grant, P. R. 1975. The classic case of character displacement. *Evolutionary Biology*, **8**, 237.

Grant, P. R. 1981. The feeding of Darwin's finches on *Tribulus cistoides* (L.) seeds. *Animal Behaviour*, **29**, 785–93.

Grant, P. R. 1986. *Ecology and evolution of Darwin's finches.* Princeton University Press, Princeton, N.J.

Grant, P. R. and Abbott, I. 1980. Interspecific competition, island biogeography and null hypotheses. *Evolution*, **34**, 332–41.

Grant, P. R. and Price, T. 1981. Population variation as an ecological genetics problem. *American Zoologist*, **21**, 795–811.

Grant, P. R. and Schluter, D. 1984. Interspecific competition inferred from patterns of guild structure. In *Ecological communities: conceptual issues and the evidence* (ed. D. R. Strong, D. S. Simberloff, L. G. Abele, and A. B. Thistle), pp.. 201–33. Princeton University Press, Princeton, N.J.

Grant, P. R., Abbott, I., Schluter, D., Curry, R. L., and Abbott, L. K. 1985. Variation in the size and shape of Darwin's finches. *Biological Journal of the Linnean Society*, **25**, 1–39.

Grant, P. R. and Grant, B. R. 1992. Hybridization of bird species. *Science (Washington, D.C.)*, **256**, 193–7.

Grant, P. R. and Grant, B. R. 1994. Phenotypic and genetic effects of hybridization in Darwin's finches. *Evolution*, **48**, 297–316.

Grant, P. R. and Grant, B. R. 1995. Predicting microevolutionary responses to directional selection on heritable variation. *Evolution*, **49**, 241–51.

Grant, P. R. and Grant, B. R. 1996. Speciation and hybridization in island birds. *Philosophical Transactions of the Royal Society of London B, Biological Sciences*, **351**, 765–72.

Grant, P. R. and Grant, B. R. 1997. Genetics and the origin of bird species. *Proceedings of the National Academy of Sciences, USA*, **94**, 7768–75.

Grant, V. 1949. Pollinating systems as isolating mechanisms in angiosperms. *Evolution*, **3**, 82–97.

Grant, V. 1952. Isolation and hybridization between *Aquilegia formosa* and *A. pubescens*. *Aliso*, **2**, 341–60.

Grant, V. and Grant, K. A., 1965. *Flower pollination in the Phlox family.* Columbia University Press, New York.
Grant, V. 1981. *Plant speciation.* Columbia University Press, New York.
Grant, V. 1989. The theory of speciational trends. *American Naturalist,* **133,** 604–12.
Greenwood, P. H. 1974. The cichlid fishes of Lake Victoria, East Africa: the biology and evolution of a species flock. *Bulletin of the British Museum of Natural History Zoology Series,* **6,** 1–134.
Greenwood, P. H. 1984. African cichlids and evolutionary theories. In *Evolution of fish species flocks* (ed. A. A. Echelle and I. Kornfield), pp. 141–54. University of Maine Press, Orono, Maine.
Grinnell, J. 1917. The niche-relationships of the California thrasher. *Auk,* **34,** 427–33.
Grinnell, J. 1924. Geography and evolution. *Ecology,* **5,** 225–9.
Groth, J. G. 1993. Evolutionary differentiation in morphology, vocalizations, and allozymes among nomadic sibling species in the North American red crossbill (*Loxia curvirostra*) complex. University of California Publication in Zoology, Berkeley.
Grudemo, J. and Johannesson, K. 1999. Size of mudsnails, *Hydrobia ulvae* (Pennant) and *H. ventrosa* (Montagu), in allopatry and sympatry: conclusions from field distributions and laboratory growth experiments. *Journal of Experimental Marine Biology and Ecology,* **239,** 167–81.
Gurevitch, J., Morrison, J. A., and Hedges L. V. 2000. The interaction between competition and predation: a meta-analysis of field experiments. *American Naturalist,* **155,** 435–53.
Gurevitch, J., Morrow, L. L., Wallace, A., and Walsh, J. S. 1992. A meta-analysis of competition in field experiments. *American Naturalist,* **140,** 539–72.
Gustafsson, L. 1988. Foraging behaviour of individual coal tits, *Parus ater,* in relation to their age, sex and morphology. *Animal Behaviour,* **36,** 696–704.
Hairston, H. G., Smith, F. E., and Slobodkin, L. B. 1960. Community structure, population control, and competition. *American Naturalist,* **94,** 421–5.
Hansen, T. F. 1997. Stabilizing selection and the comparative analysis of adaptation. *Evolution,* **51,** 1341–51.
Hansen, T. F., Armbruster, W. S., and Antonsen, L. 2000. Comparative analysis of character displacement. *American Naturalist,* **156** *(Supplement),* in press.
Hapeman, J. R. and Inouye, J. R. 1997. Plant-pollinator interactions and floral radiation in *Platanthera* (Orchidaceae). In *Molecular evolution and adaptive radiation* (ed. T. J. Givnish and K. J. Sytsma), pp. 433–454. Cambridge University Press, Cambridge.
Harrison, M. K. and Crespi, B. J. 1999. A phylogenetic test of ecological adaptation in *Cancer* crabs. *Evolution,* **53,** 961–5.
Harvey, P. H. and Pagel, M. D. 1991. *The comparative method in evolutionary biology.* Oxford University Press, Oxford.
Harvey, P. H., May, R. M., and Nee, S. 1994. Phylogenies without fossils. *Evolution,* **48,** 523–9.
Harvey, P. H. and Rambaut, A. 2000. Comparative analyses for adaptive radiations. *Philosophical Transactions of the Royal Society of London B, Biological Sciences,* in press.
Hass, C. A. 1991. Evolution and biogeography of West Indian *Sphaerodactylus* (Sauria: Gekkonidae): a molecular approach. *Journal of Zoology (London),* **225,** 525–62.
Hass, C. A. and Hedges, S. B. 1991. Albumis evolution in West Indian frogs of the genus *Eleutherodactylus* (Leptodactylidae). Caribbean biogeography and a calibration of the albumin immunological clock. *Journal of Zoology (London),* **225,** 413–26.

Hass, C. A., Hedges, S. B., and Maxson, L. R. 1993. Molecular insights into the relationships and biogeography of West Indian Anoline lizards. *Biochemical Systematics and Ecology*, **21**, 97–114.

Hastie, T. and Tibshirani, R. 1990. *Generalized additive models*. Chapman & Hall, New York.

Hatfield, T. and Schluter, D. 1999. Ecological speciation in sticklebacks: environment-dependent hybrid fitness. *Evolution*, **53**, 866–73.

Hayes, J. F. and Hill, W. G. 1981. Modification of estimates of parameters in the construction of genetic selection indices ('bending'). *Biometrics*, **37**, 483–93.

Heard, S. B. and Hauser, D. L. 1995. Key evolutionary innovations and their ecological mechanisms. *Historical Biology*, **10**, 151–73.

Hebert, P. D. N. and Emery, C. J. 1990. The adaptive significance of cuticular pigmentation in *Daphnia*. *Functional Ecology*, **4**, 703–10.

Heed, W. B. 1968. Ecology of the Hawaiian *Drosophila*. *University of Texas Publications*, **6818**, 387–419.

Heilbuth, J. C. 2000. Species richness in dioecious clades: lower speciation or higher extinction? *American Naturalist*, in press.

Hewitt, G. M. 1989. The subdivision of species by hybrid zones. In *Speciation and its consequences* (ed. D. Otte and J. A. Endler), pp. 85–110. Sinauer, Sunderland, Mass.

Helenurm, K. and Ganders, F. R. 1985. Adaptive radiation and genetic differentiation in Hawaiian *Bidens*. *Evolution*, **39**, 753–65.

Helling, R. B., Vargas, C. N., and Adams, J. 1987. Evolution of *Escherichia coli* during growth in a constant environment. *Genetics*, **116**, 349–58.

Hertel, F. 1994. Diversity in body size and feeding morphology within past and present vulture assemblages. *Ecology*, **75**, 1074–84.

Heske, E. J., Brown, J. H., and Mistry, S. 1994. Long-term experimental study of a Chihuahuan Desert rodent community: 13 years of competition. *Ecology*, **75**, 438–45.

Heslop-Harrison, J. 1964. Forty years of genecology. *Advances in Ecological Research*, **2**, 159–247.

Hewitt, G. M. 1989. The subdivision of species by hybrid zones. In *Speciation and its consequences* (ed. D. Otte and J. A. Endler), pp. 85–110. Sinauer, Sunderland, Mass.

Hiesey, W. H., Nobs, M. A., and Björkman, O. 1971. Experimental studies on the nature of species. V. Biosystematics, genetics, and physiological ecology of the Erythranthe section of *Mimulus*. *Carnegie Institution of Washington Publication*, **628**, 1–213.

Higashi, M., Takimoto, G. and Yamamura, N. 1999. Sympatric speciation by sexual selection. *Nature (London)*, **402**, 523–6.

Hodges, S. A. 1997. Rapid radiation due to a key innovation in colombines (Ranunculaceae: *Aquilegia*). In *Molecular evolution and adaptive radiation* (ed. T. J. Givnish and K. J. Sytsma), pp. 391–406. Cambridge University Press, Cambridge.

Hodges, S. A. and Arnold, M. L. 1994a. Columbines: a geographically widespread species flock. *Proceedings of the National Academy of Sciences, USA*, **91**, 5129–32.

Hodges, S. A. and Arnold, M. L. 1994b. Floral and ecological isolation between *Aquilegia formosa* and *A. pubescens*. *Proceedings of the National Academy of Sciences, USA*, **91**, 2493–6.

Hodges, S. A. and Arnold, M. L. 1995. Spurring plant diversification: are floral nectar spurs a key innovation. *Proceedings of the Royal Society of London B, Biological Sciences*, **262**, 343–8.

Holland, B. and Rice, W. R. 1998. Chase-away selection: antagonistic seduction versus resistance. *Evolution*, **52**, 1–7.
Holland, B. and Rice, W. R. 1999. Experimental removal of sexual selection reverses intersexual antagonistic coevolution and removes a reproductive load. *Proceedings of the National Academy of Sciences, USA*, **96**, 5083–8.
Holloway, J. D. and Hebert, P. D. N. 1979. Ecological and taxonomic trends in macrolepidopteran host plant selection. *Biological Journal of the Linnean Society*, **11**, 229–51.
Holt, R. D. 1977. Predation, apparent competition, and the structure of prey communities. *Theoretical Population Biology*, **12**, 197–229.
Holt, R. D. 1987. Prey communities in patchy environments. *Oikos*, **50**, 276–90.
Holt, R. D. and Gaines, M. S. 1992. Analysis of adaptation in heterogeneous landscapes: implications for the evolution of fundamental niches. *Evolutionary Ecology*, **6**, 433–47.
Holt, R. D., Grover, J., and Tilman, D. 1994. Simple rules for interspecific dominance in systems with exploitation and apparent competition. *American Naturalist*, **144**, 741–71.
Holt, R. D. and Lawton, J. H. 1994. The ecological consequences of shared natural enemies. *Annual Review of Ecology and Systematics*, **25**, 495–520.
Holt, R. D. and Polis, G. A. 1997. A theoretical framework for intraguild predation. *American Naturalist*, **149**, 745–64.
Houde, A. E. 1997. *Sex, color, and mate choice in guppies*. Princeton University Press, Princeton, N.J.
Houle, D. 1991. Genetic covariance of fitness correlates: what genetic correlations are made of and why it matters. *Evolution*, **45**, 630–48.
Houle, D. 1992. Comparing evolvability and variability of quantitative traits. *Genetics*, **130**, 195–204.
Howard, D. J. 1993. Reinforcement: origin, dynamics, and fate of an evolutionary hypothesis. In *Hybrid zones and the evolutionary process* (ed. R. G. Harrison), pp. 46–69. Oxford University Press, Oxford.
Huey, R. B., Pianka E. R., Egan, M. E., Coons, L. W. 1974. Ecological shifts in sympatry: Kalahari fossorial lizards (*Typhlosaurus*). *Ecology*, **55**, 304–16.
Hunter, J. P. 1998. Key innovations and the ecology of macroevolution. *Trends in Ecology & Evolution*, **13**, 31–5.
Hunter, J. P. and Jernvall, J. 1995. The hypocone as a key innovation in mammalian evolution. *Proceedings of the National Academy of Sciences, USA*, **92**, 10718–22.
Hunter, M. D. and Price, P. W. 1992. Playing chutes and ladders: heterogeneity and the relative roles of bottom-up and top-down forces in natural communities. *Ecology*, **73**, 724–32.
Hurlbert, S. H. 1984. Pseudoreplication and the design of ecological field experiments. *Ecological Monographs*, **54**, 187–211.
Husband, B. C. and Barrett, S. C. H. 1993. Multiple origins of self-fertilization in tristylous *Eichornia paniculata* (Pontederiaceae): inferences from style morph and isozyme variation. *Journal of Evolutionary Biology*, **6**, 591–608.
Huxley, J. 1942. *Evolution, the modern synthesis*. Allen & Unwin, London, UK.
Hynes, R. A., Ferguson, A., and McCann, M. A. 1996. Variation in mitochondrial DNA and post-glacial colonization of north western Europe by brown trout. *Journal of Fish Biology*, **48**, 54–67.
Irschick, D. J., Vitt, L. J., Zani, P. A., and Losos, J. B. 1997. A comparison of evolutionary radiations in mainland and Caribbean *Anolis* lizards. *Ecology*, **78**, 2191–203.

Iwasa, Y. and Pomiankowski, A. 1995. Continual change in mate preferences. *Nature (London)*, **377**, 420–2.
Jablonski, D. 1986. Evolutionary consequences of mass extinctions. In *Patterns and processes in the history of life* (ed. D. M. Raup and D. Jablonski), pp. 313–29. Springer-Verlag, Berlin, Germany.
Jablonski, D. 1989. The biology of mass extinctions: a palaeontological perspective. *Philosophical Transactions of the Royal Society of London B, Biological Sciences*, **325**, 357–68.
Jablonski, D. and Sepkoski, J. J., Jr. 1996. Paleobiology, community ecology, and scales of ecological pattern. *Ecology*, **77**, 1367–78.
Jackman, T., Losos, J. B., Larson, A., and de Queiroz, K. 1997. Phylogenetic studies of convergent adaptive radiations in Caribbean *Anolis* lizards. In *Molecular evolution and adaptive radiation* (ed. T. J. Givnish and K. J. Sytsma), pp. 535–57. Cambridge University Press, Cambridge.
Jaenike, J. 1990. Host specialization in phytophagous insects. *Annual Review of Ecology and Systematics*, **21**, 243–73.
Janson, K. 1983. Selection and migration in two distinct phenotypes of *Littorina saxatilis* in Sweden. *Oecologia*, **59**, 58–61.
Jeffries, M. J. and Lawton, J. H. 1984. Enemy free space and the structure of ecological communities. *Biological Journal of the Linnean Society*, **23**, 269–86.
Jernigan, R. W., Culver, D. C., and Fong, D. W. 1994. The dual role of selection and evolutionary history as reflected in genetic correlations. *Evolution*, **48**, 587–96.
Jernvall, J., Hunter, J. P. and Fortelius, M. 1996. Molar tooth diversity, disparity, and ecology in Cenezoic ungulate radiations. *Science (Washington, D.C.)*, **274**, 1489–92.
Johansson, M. E. 1994. Life history differences between central and marginal populations of the clonal aquatic plant *Ranunculus lingua*: a reciprocal transplant experiment. *Oikos*, **70**, 65–72.
Johns, G. C. and Avise, J. C. 1998. Tests for ancient species flocks based on molecular phylogenetic appraisals of *Sebastes* rockfishes and other marine fishes. *Evolution*, **52**, 1135–46.
Johnson, M. S., Murray, J., and Clarke, B. 1993. The ecological genetics and adaptive radiation of *Partula* on Moorea. *Oxford Surveys in Evolutionary Biology*, **9**, 167–238.
Johnson, N. K., Marten, J. A., and Ralph, C. J. 1991. Genetic evidence for the origin and relationships of the Hawaiian honeycreepers (Aves: Fringillidae). *Copeia*, **91**, 379–96.
Jones, C. D. 1998. The genetic basis of *Drosophila sechellia*'s resistance to a host plant toxin. *Genetics*, **149**, 1899–1908.
Jones, M. 1997. Character displacement in Australian dasyurid carnivores: size relationships and prey size patterns. *Ecology*, **78**, 2569–87.
Jones, R., Culver, D. C., and Kane, T. C. 1992. Are parallel morphologies of cave organisms the result of similar selection pressures? *Evolution*, **46**, 353–65.
Jordan, N. 1991. Multivariate analysis of selection in experimental populations derived from hybridization of two ecotypes of the annual plant *Diodea teres* W. (Rubiaceae). *Evolution*, **45**, 1760–72.
Jordan, N. 1992. Path analysis of local adaptation in two ecotypes of the anual plant *Diodia teres* Walt. (Rubiaceae). *American Naturalist*, **140**, 149–65.
Joshi, A. and Thompson, J. N. 1995. Trade-offs and the evolution of host specialization. *Evolutionary Ecology*, **9**, 82–92.

Juliano, S. A. and Lawton, J. H. 1990*a*. The relationship between competition and morphology. I. Morphological patterns among co-occurring dyticid beetles. *Journal of Animal Ecology*, **59**, 403–19.

Juliano, S. A. and Lawton, J. H. 1990*b*. The relationship between competition and morphology. II. Experiments on co-occurring dyticid beetles. *Journal of Animal Ecology*, **59**, 831–48.

Kambysellis, M. P. and Craddock, E. M. 1997. Ecological and reproductive shifts in the diversification of the endemic Hawaiian *Drosophila*. In *Molecular evolution and adaptive radiation* (ed. T. J. Givnish and K. J. Sytsma), pp. 475–509. Cambridge University Press, Cambridge.

Kambysellis, M. P., Ho, K-F., Craddock, E. M., Piano, F., Parisi, M. and Cohen, J. 1995. Pattern of ecological shifts in the diversification of Hawaiian *Drosophila* inferred from a molecular phylogeny. *Current Biology*, **5**, 1129–39.

Kapan, D. 1998. Divergent natural selection and Müllerian mimicry in polymorphic *Heliconius cydno* (Lepidoptera: Nymphalidae). Ph.D. Thesis, University of British Columbia, Canada.

Karr, J. R. and James, F. C. 1975. Ecomorphological configurations and convergent evolution. In *Ecology and evolution of communities* (ed. M. L. Cody and J. M. Diamond), pp. 258–91. Harvard University Press, Cambridge, Mass.

Kawano, K. 1995. Habitat shift and phenotypic character displacement of two closely related rhinocerus beetle species (Coleoptera: Scarabaeidae). *Annals of the Entomological Society of America*, **88**, 641–52.

Kawecki, T. J. 1998. Red queen meets Santa Rosalia: arms races and the evolution of host specialization in organisms with parasitic lifestyles. *American Naturalist*, **152**, 635–51.

Kawecki, T. J. and Abrams, P. A. 1999. Character displacement mediated by the accumulation of mutations affecting resource consumption abilities. *Evolutionary Ecology Research*, **1**, 173–88.

Kelley, S. T. and Farrell, B. D. 1998. Is specialization a dead end? The phylogeny of host use in *Dendroctonus* bark beetles (Scolytidea). *Evolution*, **52**, 1731–43.

Kellogg, D. E. 1975. Character displacement in the radiolarian genus, *Eucyrtidium*. *Evolution*, **29**, 736–49.

Kieser, J. A. 1995. Gnathomandibular morphology and character displacement in the bat-eared fox. *Journal of Mammalogy*, **76**, 542–50.

Kiester, A. R., Lande, R., and Schemske, D. W. 1984. Models of coevolution and speciation in plants and their pollinators. *American Naturalist*, **124**, 220–43.

Kilias, G., Alahiotis, S. N. and Pelecanos, M. 1980. A multifactorial genetic investigation of speciation theory using *Drosophila melanogaster*. *Evolution*, **34**, 730–7.

Kiltie, R. A. 1984. Size ratios among sympatric neotropical cats. *Oecologia*, **61**, 411–6.

Kim, H.-G., Keeley, S. C., Vroom, P. S., and Jansen, R. K. 1998. Molecular evidence for an African origin of the Hawaiian endemic *Hesperomannia (Asteraceae)*. *Proceedings of the National Academy of Sciences, USA*, **95**, 15440–5.

Kingsolver, J. G. 1988. Thermoregulation, flight, and the evolution of wing pattern in pierid butterflies: the topography of adaptive landscapes. *American Zoologist*, **28**, 899–912.

Kirkpatrick, M. 1982*a*. Sexual selection and the evolution of female choice. *Evolution*, **36**, 1–12.

Kirkpatrick, M. 1982*b*. Quantum evolution and punctuated equilibrium in continuous genetic characters. *American Naturalist*, **119**, 833–48.

Kirkpatrick, M. and Lofsvold, D. 1992. Measuring selection and constraint in the evolution of growth. *Evolution*, **46**, 954–71.

Kirkpatrick, M. and Barton, N. H. 1997. Evolution of a species' range. *American Naturalist*, **150**, 1–23.
Kirkpatrick, M. and Servedio, M. R. 1999. The reinforcement of mating preferences on an island. *Genetics*, **151**, 865–84.
Klein, N. K. and Payne, R. B. 1998. Evolutionary associations of brood parasitic finches (*Vidua*) and their host species: analyses of mitochondrial DNA restriction sites. *Evolution*, **52**, 566–82.
Klinka, J. and Zink, R. M. 1997. The importance of recent ice ages in speciation: a failed paradigm. *Science (Washington, D.C.)*, **277**, 1666–69.
Kluge, A. R. and Kerfoot, W. C. 1973. The predictability and regularity of character divergence. *American Naturalist*, **107**, 426–42.
Kocher, T. D., Conroy, J. A., McKaye, K. R., and Stauffer, J. R. 1993. Similar morphologies of cichlid fish in Lakes Tanganyika and Malawi are due to convergence. *Molecular Biology and Evolution*, **2**, 158–65.
Kondrashov, A. S. and Kondrashov, F. A. 1999. Interactions among quantitative traits in the course of sympatric speciation. *Nature (London)*, **400**, 351–4.
Kondrashov, A. S. and Turelli, M. 1992. Deleterious mutations, apparent stabilizing selection and the maintenance of quantitative variation. *Genetics*, **132**, 603–18.
Krebs, R. A. 1990. Courtship behavior and control of reproductive isolation in *Drosophila mojavensis*: Genetic analysis of population hybrids. *Behavior Genetics*, **20**, 535–43.
Lachance, S. and Magnan, P. 1990. Performance of domestic, hybrid, and wild strains of brook trout, *Salvelinus fontinalis*, after stocking: the impact of intra- and interspecific competition. *Canadian Journal of Fisheries and Aquatic Sciences*, **47**, 2278–84.
Lack, D. 1947. *Darwin's finches*. Cambridge University Press, Cambridge.
Lakovaara, S., Saura, A., and Falk, C. 1972. Genetic distance and evolutionary relationships in the *Drosophila obscura* group. *Evolution*, **26**, 177–84.
Lande, R. 1976. Natural selection and random genetic drift in phenotypic evolution. *Evolution*, **30**, 314–34.
Lande, R. 1977. Statistical tests for natural selection on quantitative characters. *Evolution*, **31**, 442–4.
Lande, R. 1979. Quantitative genetic analysis of multivariate evolution, applied to brain:body size allometry. *Evolution*, **33**, 402–16.
Lande, R. 1980. The genetic covariance between characters maintained by pleiotropic mutations. *Genetics*, **94**, 309–20.
Lande, R. 1981. Models of speciation by sexual selection on polygenic traits. *Proceedings of the National Academy of Sciences, USA*, **78**, 3721–5.
Lande, R. 1982. Rapid origin of sexual isolation and character divergence in a cline. *Evolution*, **36**, 213–23.
Lande, R. 1985. Expected time for random genetic drift of a population between stable phenotypic states. *Proceedings of the National Academy of Sciences, USA*, **82**, 7641–5.
Lande, R. 1992. Neutral theory of quantitative genetic variance in an island model with local extinction and colonization. *Evolution*, **46**, 381–9.
Lande, R. and Arnold, S. J. 1983. The measurement of selection on correlated characters. *Evolution*, **37**, 1210–26.
Lanyon, S. M. 1992. Interspecific brood parasitism in blackbirds (Icterinae): a phylogenetic perspective. *Science (Washington, D.C.)*, **255**, 77–9.

Lauder, G. V. 1983. Functional and morphological bases of trophic specialization in sunfishes (Teleostei, Centrarchidae). *Journal of Mammalogy*, **178**, 1–21.

Lavin, P. A. and McPhail, J. D. 1987. Morphological divergence and the organization of trophic characters among lacustrine populations of the threespine stickleback (*Gasterosteus aculeatus*). *Canadian Journal of Fisheries and Aquatic Sciences*, **44**, 1820–9.

Lawlor, L. R. and Maynard Smith, J. 1976. The coevolution and stability of competing species. *American Naturalist*, **110**, 79–99.

Leibold, M. A. 1995. The niche concept revisited: mechanistic models and community context. *Ecology*, **76**, 1371–1382.

Leibold, M. A., Chase, J. M., Shurin, J. B., and Downing, A. L. 1997. Species turnover and the regulation of trophic structure. *Annual Review of Ecology and Systematics*, **28**, 467–94.

Lenski, R. E. and Travisano, M. 1994. Dynamics of adaptation and diversification: a 10,000-generation experiment with bacterial populations. *Proceedings of the National Academy of Sciences, USA*, **91**, 6808–14.

Liebherr, J. K. and Hajek, A. E. 1990. A cladistic test of the taxon cycle and taxon pulse hypotheses. *Cladistics*, **6**, 39–59.

Liem, K. F. 1973. Evolutionary strategies and morphological innovations: cichlid pharyngeal jaws. *Systematic Zoology*, **22**, 425–41.

Liem, K. F. 1991. Functional morphology. In *Cichlid fishes: behaviour, ecology and evolution* (ed. M. H. A. Keenleyside), pp. 129–150. Chapman & Hall, New York.

Lindsey, C. C. 1981. Stocks are chameleons: plasticity in gill-rakers of coregonid fishes. *Journal of the Fisheries Research Board of Canada*, **20**, 749–67.

Liou, L. and Price, T. 1994. Speciation by reinforcement of premating isolation. *Evolution*, **48**, 1451–59.

Lomolino, M. V. 1985. Body sizes of mammals on islands: the island rule reexamined. *American Naturalist*, **125**, 310–16.

Losos, J. B. 1990a. Ecomorphology, peformance capability, and scaling of West Indian *Anolis* lizards: an evolutionary analysis. *Ecological Monographs*, **60**, 369–88.

Losos, J. B. 1990b. The evolution of form and function: morphology and locomotor performance in West Indian *Anolis* lizards. *Evolution*, **44**, 1189–1203.

Losos, J. B. 1990c. A phylogenetic analysis of character displacement in Caribbean *Anolis* lizards. *Evolution*, **44**, 1189–1203.

Losos, J. B. 1992. The evolution of convergent structure in Caribbean *Anolis* communities. *Systematic Biology*, **41**, 403–20.

Losos, J. B. 1998. Ecological and evolutionary determinants of the species-area relationship in Caribbean anoline lizards. In *Evolution on islands* (ed. P. R. Grant), pp. 210–24. Oxford University Press, Oxford.

Losos, J. B. and Sinervo, B. 1989. The effects of morphology and perch diameter on sprint performance of *Anolis* lizards. *Journal of Experimental Biology*, **145**, 23–30.

Losos, J. B., Naeem, S., and Colwell, R. K. 1989. Hutchinsonian ratios and statistical power. *Evolution*, **43**, 1820–6.

Losos, J. B. and Miles, D. B. 1994. Adaptation, constraint, and the comparative method: phylogenetic issues and methods. In *Ecological morphology* (ed. P. C. Wainwright and S. M. Reilly), pp. 60–98. Chicago University Press, Chicago.

Losos, J. B. and Irschick, D. J. 1996. The effect of perch diameter on escape performance of *Anolis* lizards: laboratory predictions and field tests. *Animal Behaviour*, **51**, 593–602.

Losos, J. B., Warheit, K. I., and Schoener, T. W. 1997. Adaptive differentiation following experimental island colonization in *Anolis* lizards. *Nature (London)*, **387**, 70–3.

Losos, J. B., Jackman, T. R., Larson, A., de Queiroz, K., and Rodríguez-Schettino, L. 1998. Contingency and determinism in replicated adaptive radiations of island lizards. *Science (Washington, D.C.)*, **279**, 2115–8.

Losos, J. B., Creer, D. A., Glossip, D., Goellner, R., Hampton, A., Roberts, G., Haskell, N., Taylor, P., and Ettling, J. 2000. Evolutionary implications of phenotypic plasticity in the hindlimb of the lizard *Anolis sagrei*. *Evolution*, **54**, 301–5.

Lovett Doust, L. 1981. Population dynamics and local specialization in a clonal perennial (*Ranunculus repens*). II. The dynamics of leaves, and a reciprocal transplant-replant experiment. *Journal of Ecology*, **69**, 757–68.

Lowrey, T. K. 1995. Phylogeny, adaptive radiation, and biogeography of Hawaiian *Tetramolopium* (Asteraceae: Astereae). In *Hawaiian biogeography* (ed. W. L. Wagner and V. A. Funk), pp. 195–219. Smithsonian Institution Press, Washington D.C.

Loy, A. and Capanna, E. 1999. A parapatric contact area between two species of moles: character displacement investigated through the geometric morphometrics of skull. *Acta Zoologica*, **44**, 151–64.

Lüscher, A., Connoly, J., and Jacquard, P. 1992. Neighbor specificity between *Lolium perenne* and *Trifolium repens* from a natural pasture. *Oecologia*, **91**, 404–9.

Lynch, M. 1988. The rate of polygenic mutation. *Genetical Research*, **51**, 137–48.

Lynch, M. 1990. The rate of morphological evolution in mammals from the standpoint of the neutral expectation. *American Naturalist*, **136**, 727–41.

Lynch, M. 1991. The genetic interpretation of inbreeding depression and outbreeding depression. *Evolution*, **45**, 622–9.

Lynch, M. and Hill, W. G. 1986. Phenotypic evolution by neutral mutation. *Evolution*, **40**, 915–35.

Lynch, M. and Walsh, B. 1998. Genetics and analysis of quantitative traits. Sinauer, Sunderland, Mass.

Lynch, M., Pfrender, M., Spitze, K., Lehman, N., Hicks, J., Allen, D., Latta, L., Ottene, M., Bogue, F., and Colbourne, J. 1998. The quantitative and molecular genetic architecture of a subdivided species. *Evolution*, **53**, 100–10.

Lynch, M., Blanchard, J., Houle, D., Kibota, T., Schultz, S., Vassilieva, L., and Willis, J. 1999. Perspective: spontaneous deleterious mutation. *Evolution*, **53**, 645–63.

MacArthur, R. W. 1972. *Geographical ecology*. Princeton University Press, Princeton, N.J.

MacArthur, R. H. and Pianka, E. R. 1966. On optimal use of a patchy environment. *Evolution*, **100**, 603–9.

Macnair, M. R. and Christie, P. 1983. Reproductive isolation as a pleiotropic effect of copper tolerance in *Mimulus guttatus*? *Heredity*, **50**, 295–302.

Macnair, M. R. and Gardner, M. 1998. The evolution of edaphic endemics. In *Endless forms: species and speciation* (ed. D. Howard), pp. 157–71. Oxford University Press, Oxford.

Maddison, W. P. and Maddison, D. R. 1992. *MacClade, version 3*. Sinauer, Sunderland, Mass.

Magnan, P. 1988. Interactions between brook charr, *Salvelinus fontinalis*, and nonsalmonid species: ecological shift, morphological shift, and their impact on zooplankton communities. *Canadian Journal of Fisheries and Aquatic Sciences*, **45**, 999–1009.

Magurran, A. E. 1998. Population differentiation without speciation. *Proceedings of the Royal Society of London B, Biological Sciences*, **353**, 275–86.

Mahmood M. S, Chippendale, P. T. and Johnson, N. A. 1998. Patterns of postzygotic isolation in frogs. *Evolution*, **52**, 1811–20.

Mallet J. 1995. A species definition for the modern synthesis. *Trends in Ecology & Evolution*, **10**, 294–9.

Mallet, J. and Barton, N. H. 1989. Strong natural selection in a warning-color hybrid zone. *Evolution*, **43**, 421–31.

Malmquist, M. G. 1985. Character displacement and biogeography of the pygmy shrews in Northern Europe. *Ecology*, **66**, 372–7.

Manly, B. F. J. 1985. *The statistics of natural selection on animal populations* Chapman & Hall, New York.

Marchetti, K. 1993. Dark habitats and bright birds illustrate the role of the environment in species divergence. *Nature (London)*, **362**, 149–52.

Marcogliese, D. J. and Cone, D. K. 1997. Food webs: a plea for parasites. *Trends in Ecology & Evolution*, **12**, 320–4.

Marten, J. A. and Johnson, N. K. 1986. Genetic relationships of North American cardueline finches. *Copeia*, **88**, 409–20.

Martin, M. M. and Harding, J. 1981. Evidence for the evolution of competition between two species of annual plants. *Evolution*, **35**, 975–87.

Martin, T. E. 1988. Processes organizing open-nesting bird assemblages: competition or nest predation? *Evolutionary Ecology*, **2**, 37–50.

Martin, T. E. 1998. Are microhabitat preferences of coexisting species under selection and adaptive? *Ecology*, **79**, 656–70.

Martins, E. P. 1994. Estimating the rate of phenotypic evolution from comparative data. *American Naturalist*, **144**, 193–209.

Martins, E. P. and Garland, T, Jr. 1991. Phylogenetic analyses of the correlated evolution of continuous characters: a simulation study. *Evolution*, **45**, 534–57.

Martins, E. P. and Hansen, T. F. 1996. Phylogenies and the comparative method: A general approach to incorporating phylogenetic information into the analysis of interspecific data. *American Naturalist*, **149**, 646–67.

Mathsoft, Inc. 1999. *S-Plus 2000 Language Reference*. Seattle, Washington.

Matsuda, H., Hori, M., and Abrams, P. A. 1996. Effects of predator-specific defense on biodiversity and community complexity in two-trophic-level communities. *Evolutionary Ecology*, **10**, 13–28.

Maynard Smith, J., Burian, R., Kauffman, S., Alberch, P., Campbell, J., Goodwin, B., Lande, R., Raup, D., and Wolpert, L. 1985. Developmental constraints and evolution. *Quarterly Review of Biology*, **60**, 265–87.

Mayr, E. 1942. *Systematics and the origin of species*. Columbia University Press, New York.

Mayr, E. 1954. Change of genetic environment and evolution. In *Evolution as a process* (ed. J. Huxley, A. C. Hardy, and E. B. Ford), pp. 157–80. Allen & Unwin, London, UK.

Mayr, E. 1963. *Animal species and evolution*. Harvard University Press, Cambridge, Mass.

McCune, A. R. 1997. How fast is speciation? Molecular, geological, and phylogenetic evidence from adaptive radiations of fishes. In *Molecular evolution and adaptive radiation* (ed. T. J. Givnish and K. J. Sytsma), pp. 585–610. Cambridge University Press, Cambridge.

McDowall, R. M. 1998. Phylogenetic relationships and ecomorphological divergence in sympatric and allopatric species of *Paragalaxias* (Teleostei: Galaxiidae) in lakes of high elevation Tasmanian lakes. *ebf*, **53**, 235–57.

McEachran, J. D. and Martin, C. O. 1977. Possible occurrence of character displacement in the sympatric skates *Raja erinacea* and *R. ocellata* (Pisces: Rajidae). *Environmental Biology of Fishes*, **2**, 121–30.

McKinnon, J. S., S. Mori and D. Schluter. Manuscript. Parallel reproductive isolation implicates divergent environments in the origin of stickleback species.

McMillan, W. O., Jiggins, C. D., and Mallet, J. 1997. What initiates speciation in passion-vine butterflies? *Proceedings of the National Academy of Sciences, USA*, **94**, 8628–33.

McPeek, M. A. 1990a. Determination of species composition in the *Enallagma* damselfly assemblages of permanent lakes. *Ecology*, **71**, 83–98.

McPeek, M. A. 1990b. Behavioral differences between *Enallagma* species (Odonata) influencing differential vulnerability to predators. *Ecology*, **71**, 1714–26.

McPeek, M. A. 1995. Morphological evolution mediated by behavior in the damselflies of two communities. *Evolution*, **49**, 749–69.

McPeek, M. A., Schrot, A. K., and Brown, J. M. 1996. Adaptation to predators in a new community: swimming performance and predator avoidance in damselflies. *Ecology*, **77**, 617–29.

McPeek, M. A. and Wellborn, G. A. 1998. Genetic variation and reproductive isolation among phenotypically divergent amphipod populations. *Limnology and Oceanography*, **43**, 1162–9.

McPhail, J. D. 1994. Speciation and the evolution of reproductive isolation in the sticklebacks (*Gasterosteus*) of southwestern British Columbia. In *Evolutionary biology of the threespine stickleback* (ed. M. A. Bell and S. A. Foster), pp. 399–437. Oxford University Press, Oxford.

McPhail, J. D. and Lindsey, C. C. 1986. Zoogeography of the freshwater fishes of Cascadia (the Columbia system and rivers north to the Stikine). In *The zoogeography of North American freshwater fishes* (ed. C. H. Hocutt and E. O. Wiley), pp. 615–37. Wiley, Chichester.

Merilä, J. 1997. Quantitative trait and allozyme divegence in the greenfinch (*Carduelis chloris*, Aves: Fringillidae). *Biological Journal of the Linnean Society*, **61**, 243–66.

Meyer, A. 1993. Phylogenetic relationships and evolutionary processes in East African cichlid fishes. *Trends in Ecology & Evolution*, **8**, 279–84.

Meyer, A., Kocher, T. D., Basasibwaki, P., and Wilson, A. C. 1990. Monophyletic origin of Lake Victoria cichlid fishes suggested by mitochondrial DNA sequences. *Nature (London)*, **347**, 550–3.

Miller, R. B. 1981. Hawkmoths and the geographic patterns of floral variation in *Aquilegia caerulea*. *Evolution*, **35**, 763–74.

Milligan, B. G. 1985. Evolutionary divergence and character displacement in two phenotypically-variable competing species. *Evolution*, **39**, 1207–22.

Mitchell-Olds, T. 1996. Pleiotropy causes long-term genetic constraints on life-history evolution in *Brassica rapa*. *Evolution*, **50**, 1849–58.

Mitchell-Olds, T. and Shaw, R. G. 1987. Regression analysis of natural selection: statistical inference and biological interpretation. *Evolution*, **41**, 1149–61.

Mitra, S., Landel, H., and Pruett-Jones, S. 1996. Species richness covaries with mating system in birds. *Auk*, **113**, 554–57.

Mittelbach, G. G. 1981. Foraging efficiency and body size: a study of optimal diet and habitat use by bluegills. *Ecology*, **62**, 1370–86.

Mittelbach, G. G. 1984. Predation and resource use in two sunfishes (Centrarchidae). *Ecology*, **65**, 499–513.

Mittelbach, G. G. and Chesson, P. L. 1987. Predation risk: indirect effects on fish populations. In *Predation: direct and indirect impacts on aquatic communities* (ed. W. C. Kerfoot and A. Sih), pp. 315–32. University Press New England, Hanover, NH, USA.

Mitter, C., Farrell, B., and Wiegmann, B. 1988. The phylogenetic study of adaptive zones: has phytophagy promoted insect diversification? *American Naturalist*, **132**, 107–28.

Miyatake, T. and Shimizu, T. 1999. Genetic correlations between life-history and behavioral traits can cause reproductive isolation. *Evolution*, **53**, 201–8.

Møller, A. P. and Cuervo, J. J. 1998. Speciation and feather ornamentation in birds. *Evolution*, **52**, 859–69.

Montgomery, S. L. 1975. Comparative breeding site ecology and the adaptive radiation of picture-winged *Drosophila*. *Proceedings of the Hawaiian Entomological Society*, **22**, 65–102.

Mooers, A. Ø. and Schluter, D. 1999. Reconstructing ancestor states using maximum likelihood: support for one and two-rate models. *Systematic Biology*, **48**, 623–33.

Mooers, A. Ø., Vamosi, S. M., and Schluter D. 1999. Using phylogenies to test macroevolutionary hypotheses of trait evolution in Cranes (Gruinae). *American Naturalist*, **154**, 249–59.

Mopper, S. and Strauss, S. Y. 1998. Genetic structure and local adaptation in natural insect populations: effects of ecology, life history, and behavior. Chapman & Hall, New York.

Moreno, E. and L. M. Carrascal. 1993. Leg morphology and feeding postures in four *Parus* species: an experimental ecomorphological approach. *Ecology*, **74**, 2037–44.

Mousseau, T. A. and Roff, D. A. 1987. Natural selection and the heritability of fitness components. *Heredity*, **59**, 181–91.

Müller, A. 1996. Host-plant specialization in Western Palearctic anthidiine bees (Hymenoptera: Apoidea: Megachilidae). *Ecological Monographs*, **66**, 235–57.

Muller, H. J. 1940. Bearings of the *Drosophila* work on systematics. In *The new systematics* (ed. J. S. Huxley), pp. 185–268. Clarendon Press, Oxford.

Munz, P. A. and Keck, D. D. 1970. *A California Flora*. University of California Press, Berkeley, CA, USA.

Murdoch, D. J. and Chow, E. D. 1994. A graphical display of large correlation matrices. Mathematical preprint #1994–09, Department of Mathematics and Statistics, Queen's University, Kingston, Canada.

Nagel, L. and Schluter, D. 1998. Body size, natural selection, and speciation in sticklebacks. *Evolution*, **52**, 209–18.

Nagy, E. S. 1997. Selection for native characters in hybrids between two locally adapted plant subspecies. *Evolution*, **51**, 1469–80.

Nagy, E. S. and Rice, K. J. 1997. Local adaptation in two subspecies of an annual plant: implications for migration and gene flow. *Evolution*, **51**, 1079–89.

Nee, S., Mooers, A. O., and Harvey, P. H. 1992. Tempo and mode of evolution revealed from molecular phylogenies. *Proceedings of the National Academy of Sciences, USA*, **89**, 8322–6.

Nee, S., May, R. M., and Harvey, P. 1994. The reconstructed evolutionary process. *Philosophical Transactions of the Royal Society of London B, Biological Sciences*, **344**, 305–11.

Nee, S., Barraclough, T. G., and Harvey, P. H. 1996. Temporal changes in biodiversity: detecting patterns and identifying causes. In *Biodiversity: a biology of numbers and difference* (ed. K. J. Gaston), pp. 230–52. Blackwell Scientific, Oxford.

Nehm, R. H. and Geary, D. H. 1994. A gradual morphologic transition during a rapid speciation event in marginellid gastropods (Neogene: Dominican Republic). *Journal of Paleontology*, **68**, 787–95.

Nei, M. 1972. Genetic distance between populations. *American Naturalist*, **106**, 283–92.

Nei, M. 1978. Estimation of average heterozygosity and genetic distance from a small number of individuals. *Genetics*, **89**, 583–90.

Newton, I. 1972. *Finches*. Collins, London.

Niewiarowski, P. H. and Roosenburg, W. 1993. Reciprocal transplant reveals sources of variation in growth rates of the lizard *Sceloporus undulatus*. *Ecology*, **74**, 1992–2002.

Niklas, K. J. 1997. *The evolutionary biology of plants*. University of Chicago Press, Chicago.

Nilsson, L. A. 1988. The evolution of flowers with deep corolla tubes. *Nature (London)*, **334**, 147–9.

Noor, M. A. 1995. Speciation driven by natural selection in *Drosophila*. *Nature (London)*, **375**, 674–5.

Noor, M. A. F. 1997. Genetics of sexual isolation and courtship dysfunction in male hybrids of *Drosophila pseudoobscura* and *Drosophila persimilis*. *Evolution*, **51**, 809–15.

Norton, S. F. 1991. Capture success and diet of cottid fishes: the role of predator morphology and attack kinematics. *Ecology*, **72**, 1807–19.

Nurminsky, D. I., Nurminskaya, M. V., De Aguiar, D., and Hartl, D. L. 1998. Selective sweep of a newly-evolved sperm-specific gene in *Drosophila*. *Nature (London)*, **396**, 572–5.

Oksanen, L., Fretwell, S. D., Arruda, J., and Niemela, P. 1981. Exploitation ecosystems in gradients of primary productivity. *American Naturalist*, **118**, 240–61.

Olson, S. L. and James, H. F. 1982. Prodromus of the fossil avifauna of the Hawaiian Islands. *Smithsonian Contributions to Zoology*, **365**, 1–59.

Orr, H. A. 1995. The population genetics of speciation: the evolution of hybrid incompatibilities. *Genetics*, **139**, 1805–13.

Orr, H. A. 1997. Haldane's rule. *Annual Review of Ecology and Systematics*, **28**, 195–218.

Orr, H. A. 1998*a*. Testing natural selection *vs*. genetic drift in phenotypic evolution using quantitative trait locus data. *Genetics*, **149**, 2099–104.

Orr, H. A. 1998*b*. The population genetics of adaptation: the distribution of factors fixed during adaptive evolution. *Evolution*, **52**, 935–49.

Orr, M. 1996. Life-history adaptation and reproductive isolation in a grasshopper hybrid zone. *Evolution*, **50**, 704–16.

Orti, G., Bell, M. A., Reimchen, T. E., and Meyer, A. 1994. Global survey of mitochondrial DNA sequences in the threespine stickleback: evidence for recent migrations. *Evolution*, **48**, 608–22.

Osenberg, C. W. and Mittelbach, G. G. 1996. The relative importance of resource limitation and predator limitation in food chains. In *Food webs* (ed. G. A. Polis and K. O. Winemiller), pp. 134–48. Chapman & Hall, New York.

O'Steen, S., Cullum, A. J. and Bennett, A. F. Manuscript. Rapid evolution of escape performance in Trinidad guppies (*Poecilia reticulata*).

Owen, D. F. 1963. Polymorphism and population density in the African land snail *Limicolaria martensiana*. *Science (Washington, D.C.)*, **140**, 666–7.

Owen, D. F. and Whiteley, D. 1989. Evidence that reflexive polymorphisms are maintained by visual selection by predators. *Oikos*, **55**, 130–3.

Pacala, S. W. and Roughgarden, J. 1982. Resource partitioning and interspecific competition in two two-species insular *Anolis* lizard communities. *Science (Washington, D.C.)*, **217**, 444–6.

Pacala, S. W. and Roughgarden, J. 1985. Population experiments with the *Anolis* lizards of St. Maarten and St. Eustacius. *Ecology*, **66**, 129–41.

Pagel, M. 1994. Detecting correlated evolution on phylogenies: ageneral method for the comparative analysis of discrete characters. *Proceedings of the Royal Society of London B, Biological Sciences*, **255**, 37–45.

Pagel, M. 1999. The maximum likelihood approach to reconstructing ancestral character states of discrete characters on phylogenies. *Systematic Biology*, **48**, 612–22.

Palomares, F. and Caro, T. M. 1999. Interspecific killing among mammalian carnivores. *American Naturalist*, **153**, 492–508.

Palumbi, S. R. 1998. Species formation and the evolution of gamete recognition loci. In *Endless forms: species and speciation* (ed. D. Howard and S. Berlocher), pp. 271–8. Oxford University Press, Oxford.

Parra, V., Loreau, M., and Jaeger, J. J. 1999. Incisor size and community structure in rodents: two tests of the role of competition. *Acta Oecologica*, **20**, 93–101.

Partridge, L. 1976. Field and laboratory observations on the foraging and feeding techniques of blue tits (*Parus caeruleus*) and coal tits (*Parus ater*) in relation to their habitats. *Animal Behaviour*, **24**, 230–40.

Patterson. 1995. Phylogenetic analysis of the Hawaiian and other Pacific species of *Scaevola* (Goodeniaceae). In *Hawaiian biogeography* (ed. W. L. Wagner and V. A. Funk), pp. 363–78. Smithsonian Institution Press, Washington D.C.

Payne, R. B. 1977. The ecology of brood parasitism in birds. *Annual Review of Ecology and Systematics*, **8**, 1–28.

Payne, R. J. H. and Krakauer, D. C. 1997. Sexual selection, space, and speciation. *Evolution*, **51**, 1–9.

Pearson, D. L. 1980. Patterns of limiting similarity in tropical forest tiger beetles (Coleoptera: Cicindelidae). *Biotropica*, **12**, 195–204.

Pearson, D. L. and Mury, E. J. 1979. Character divergence and convergence among tiger beetles (Coleoptera: Cicindelidae) *Ecology*, **60**, 557–66.

Pearson, K. 1903. Mathematical contributions to the theory of evolution. XI. On the influence of natural selection on the variability and correlation of organs. *Philosophical Transactions of the Royal Society of London A*, **200**, 1–66.

Pellmyr, O., Leebens-Mack, J., and Huth, C. J. 1996. Non-mutualistic yucca moths and their evolutionary consequences. *Nature (London)*, **380**, 155–6.

Pellmyr, O. and Leebens-Mack, J. 2000. Adaptive radiation in yucca moths and the reversal of mutualism. *American Naturalist*, **156** *(Supplement)*, in press.

Peterson, R. T. and McKenny, M. 1968. *A field guide to wildflowers of northeastern and north-central North America*. Houghton Mifflin, Boston, Mass.

Petren, K., Grant, B. R., and Grant, P. R. 1999. A phylogeny of Darwin's finches based on microsatellite DNA length variation. *Proceedings of the Royal Society of London B, Biological Sciences*, **266**, 321–30.

Pfennig, D. W. 1992. Polyphenism in spadefoot toad tadpoles as a locally adjusted evolutionarily stable strategy. *Evolution*, **46**, 1408–20.

Pfennig, D. W. and P. J. Murphy. 2000. Character displacement in polyphenic tadpoles. *Evolution*, **54**, in press.

Phillips, P. C. and Arnold, S. J. 1989. Visualizing multivariate selection. *Evolution*, **43**, 1209–22.
Pianka, E. R. 1969. Sympatry of desert lizards (*Ctenotus*) in Western Australia. *Ecology*, **50**, 1012–30.
Pianka, E. R. 1986. *Ecology and natural history of desert lizards*. Princeton University Press, Princeton, N.J.
Pianka, E. R. 1998. Phylogenetic analysis of a major adaptive radiation [Online]. Available at: http://uts.cc.utexas.edu/~varanus/ctenotus.html [1998, June 8].
Polis, G. A., Myers, C. A., and Holt, R. D. 1989. The ecology and evolution of intraguild predation: potential competitors that eat each other. *Annual Review of Ecology and Systematics*, **20**, 297–330.
Pomiankowski, A. and Iwasa, Y. 1998. Runaway ornament diversity caused by Fisherian sexual selection. *Proceedings of the National Academy of Sciences, USA*, **95**, 5106–11.
Pöysä, H., Elmberg, J., Nummi, P., and Sjöberg, K. 1994. Species composition of dabbling duck assemblages: ecomorphological patterns compared with null models. *Oecologia*, **98**, 193–200.
Pregill, G. K. and Olson, S. L. 1981. Zoogeography of West Indian vertebrates in relation to Pleistocene climate cycles. *Annual Review of Ecology and Systematics*, **12**, 75–98.
Price, D. K. and Boake, C. R. B. 1995. Behavioral reproductive isolation in *Drosophila silvestris*, *D. heteroneura*, and their F1 hybrids (Diptera: Drosophilae). *Journal of Insect Behavior*, **8**, 595–616.
Price, M. V. and Waser, N. M. 1979. Pollen dispersal and optimal outcrossing in *Delphinium nelsoni*. *Nature (London)*, **277**, 294–7.
Price, T. 1987. Diet variation in a population of Darwin's finches. *Ecology*, **68**, 1015–28.
Price, T. 1991. Morphology and ecology of breeding warblers along an altitudinal gradient in Kashmir, India. *Journal of Animal Ecology*, **60**, 643–64.
Price, T. 1997. Correlated evolution and independent contrasts. *Philosophical Transactions of the Royal Society of London B, Biological Sciences*, **352**, 519–29.
Price, T. 1998. Sexual selection and natural selection in bird speciation. *Philosophical Transactions of the Royal Society of London B, Biological Sciences*, **353**, 1–12.
Price, T. and Grant, P. R. 1984. Life history traits and natural selection for small body size in a population of Darwin's finches. *Evolution*, **38**, 483–94.
Price, T. D., Grant P. R., Gibbs H. L., and Boag P. T. 1984*a*. Recurrent patterns of natural selection in a population of Darwin's finches. *Nature (London)*, **309**, 787–9.
Price, T. D., Grant, P. R., and Boag, P. T. 1984*b*. Genetic changes in the morphological differentiation of Darwin's ground finches. In *Population biology and Evolution* (ed. K. Wohrmann and V. Loechske), pp. 49–66. Springer-Verlag, Berlin.
Price, T. D., Kirkpatrick, M., and Arnold, S. J. 1988. Directional selection and the evolution of breeding date in birds. *Science (Washington, D.C.)*, **240**, 798–9.
Price, T. and Liou, L. 1989. Seletion in clutch size in birds. *American Naturalist*, **134**, 950–9.
Price, T. D. and Schluter, D. 1991. On the low heritability of life history traits. *Evolution*, **45**, 853–61.
Price, T., Turelli, M., and Slatkin, M. 1993. Peak shifts produced by correlated response to selection. *Evolution*, **47**, 280–90.
Price, T., Gibbs, H. L., de Sousa, L., and Richman, A. D. 1998. Different timings of the adaptive radiations of North American and Asian warblers. *Proceedings of the Royal Society of London B, Biological Sciences*, **26**, 1969–75.

Price, T., Lovette, I., Bermingham, H. L., Gibbs, H. L., and Richman, A. D. 2000. The imprint of history on communities of North American and Asian warbers. *American Naturalist*, in press.

Primack, R. B. and Kang, H. 1989. Measuring fitness and natural selection in wild plant populations. *Annual Review of Ecology and Systematics*, **20**, 367–96.

Pritchard, J. R. and Schluter, D. Manuscript. Declining competition during character displacement: summoning the ghost of competition past.

Proctor, H. C. 1992. Sensory exploitation and the evolution of male mating behavior: a cladistic test using water mites (Acari, Parasitengona) *Animal Behaviour*, **44**, 745–52.

Provine, W. B. 1986. *Sewall Wright and evolutionary biology*. Chicago University Press, Chicago.

Racz, G. and Demeter, A. 1999. Character displacement in mandible shape and size in two species of water shrews (*Neomys*, Mammalia, Insectivora). *American Zoologist*, **44**, 165–75.

Radtkey, R. R. 1996. Adaptive radiation of day-geckos (*Phelsuma*) in the Seychelles archipelago: a phylogenetic analysis. *Evolution*, **50**, 604–23.

Radtkey, R. R., Fallon, S. M., and Case, T. J. 1997. Character displacement in some *Cnemidophorus* lizards revisited: a phylogenetic analysis. *Proceedings of the National Academy of Sciences, USA*, **94**, 9740–5.

Rainey, P. B. and Travisano, M. 1998. Adaptive radiation in a heterogeneous environment. *Nature (London)*, **394**, 69–72.

Ramsey, J. and Schemske, D. W. 1998. Pathways, mechanisms, and rates of polyploid formation in flowering plants. *Annual Review of Ecology and Systematics*, **29**, 467–501.

Ratcliffe, L. M. and Grant, P. R. 1983. Species recognition in Darwin's finches (*Geospiza*, Gould). I. Discrimination by morphological cues. *Animal Behaviour*, **31**, 1139–53.

Rathcke, B. 1983. Competition and facilitation among plants for pollinators. In *Pollination biology* (ed. L. Real), pp. 305–29. Academic Press, New York.

Rausher, M. D. 1984. Trade-offs in performance on different hosts: evidence from within- and between-site variation in the beetle *Deloyala guttata*. *Evolution*, **38**, 582–95.

Rausher, M. D. 1988. Is coevolution dead? *Ecology*, **69**, 898–901.

Rausher, M. D. 1992. The measurement of selection on quantitative traits biased due to environmental covariances between traits and fitness *Evolution*, **46**, 616–26.

Ree, R. H. and Donoghue, M. J. 1999. Inferring rates of change in flower symmetry in asterid angiosperms. *Systematic Biology*, **48**, 633–41.

Reeve, H. K. and Sherman, P. W. 1993. Adaptation and the goals of evolutionary research. *Quarterly Review of Biology*, **68**, 1–32.

Repasky, R. R. and Schluter, D. 1996. Habitat distributions of sparrows: foraging success in a transplant experiment. *Ecology*, **77**, 452–60.

Reznick, D. N., Shaw, F. H., Rodd, F. H., and Shaw, R. G. 1997. Evaluation of the rate of evolution in natural populations of guppies (*Poecilia reticulata*). *Science (Washington, D.C.)*, **275**, 1934–37.

Rice, W. R. 1996. Sexually antagonistic male adaptation triggered by experimental arrest of female evolution. *Nature (London)*, **381**, 232–4.

Rice, W. R. 1998. Intergenomic conflict, interlocus antagonistic coevolution, and the evolution of reproductive isolation. In *Endless forms: species and speciation* (ed. D. Howard and S. Berlocher), pp. 261–70. Oxford University Press, Oxford.

Rice, W. R. and Hostert, E. E. 1993. Laboratory experiments on speciation: what have we learned in 40 years? *Evolution*, **47**, 1637–53.
Rice, W. R. and Holland, B. 1997. The enemies within: interlocus contest evolution (ICE), and the interspecific Red Queen. *Behavioural Ecology and Sociobiology*, **41**, 1–10.
Richman, A. D. and Price, T. 1992. Evolution of ecological differences in the Old World leaf warblers. *Nature (London)*, **355**, 817–21.
Ricklefs, R. E. and O'Rourke, K. 1975. Aspect diversity in moths: a temperate–tropical comparison. *Evolution*, **29**, 313–24.
Ricklefs, R. E. and Cox, G. W. 1972. Taxon cycles in the West Indian avifauna. *American Naturalist*, **106**, 195–219.
Ricklefs, R. E. and Renner, S. S. 1994. Species richness within families of flowering plants. *Evolution*, **48**, 1619–36.
Ricklefs, R. E. and Starck, J. M. 1996. Applications of phylogenetically independent contrasts—a mixed progress report. *Oikos*, **77**, 167–72.
Ricklefs, R. E. and Bermingham, E. 1999. Taxon cycles in the Lesser Antilles avifauna. *Ostrich*, **70**, 49–59.
Rieseberg, L. H. 1997. Hybrid origins of plant species. *Annual Review of Ecology and Systematics*, **28**, 359–89.
Rieseberg, L. H. and Wendel, J. F. 1993. Introgression and its consequences in plants. In *Hybrid zones and the evolutionary process* (ed. R. G. Harrison), pp. 70–109. Oxford University Press, Oxford.
Ritland, C., and Ritland, K. 1989. Variation of sex allocation among 8 taxa of the *Mimulus guttatus* species complex (Scrophulariaceae). *American Journal of Botany*, **76**, 1731–9.
Robichaux, R. H. 1984. Variation in the tissue water relations of two sympatric Hawaiian *Dubautia* species and their natural hybrid. *Oecologia*, **65**, 75–81.
Robichaux, R. H. and Canfield, J. E. 1985. Tissue elastic properties of eight Hawaiian *Dubautia* species that differ in habitat and diploid chromosome number. *Oecologia*, **66**, 77–80.
Robichaux, R. H., Carr, G. D., Liebman, M., and Pearcy, R. W. 1990. Adaptive radiation of the Hawaiian silversword alliance (Compositae–Madiinae): ecological, morphological, and physiological diversity. *Annals of the Missouri Botanical Garden*, **77**, 64–72.
Robinson, B. W., Wilson, D. S., Margosian, A. S., and Lotito, P. T. 1993. Ecological and morphological differentiation of pumpkinseed sunfish in lakes without bluegill sunfish. *Evolutionary Ecology*, **7**, 451–64.
Robinson, B. W. and Wilson, D. S. 1994. Character release and displacement in fishes: a neglected literature. *American Naturalist*, **144**, 596–627.
Robinson, B. W., Wilson, D. S., and Shea, G. O. 1996. Trade-offs of ecological specialization: an intraspecific comparison of pumpkinseed sunfish phenotypes. *Ecology*, **77**, 170–8.
Robinson, B. W. and Wilson, D. S. 1996. Genetic variation and phenotypic plasticity in a trophically polymorphic population of pumpkinseed sunfish (*Lepomis gibbosus*). *Evolutionary Ecology*, **10**, 631–52.
Robinson, B. W. and Wilson, D. S. 2000. A pleuralistic analysis of character release in pumpkinseed sunfish (*Lepomis gibbosus*). *Ecology*, **81**, in press.
Robinson, B. W. and Schluter, D. 2000. Natural selection and the evolution of adaptive genetic variation in northern freshwater fishes. In *Adaptive genetic variation* (ed. T. A. Mousseau, B. Sinervo, and J. Endler), pp. 65–94. Oxford University Press, Oxford.

Robinson, S. K. and Terborgh, J. 1995. Interspecific aggression and habitat selection by Amazonian birds. *Journal of Animal Ecology*, **64**, 1–11.
Roff, D. A. 1997. *Evolutionary quantitative genetics*. Chapman & Hall, New York.
Roff, D. A. and Mousseau, T. A. 1987. Quantitative genetics and fitness: lessons from *Drosophila*. *Heredity*, **58**, 103–18.
Rohlf, F. J., Gilmartin, A. J., and Hart, G. 1983. The Kluge-Kerfoot phenomenon—a statistical artifact. *Evolution*, **37**, 180–202.
Rolán-Alvarez, E., Johannesson, K., and Ekendahl, A. 1995. Frequency- and density-dependent sexual selection in natural populations of Galician *Littorina saxatilis* Olivi. *Hydrobiologia*, **309**, 167–72.
Rolán-Alvarez, E., Johannesson, K., and Erlandsson, J. 1997. The maintenance of a cline in the marine snail *Littorina saxatilis*: the role of home site advantage and hybrid fitness. *Evolution*, **51**, 1838–47.
Rosenzweig, M. L. 1978. Competitive speciation. *Biological Journal of the Linnean Society*, **10**, 275–89.
Rosenzweig, M. L. 1981. A theory of habitat selection. *Ecology*, **62**, 327–35.
Rosenzweig, R. F. Sharp, R. R., Treves, D. S., and Adams, J. 1994. Microbial evolution in a simple unstructured environment: genetic differentiation in *Escherichia coli*. *Genetics*, **137**, 903–17.
Ross, H. H. 1972. The origin of species diversity in ecological communities. *Taxon*, **21**, 253–9.
Rossi, L., Basset, A., and Nobile, L. 1983. A coadapted trophic niche in two species of crustacea (Isopoda): *Acellus aquaticus* (L.) and *Proacellus coxalis* Dolff. *Evolution*, **37**, 810–20.
Rothstein, S. I., Patten, M. A., and Fleischer, R. C. Manuscript. Phylogeny, specialization and parasite-host coevolution.
Roughgarden, J. 1972. Evolution of niche width. *American Naturalist*, **106**, 683–718.
Roughgarden, J. 1976. Resource partitioning among competing species—coevolutionary approach. *Theoretical Population Biology*, **9**, 388–424.
Roughgarden, J., Heckel, D., and Fuentes, E. R. 1983. Coevolutionary theory and the biogeography and community structure of *Anolis*. In *Lizard ecology* (ed. R. B. Huey, E. R. Pianka, and T. W. Schoener), pp. 371–410. Harvard University Press, Cambridge, Mass.
Rowe, L. and Houle, D. 1996. The lek paradox and the capture of genetic variance by condition dependent traits. *Proceedings of the Royal Society of London B, Biological Sciences*, **263**, 1415–21.
Roy, K. and Foote, M. 1997. Morphological approaches to measuring biodiversity. *Trends in Ecology & Evolution*, **12**, 277–81.
Rummel, J. D. and Roughgarden, J. 1985. Effects of reduced perch height separation on competition between two *Anolis* lizards. *Ecology*, **66**, 430–44.
Rundle, H. D. and Schluter, D. 1998. Reinforcement of stickleback mate preferences: sympatry breeds contempt. *Evolution*, **52**, 200–8.
Rundle, H. D., Mooers, A. O. and Whitlock, M. C. 1999. Single founder-flush events and the evolution of reproductive isolation. *Evolution*, **52**, 1850–5.
Rundle, H. D., Nagel, L., Boughman, J. W., and Schluter, D. 2000. Natural selection and parallel speciation in sticklebacks. *Science (Washington, D.C.)*, **287**, 306–8.
Ryan, M. J. 1998. Sexual selection, receiver biases, and the evolution of sex differences. *Science (Washington, D.C.)*, **281**, 1999–2003.

Ryan, M. J. and Rand, A. S. 1993. Species recognition and sexual selection as a unitary problem in animal communication. *Evolution*, **47**, 647–57.

Sætre, G.-P., Moum, T., Bureš, S., Král, M., Adamjan, M., and Moreno, J. 1997. A sexually selected character displacement in flycatchers reinforces premating isolation. *Nature (London)*, **387**, 589–92.

Sakai, A. K., Weller, S. G., Wagner, W. L., Soltis, P. S., and Soltis, D. E. 1997. Phylogenetic persective on the evolution of dioecy: adaptive radiation in the endemic Hawaiian genera *Schiedea* and *Alsinidendron* (Caryophyllaceae: Alsinoideae). In *Molecular evolution and adaptive radiation* (ed. T. J. Givnish and K. J. Sytsma), pp. 455–73. Cambridge University Press, Cambridge.

Saloniemi, I. 1993. An environmental explanation for the character displacement pattern in *Hydrobia* snails. *Oikos*, **67**, 75–80.

Sanderson, M. J. 1998. Reappraising adaptive radiation. *American Journal of Botany*, **85**, 1650–5.

Sanderson, M. J. and Donoghue, M. J. 1994. Shifts in diversification rate with the origin of angiosperms. *Science (Washington, D.C.)*, **264**, 1590–3.

Sandoval, C. P. 1994. The effects of the relative geographical scales of gene flow and selection on morph frequencies in the walking-stick *Timema cristinae*. *Evolution*, **48**, 1866–79.

Sato, A., O'Huigin, C., Figueroa, F., Grant, P. R., Grant, B. R., Tichy, H., and Klein, J. 1999. Phylogeny of Darwin's finches as revealed by mtDNA sequences. *Proceedings of the National Academy of Sciences, USA*, **96**, 5101–6.

Schemske, D. W. and Bradshaw, H. D., Jr. 1999. Pollinator preference and the evolution of floral traits in monkeyflowers (*Mimulus*). *Proceedings of the National Academy of Sciences, USA*, **96**, 11910–5.

Schindel, D. E. and Gould, S. J. 1977. Biological interaction between fossil species: character displacement in Bermudian land snails. *Paleobiology*, **3**, 259–69.

Schliewen, U. K., Tautz, D., and Pääbo, S. 1994. Sympatric speciation suggested by monophyly of crater lake cichlids. *Nature (London)*, **368**, 629–32.

Schluter, D. 1982. Seed and patch selection by Galápagos ground finches: relation to foraging efficiency and food supply. *Ecology*, **63**, 1106–20.

Schluter, D. 1984. Morphological and phylogenetic relations among the Darwin's finches. *Evolution*, **38**, 921–30.

Schluter, D. 1986. Tests for similarity and convergence of finch communities. *Ecology*, **67**, 1073–85.

Schluter, D. 1988a. Estimating the form of natural selection on a quantitative trait. *Evolution*, **42**, 849–61.

Schluter, D. 1988b. The evolution of finch communities on islands and continents: Kenya vs. Galápagos. *Ecological Monographs*, **58**, 229–49.

Schluter, D. 1988c. Character displacement and the adaptive divergence of finches on islands and continents. *American Naturalist*, **131**, 799–824.

Schluter, D. 1990. Species-for-species matching. *American Naturalist*, **136**, 560–8.

Schluter, D. 1993. Adaptive radiation in sticklebacks: size, shape, and habitat use efficiency. *Ecology*, **74**, 699–709.

Schluter, D. 1994. Experimental evidence that competition promotes divergence in adaptive radiation. *Science (Washington, D.C.)*, **266**, 798–801.

Schluter, D. 1995. Adaptive radiation in sticklebacks: trade-offs in feeding performance and growth. *Ecology*, **76**, 82–90.

Schluter, D. 1996a. Ecological causes of adaptive radiation. *American Naturalist*, **148** *(Supplement)*, S40–S64.

Schluter, D. 1996b. Ecological speciation in postglacial fishes. *Philosophical Transactions of the Royal Society of London B, Biological Sciences*, **351**, 807–14.

Schluter, D. 1996c. Adaptive radiation along genetic lines of least resistance. *Evolution*, **50**, 1766–74.

Schluter, D. 1997. Ecological speciation in postglacial fishes. In *Evolution on islands* (ed. P. R. Grant), pp. 114–29. Oxford University Press, Oxford.

Schluter, D. 1998. Ecological causes of speciation. In *Endless forms: species and speciation* (ed. D. Howard and S. Berlocher), pp. 114–29. Oxford University Press, Oxford.

Schluter, D. and Grant, P. R. 1984. Determinants of morphological patterns in communities of Darwin's finches. *American Naturalist*, **123**, 175–96.

Schluter, D., Price, T. D., and Grant, P. R. 1985. Ecological character displacement in Darwin's finches. *Science (Washington, D.C.)*, **227**, 1056–9.

Schluter, D., Price, T. D., and Rowe, L. 1991. Conflicting selection pressures and life history trade-offs. *Proceedings of the Royal Society of London B, Biological Sciences*, **246**, 11–17.

Schluter, D. and McPhail, J. D. 1992. Ecological character displacement and speciation in sticklebacks. *American Naturalist*, **140**, 85–108.

Schluter, D. and McPhail, J. D. 1993. Character displacement and replicate adaptive radiation. *Trends in Ecology & Evolution*, **8**, 197–200.

Schluter, D. and Ricklefs, R. E. 1993. Convergence and the regional component of species diversity. In *Species diversity in ecological communities: historical and geographical perspectives* (eds. R. E. Ricklefs and D. Schluter), pp. 230–40. Chicago University Press, Chicago.

Schluter, D. and Price, T. 1993. Honesty, perception and population divergence in sexually selected traits. *Proceedings of the Royal Society of London B, Biological Sciences*, **253**, 117–22.

Schluter, D. and Nychka, D. 1994. Exploring fitness surfaces. *American Naturalist*, **143**, 597–616.

Schluter, D and Nagel, L. M. 1995. Parallel speciation by natural selection. *American Naturalist*, **146**, 292–301.

Schluter, D., Price, T., Mooers, A. Ø., and Ludwig, D. 1997. Likelihood of ancestor states in adaptive radiation. *Evolution*, **51**, 1699–711.

Schmidt, K. P. and Levin, D. A. 1985. The comparative demography of reciprocally sown populations of *Phlox drummondii* Hook. I. Survivorships, fecundities, and finite rates of increase. *Evolution*, **39**, 396–404.

Schoener, T. W. 1970. Size patterns in West Indian *Anolis* lizards. II. Correlations with the size of particular sympatric species—displacement and convergence. *American Naturalist*, **104**, 155–74.

Schoener, T. W. 1983. Field experiments on interspecific competition. *American Naturalist*, **122**, 240–85.

Schoener, T. W. 1984. Size differences among sympatric, bird-eating hawks: a worldwide survey. In *Ecological communities: conceptual issues and the evidence* (ed. D. R. Strong,

D. S. Simberloff, L. G. Abele, and A. B. Thistle), pp. 254–81. Princeton University Press, Princeton, N.J.
Schoener, T. W. 1989. The ecological niche. In *Ecological concepts* (ed. J. M. Cherrett), pp. 79–113. Blackwell Scientific, Oxford.
Scriber, M. J., Lederhouse, R. C., and Dowell, R. V. 1995. Hybridization studies with North American swallowtails. In *Swallowtail butterflies: their ecology and evolutionary biology* (ed. J. M. Scriber, Y. Tsubaki, and R. C. Lederhouse), pp. 269–81. Scientific Publishers, Gainsville, Florida.
Seehausen, O. and Bouton, N. 1997. Microdistribution and fluctuations in niche overlap in a rocky shore cichlid community in Lake Victoria. *Ecology of Freshwater Fish*, **6**, 161–73.
Seehausen, O., van Alphen, J. J. M., and Witte, F. 1997. Cichlid fish diversity threatened by eutrophication that curbs sexual selection. *Science (Washington, D.C.)*, **277**, 1808–11.
Seehausen, O. Lippitsch, E., Bouton, N., and Zwennes, H. 1998. Mbipi, the rock-dwelling cichlids of Lake Victoria: Description of three new genera and fifteen new species (Teleostei). *Ichthyological Exploration of Freshwaters*, **9**, 129–228.
Seehausen, O., van Alphen, J. J. M., and Lande, R. 1999. Color polymorphism and sex ratio distortion in a cichlid fish as an incipient stage in sympatric speciation by sexual selection. *Ecology Letters*, **2**, 367–78.
Sepkoski, J. J., Jr. 1979. A kinetic model of Phanerozoic taxonomic diversity. II. Early Phanerozoic families and multiple equilibria. *Paleobiology*, **5**, 222–51.
Sepkoski, J. J., Jr. 1984. A kinetic model of Phanerozoic taxonomic diversity. III. Post-Paleozoic families and mass extinctions. *Paleobiology*, **10**, 246–67.
Sepkoski, J. J., Jr. 1988. Alpha, beta, or gamma—where does all the diversity go? *Paleobiology*, **14**, 221–34.
Sepkoski, J. J., Jr. 1996. Competition in macroevolution: the double wedge revisited. In *Evolutionary paleobiology* (ed. D. Jablonski, D. H. Erwin, and J. H. Lipps), pp. 211–55. Chicago University Press, Chicago.
Sequeira, A. S., Lanteri, A. A., Scataglini, M. A., Confalonieri, V. A., and Farrell, B. D. Manuscript. *Are flightless* Galapaganus *weevils older than the Galápagos islands they inhabit?*
Service, P. M. and Rose, M. R. 1985. Genetic covariation among life history components: the effect of novel environments. *Evolution*, **39**, 943–5.
Shaw, F. H., Shaw, R. G., Wilkinson, G. S., and Turelli, M. 1995. Changes in genetic variances and covariances: **G** whiz! *Evolution*, **49**, 1260–67.
Shields, W. M. 1982. *Philopatry, inbreeding, and the evolution of sex.* State University of New York Press, Albany, New York.
Simberloff, D. and Boecklen, W. 1981. Santa Rosalia reconsidered: size ratios and competition. *Evolution*, **35**, 1206–28.
Simpson, G. G. 1944. *Tempo and mode in evolution.* Columbia University Press, New York, New York.
Simpson, G. G. 1953. *The major features of evolution.* Columbia University Press, New York, New York.
Skelly, D. K. 1995. A behavioural trade-off and its consequences for the distribution of *Pseudacris* treefrog larvae. *Ecology*, **76**, 150–64.
Skelton, P. W. 1993. Adaptive radiation: definition and diagnostic tests. In *Evolutionary patterns and processes* (ed. D. R. Lees and D. Edwards), pp. 45–58. Academic Press, New York.

Skúlason, S., Noakes, D. L. G., and Snorrason, S. S. 1989. Ontogeny of trophic morphology in 4 sympatric morphs of Arctic charr *Salvelinus alpinus* in Thingvallavatn, Iceland. *Biological Journal of the Linnean Society*, **38**, 281–301.

Slatkin, M. 1979. Frequency- and density-dependent selection on a quantitative character. *Genetics*, **93**, 755–71.

Slatkin, M. 1980. Ecological character displacement. *Ecology*, **61**, 163–77.

Slowinski, J. B. and Guyer, C. 1989. Testing the stochasticity of patterns of organismal diversity: an improved null model. *American Naturalist*, **134**, 907–21.

Slowinski, J. B. and Guyer, C. 1993. Testing whether certain traits have caused amplified diversification: an improved method based on a model of random speciation and extinction. *American Naturalist*, **142**, 1019–24.

Smith, D. C. and van Buskirk, J. 1995. Phenotypic design, plasticity, and ecological performance in two tadpole species. *American Naturalist*, **145**, 211–33.

Smith, G. R. and Todd, T. N. 1984. Evolution of species flocks of fishes in north temperate lakes. In *Evolution of fish species flocks* (ed. A. A. Echelle and I. Kornfield), pp. 45–68. University of Maine Press, Orono, Maine.

Smith, T. B. 1987. Bill size polymorphism and intraspecific niche utilization in an African finch *Nature (London)*, **329**, 717–19.

Smith, T. B. 1993. Disruptive selection and the genetic basis of bill size polymorphism in the African finch, *Pyrenestes*. *Nature (London)*, **363**, 618–20.

Sorci, G. and Clobert, J. 1999. Natural selection on hatchling body size and mass in two environments in the common lizard (*Lacerta vivipera*). *Ecology Research*, **1**, 303–16.

Southwood, T. R. E. 1972. The insect/plant relationship—an evolutionary perspective. In *Insect/plant relationships* (ed. H. F. van Emden), pp. 3–30. Blackwell Scientific, Oxford.

Spicer, G. S. 1993. Morphological evolution of the *Drosophila virilis* species group as assessed by rate tests for natural selection on quantitative characters. *Evolution*, **47**, 1240–54.

Spieth, H. T. 1974. Mating behavior and the evolution of the Hawaiian *Drosophila*. In *Genetic mechanisms of speciation in insects* (ed. M. J. D. White), pp. 94–101. Reidel, Boston, Mass.

Spiller, D. A. and Schoener, T. W. 1996. Food web dynamics on some small subtropical islands: effects of top and intermediate predators. In *Food webs: integration of patterns and dynamics* (ed. G. A. Polis and K. O. Winemiller), pp. 365–411. Chapman & Hall, New York.

Spitze, K. 1993. Population structure in *Daphnia obtusa*: quantitative genetic and allozymic variation. *Genetics*, **135**, 367–74.

Spring, J. 1997. Vertebrate evolution by interspecific hybridization—are we polyploid? *FEBS Letters*, **400**, 2–8.

Stanley, S. M. 1979. *Macroevolution: pattern and process*. W. H. Freeman, San Francisco.

Stanton, M. and Young, H. J. 1994. Selecting for floral character associations in wild radish. *Journal of Evolutionary Biology*, **7**, 271–85.

Stebbins, G. L. 1950. *Variation and evolution in plants*. Columbia University Press, New York.

Stebbins, G. L. 1970. Adaptive radiation of reproductive characteristics in angiosperms. I. Pollination mechanisms. *Annual Review of Ecology and Systematics*, **1**, 307–326.

Stebbins, G. L. 1971. Adaptive radiation of reproductive characteristics in angiosperms, II. Seeds and seedlings. *Annual Review of Ecology and Systematics*, **2**, 237–60.

Stephens, D. W. and Krebs, J. R. 1986. *Foraging theory*. Princeton University Press, Princeton, N.J.

Stern, D. L. and Grant, P. R. 1996. A phylogenetic reanalysis of allozyme variation among populations of Galápagos finches. *Zoological Journal of the Linnean Society*, **118**, 119–34.
Stiles, F. G. 1977. Coadapted competitors: the flowering seasons of hummingbird-pollinated plants in a tropical forest. *Science (Washington, D.C.)*, **198**, 1177–8.
Stone, G., Willmer, P., and Nee, S. 1996. Daily partitioning of pollinators in an African *Acacia* community. *Proceedings of the Royal Society of London B, Biological Sciences*, **263**, 1389–93.
Stratton, G. E. and Uetz, G. W. 1986. The inheritance of courtship behavior and its role as a reproductive isolating mechanism in two species of *Schizocosa* wolf spiders (Araneae; Lycosidae). *Evolution*, **40**, 129–41.
Strong, D. R., Jr. 1992. Are trophic cascades all wet? Differentiation and donor-control in speciose ecosystems. *Ecology*, **73**, 747–54.
Strong, D. R., Jr., Szyska, L. A., and Simberloff, D. S. 1979. Tests of community-wide character displacement against null hypotheses. *Evolution*, **33**, 897–913.
Strong, D. R., Jr., Lawton, J. H., and Southwood, R. 1984. Insects on plants: community patterns and mechanisms. Harvard University Press, Cambridge, Mass.
Sturmbauer, C. and Meyer, A. 1992. Genetic divergence, speciation and morphological stasis in a lineage of African cichlid fishes. *Nature (London)*, **358**, 578–81.
Sugihara, G. 1980. Minimal community structure: an explanation of species abundance patterns. *American Naturalist*, **116**, 770–87.
Suhonen, J., Alatalo, R. V., and Gustafsson. L. 1994. Evolution of foraging ecology in Fennoscandian tits (*Parus spp.*) *Proceedings of the Royal Society of London B, Biological Sciences*, **258**, 127–31.
Svärdson, G. 1979. Speciation of Scandinavian *Coregonus*. *Report from the Institute of the Freshwater Research, Drottningholm*, **57**, 1–95.
Swarth, H. S. 1931. The avifauna of the Galápagos Islands. *Occasional Papers of the California Academy of Sciences*, **18**, 1–299.
Taper, M. L. and Case, T. J. 1985. Quantitative genetic models for the coevolution of character displacement. *Ecology*, **66**, 355–71.
Taper, M. L. and Case, T. J. 1992*a*. Models of character displacement and the theoretical robustness of taxon cycles. *Evolution*, **46**, 317–33.
Taper, M. L. and Case, T. J. 1992*b*. Coevolution among competitors. *Oxford Surveys in Evolutionary Biology*, **8**, 63–109.
Taylor, E. B., McPhail, J. D., and Schluter, D. 1997. History of ecological selection in sticklebacks: uniting experimental and phylogenetic approaches. In *Molecular evolution and adaptive radiation* (ed. T. J. Givnish and K. J. Sytsma), pp. 511–34. Cambridge University Press, Cambridge.
Taylor, E. B. and McPhail, J. D. Manuscript. *Historical contingency and ecological determinism interact to prime speciation in sticklebacks*, Gasterosteus.
Templeton, A. R. 1989. The meaning of species and speciation: a genetic perspective. In *Speciation and its consequences* (ed. D. Otte and J. A. Endler), pp. 3–27. Sinauer, Sunderland, Mass.
Terborgh, J. and Weske, J. S. 1975. The role of competition in the distribution of Andean birds. *Ecology*, **56**, 562–76.

Thomas, S. and Singh, R. S. 1992. A comprehensive study of genic variation in natural populations of *Drosophila melanogaster* VII. Varying rates of genic divergence as revealed by two-dimensional electrophoresis. *Molecular Biology and Evolution*, **9**, 507–25.

Thompson, J. N., Wehling, W., and Podolsky, R. 1990. Evolutionary genetics of host use in swallowtails. *Nature (London)*, **344**, 148–50.

Thompson, J. N. 1994. *The coevolutionary process*. Chicago University Press, Chicago.

Trewevas, E., Green, J., and Corbet, S. A. 1972. Ecological studies on crater lakes in West Cameroon fishes of Barombi Mbo. *Journal of Zoology (London)*, **167**, 41–95.

True, J. R., Liu, J., Stam, L. F., Zeng, Z.-B., and Laurie, C. C. 1997. Quantitative genetic analysis of divergence in male secondary sexual traits between *Drosophila simulans* and *Drosophila mauritiana*. *Evolution*, **51**, 816–32.

Turelli, M. 1988. Phenotypic evolution, constant covariances, and the maintenance of additive variance *Evolution*, **42**, 1342–7.

Turelli, M., Gillespie, J. H., and Lande, R. 1988. Rate tests for selection on quantitative characters during macroevolution and microevolution. *Evolution*, **42**, 1085–9.

Turelli, M. and Barton, N. H. 1994. Genetic and statistical analyses of strong selection on polygenic traits: what, me normal? *Genetics*, **138**, 913–41.

Turkington, R. 1989. The growth, distribution and neighbour relationships of *Trifolium repens* in a permanent pasture. V. The coevolution of competitors. *Journal of Ecology*, **77**, 717–33.

Turner, G. F. 1994. Speciation mechanisms in Lake Malawi cichlids. *Archiv fur Hydrobiologie*, **44**, 139–60.

Turner, G. F. and Burrows, M. T. 1995. A model of sympatric speciation by sexual selection. *Proceedings of the Royal Society of London B, Biological Sciences*, **260**, 287–92.

Turner, J. R. G. 1976. Adaptive radiation and covergence in subdivisions of the butterfly genus *Heliconius* (Lepidoptera: Nymphalidae). *Zoological Journal of the Linnean Society*, **58**, 297–308.

Valentine, J. W. 1986. Fossil record of the origin of Baupläne and its implications. In *Patterns and processes in the history of life* (ed. D. M. Raup and D. Jablonski), pp. 209–22. Springer-Verlag, Berlin, Germany.

Vamosi, S. M. and Schluter, D. 1999. Sexual selection against hybrids between sympatric stickleback species: evidence from a field experiment. *Evolution*, **53**, 874–9.

Van Buskirk, J., McCollum, S. A., and Werner, E. E. 1997. Natural selection for environmentally induced phenotypes in tadpoles. *Evolution*, **51**, 1983–92.

van Doorn, G. S., Noest, A. J. and Hogeweg, P. 1998. Sympatric speciation and extinction driven by environment dependent sexual selection. *Proceedings of the Royal Society of London B, Biological Sciences*, **265**, 1915–19.

van Noordwijk, A. J., van Balen, J. H., and Scharloo, W. 1988. Heritability of body size in a natural population of the great tit and its relation to age and environmental conditions during growth. *Genetical Research*, **51**, 149–62.

van Tienderen, P. H. and van der Toorn, J. 1991*a*. Genetic differentiation between populations of *Plantago lanceolata*. I. Local adaptation in three contrasting habitats. *Journal of Ecology*, **79**, 27–42.

van Tienderen, P. H. and van der Toorn, J. 1991*b*. Genetic differentiation between populations of *Plantago lanceolata*. II. Phenotypic selection in a transplant experiment in three contrasting habitats. *Journal of Ecology*, **79**, 43–59.

Van Valen, L. M. 1965. Morphological variation and the width of the ecological niche. *American Naturalist*, **99**, 377–90.
Van Valen, L. M. 1973. A new evolutionary law. *Evolutionary Theory*, **1**, 1–30.
Vermeij, G. J. 1974. Adaptation, versatility, and evolution. *Systematic Zoology*, **22**, 466–77.
Vermeij, G. J. 1987. *Evolution and escalation: an ecological history of life*. Princeton University Press, Princeton, N.J.
Vermeij, G. J. 1994. The evolutionary interaction among species: selection, escalation, and coevolution. *Annual Review of Ecology and Systematics*, **25**, 219–36.
Vermeij, G. J. 1995. Economics, volcanoes, and Phanerozoic revolutions. *Paleobiology*, **21**, 125–52.
Via, S. 1991. The genetic structure of host plant adaptation in a spatial patchwork: demographic variability among reciprocally transplanted pea aphid clones. *Evolution*, **45**, 827–52.
Via, S. and Lande, R. 1985. Genotype-environment interaction and the evolution of phenotypic plasticity. *Evolution*, **39**, 505–22.
Vogler, A. P. and Goldstein, P. Z. 1997. Adaptation, cladogenesis, and the evolution of habitat association in North American tiger beetles: a phylogenetic perspective. In *Molecular evolution and adaptive radiation* (ed. T. J. Givnish and K. J. Sytsma), pp. 353–73. Cambridge University Press, Cambridge.
Vulić, M., Lenski, R. E. and Radman, M. 2000. Mutation, recombination and incipient speciation of bacteria in the laboratory. *Proceedings of the National Academy of Sciences, USA*, in press.
Wade, M. J. and Goodnight, C. J. 1998. The theories of Fisher and Wright in the context of metapopulations: When nature does many small experiments. *Evolution*, **52**, 1537–53.
Wagner, W. L., Herbst, D. R., and Sohmer, S. H. 1990. *Manual of the flowering plants of Hawaii*. University of Hawaii Press, Honolulu.
Wagner, W. L., Weller, S. G., and Sakai, A. K. 1995. Phylogeny and biogeography in *Schiedea* and *Alsinidendron* (Caryophyllaceae). In *Hawaiian biogeography* (ed. W. W. Wagner and V. A. Funk), pp. 221–58. Smithsonian Institution Press, Washington D.C.
Wainwright, P. C. 1994. Functional morphology as a tool in ecological research. In *Ecological morphology* (ed. P. C. Wainwright and S. M. Reilly), pp. 42–59. Chicago University Press, Chicago.
Wainwright, P. C. and Lauder, G. V. 1992. The evolution of feeding biology in sunfishes (Centrarchidae). In *Systematics, historical ecology, and North American freshwater fishes* (ed. R. L. Mayden), pp. 472–91. Stanford University Press, Stanford, California.
Waldmann, P. and Andersson, S. 1998. Comparison of quantitative genetic variation and allozyme diversity within and between populations of *Scabiosa canescens* and *S. columbaria*. *Heredity*, **81**, 79–86.
Walker, J. A. 1997. Ecological morphology of lacustrine threespine stickleback *Gasterosteus aculeatus* L. (Gasterosteidae) body shape. *Biological Journal of the Linnean Society*, **61**, 3–50.
Wang, H., McArthur, E. D., Sanderson, S. C., Graham, J. H., and Freeman, D. C. 1997. Narrow hybrid zone between two subspecies of big sagebrush (*Artemisia tridentata*: Asteraceae). IV. Reciprocal transplant experiment. *Evolution*, **51**, 95–102.
Waser, N. M. 1983. Competition for pollination and floral character differences among sympatric plant species: a review of the evidence. In *Handbook of experimental pollination ecology* (ed. C. E. Jones and R. J. Little), pp. 277–93. Van Nostrand Reinhold, New York.

Waser, N. M. 1993. Population structure, optimal outbreeding, and assortative mating in angiosperms. In *The natural history of inbreeding and outbreeding* (ed. N. H. Thornhill), pp. 173–99. Chicago University Press, Chicago.

Waser, N. M. and Real, L. A. 1979. Effective mutualism between sequentially flowering plant species. *Nature (London)*, **281**, 670–2.

Weber, K. E. 1992. How small are the smallest selectable domains of form? *Genetics*, **130**, 345–53.

Wellborn, G. A. 1994. Size-biased predation and prey life histories: a comparative study of freshwater amphipod populations. *Ecology*, **75**, 2104–17.

Wellborn, G. A., Skelly, D. K., and Werner, E. E. 1997. Mechanisms creating community structure across a freshwater habitat gradient. *Annual Review of Ecology and Systematics*, **27**, 337–63.

Weller, S. G., Sakai, A. K., Wagner, W. L., and Herbst, D. R. 1990. Evolution of dioecy in *Schiedea* (Caryophyllaceae: Alsinoideae) in the Hawaiian Islands: biogeographical and ecological factors. *Systematic Botany*, **15**, 266–76.

Werdelin, L. 1996. Community-wide character displacement in Miocene hyaenas. *Lethaia*, **29**, 97–106.

Werner, E. E. 1977. Species packing and niche complementarity in three sunfishes. *American Naturalist*, **111**, 553–78.

Werner, E. E. and Hall, D. S. 1976. Niche shift in sunfishes: experimental evidence and significance. *Science (Washington, D.C.)*, **191**, 404–6.

Werner, E. E. and Hall, D. S. 1977. Competition and habitat shift in two sunfishes (Centrarchidae). *Ecology*, **58**, 869–76.

Werner, E. E. and Hall, D. S. 1979. Foraging efficiency and habitat switching in competing sunfishes. *Ecology*, **60**, 256–64.

West-Eberhard, M. J. 1983. Sexual selection, social competition, and speciation. *Quarterly Review of Biology*, **58**, 155–83.

West-Eberhard, M. J. 1989. Phenotypic plasticity and the origins of diversity. *Annual Review of Ecology and Systematics*, **20**, 249–78.

Westman, E., Persson, L., and Christensen, B. Manuscript. *Character displacement in whitefish (Coregonus sp.) as a result of competition with efficient planktivores.*

Westneat, M. W. 1995. Phylogenetic systematics and biomechanics in ecomorphology. *Environmental Biology of Fishes*, **44**, 263–83.

Whitlock, M. C. 1995. Variance-induced peak shifts. *Evolution*, **49**, 252–9.

Whitlock, M. C. 1996. The red queen beats the jack-of-all-trades: the limitations on the evolution of phenotypic plasticity and niche breadth. *American Naturalist*, **148** *(Supplement)*, S65–S77.

Whitlock, M. C. 1997. Founder effects and peak shifts without genetic drift: adaptive peak shifts occur easily when environments fluctuate slightly. *Evolution*, **51**, 1044–8.

Whitlock, M. C., Phillips, P. C., Moore, F. B.-G., and Tonsor, S. J. 1995. Multiple fitness peaks and epistasis. *Annual Review of Ecology and Systematics*, **26**, 601–29.

Whittaker, R. H. 1977. Evolution of species diversity in land communities. *Evolutionary Biology*, **10**, 1–67.

Wiens, J. A. 1977. On competition and variable environments. *American Scientist*, **65**, 590–7.

Wiggins, I. L. and Porter, D. M. 1971. Flora of the Galapagos Islands. Stanford University Press, Stanford, California.

Wilkinson, G. S., Fowler, K., and Partridge, L. 1990. Resistance of genetic correlation structure to directional selection in *Drosophila melanogaster*. *Evolution*, **44**, 1990–2003.
Williams, E. E. 1972. The origin of faunas. Evolution of lizard congeners in a complex island fauna: a trial analysis. *Evolutionary Biology*, **6**, 47–89.
Williams, E. E. 1983. Ecomorphs, faunas, island size, and diverse end points in island radiations of *Anolis*. In *Lizard ecology: studies of a model organism* (ed. R. B. Huey, E. R. Pianka, and T. W. Schoener), pp. 326–70. Harvard University Press, Cambridge, Mass.
Williams, S. M. and Sarkar, S. 1994. Assortative mating and the adaptive landscape. *Evolution*, **48**, 868–75.
Willis, J. H., Coyne, J. A., and Kirkpatrick, M. 1991. Can one predict the evolution of quantitative characters without genetics? *Evolution*, **45**, 441–4.
Wilson, D. S. and Turelli, M. 1986. Stable underdominance and the evolutionary invasion of empty niches. *American Naturalist*, **127**, 835–50.
Wilson, E. O. 1959. Adaptive shift and dispersal in a tropical ant fauna. *Evolution*, **13**, 122–44.
Wilson, E. O. 1961. The nature of the taxon cycle in the Melanesian ant fauna. *American Naturalist*, **95**, 169–93.
Wilson, E. O. 1971. *The insect societies*. Harvard University Press, Cambridge, Mass.
Wilson, P. 1995. Selection for pollination success and the mechanical fit of *Impatiens* flowers around bumblebee bodies. *Biological Journal of the Linnean Society*, **55**, 355–83.
Wollenberg, K., Arnold, J., and Avise, J. C. 1996. Recognizing the forest for the trees: testing temporal patterns of cladogenesis using a null model of stochastic diversification. *Molecular Biology and Evolution*, **13**, 833–49.
Wood, C. C. and Foote, C. J. 1990. Genetic differences in the early development and growth of sympatric sockeye salmon and kokanee (*Oncorhynchus nerka*) and their hybrids. *Canadian Journal of Fisheries and Aquatic Sciences*, **47**, 2250–60.
Wood, T. K. 1993. Speciation of the *Enchenopa binotata* complex (Insecta: Homoptera: Membracidae). In *Evolutionary patterns and processes* (ed. D. R. Lees and D. Edwards), pp. 299–317. Academic Press, New York.
Wright, S. 1931. Evolution in Mendelian populations. *Genetics*, **16**, 97–159.
Wright, S. 1932. The roles of mutation, inbreeding, crossbreeding and selection in evolution. *Proceedings of the 6th International Congress of Genetics*, **1**, 356–66.
Wright, S. 1940. The statistical consequences of Mendelian heredity in relation to speciation. In *The new systematics* (ed. J. S. Huxley), pp. 161–83. Clarendon Press, Oxford.
Wright, S. 1945. Tempo and mode in evolution: a critical review. *Ecology*, **26**, 415–19.
Wright, S. 1959. Physiological genetics, ecology of populations, and natural selection. *Perspectives in Biology and Medicine*, **3**, 107–51.
Wu, C.-I. and Davis, A. W. 1993. Evolution of postmating reproductive isolation: the composite nature of Haldane's rule and its genetic basis. *American Naturalist*, **142**, 187–212.
Wu, C.-I., Johnson, N. A., and Palopoli, M. F. 1996. Haldane's rule and its legacy: Why are there so many sterile males? *Trends in Ecology & Evolution*, **11**, 411–13.
Wyatt, R. 1988. Phylogenetic aspects of the evolution of self-pollination. In *Plant evolutionary biology* (eds. L. D. Gottleib and S. K. Jain), pp. 109–31. Chapman & Hall, New York.
Yang, S. Y. and Patton, J. L. 1981. Genic variability and differentiation in the Galapagos finches. *Auk*, **98**, 230–42.
Yom-Tov, Y. 1991. Character displacement in the psammophile Gerbillidae of Israel. *Oikos*, **60**, 173–9.

Yom-Tov, Y. 1993a. Character displacement among the insectivorous bats of the Dead Sea area. *Journal of Zoology (London)*, **45**, 347–56.

Yom-Tov, Y. 1993b. Size variation in *Rhabdomys pumilio*: a case of character release? *Zeitschrift für Säugetierkunde*, **58**, 48–53.

Young, N. D. 1996. An analysis of the causes of genetic isolation in two Pacific Coast *Iris* hybrid zones. *Canadian Journal of Botany*, **74**, 2006–13.

Zeng, Z.-B. 1988. Long-term correlated response, interpopulation covariation, and interspecific allometry. *Evolution*, **42**, 363–74.

Zink, R. M. 1982. Patterns of genic and morphologic variation among sparrows in the genera *Zonotrichia, Melospiza, Junco*, and *Passerella*. *Auk*, **99**, 632–49.

Zink, R. M. 1991. Concluding remarks: Modern biochemical approaches to avian systematics. *Acta XX Congressus Internationalis Ornithologici*, pp. 629–36. New Zealand Ornithological Congress Trust Board, Wellington, New Zealand.

Zink, R. M. and Avise, J. C. 1990. Patterns of mitochondrial DNA and allozyme evolution in the avian genus *Ammodramus*. *Systematic Zoology*, **39**, 148–61.

Zink, R. M. and Slowinski, J. B. 1995. Evidence from molecular systematics for decreased avian diversification in the Pleistocene epoch. *Proceedings of the National Academy of Sciences, USA*, **92**, 5832–5.

INDEX

adaptation, 11
    comparative method, 11–12
    correlated random walk, 12
    current utility, 11
    historical concept, 11
    vs. adaptive radiation, 31–2
adaptive landscape, 19, 67, 72, 86–90, 105
    definition, 88
    estimates from nature, 111–15
    frequency dependent selection, 126–8
    peak shifts, 69, 115–22, 220
adaptive peak. *See* adaptive landscape
adaptive radiation
    definition, 2, 10–14
    detecting, 10–35
    ecological theory, 65–83
    end of, 63
    examples, 20–31
    narrow sense, 14
    replicated, 55–9, 64
    speciation-driven, 82
    taxonomic levels, 8, 14
adaptive ridge. *See* genetic drift
adaptive valley. *See* adaptive landscape
adaptive zone, 70, 163
    definition, 19
    progressive occupation, 7
amphibians
    Ambystoma, 230
    frogs, 55
    Plethodon, 143
    Pseudacris, 106
    Spea, 143
amphipods
    Gammarus, 108
    Hyalella, 205
Anolis. See lizards
apparent competition. *See* predation
Aquilegia. See plants

bacteria, 175, 193
    E. coli, 73

birds, 61, 165, 179, 222
    Accipiter – hawks, 130, 156
    Amazon, 156
    Amphispiza, 99
    cowbirds, 48
    Dendroica – New World warblers, 59, 62, 63
    Ficedula – flycatchers, 190, 210, 225
    finches, 93, 170
    Galápagos finches, 1, 2, 22–3, 43, 51, 66–9, 74, 82, 86, 106, 111–13, 117–18, 123, 146–9, 159, 165, 169–71, 176, 180, 198, 201–14, 222, 225, 231–4
    Hawaiian honeycreepers, 69, 165, 169–70
    Loxia – crossbills, 105–6, 180
    New Guinea, 49–51
    Parus, 156
    Phylloscopus – Old World warblers, 33, 59, 62, 63
    Pyrenestes, 105
    Sericornis, 43
    Sitta – nuthatches, 74, 143
    West Indies, 54

character displacement, 69, 74, 124–52, 161
    between distant taxa, 164–9, 179
    constant size ratios. *See* character displacement – trait over-dispersion
    criteria, 131–43
    definition, 69
    environmental variability, 74, 161
    exaggerated divergence in sympatry, 130, 132–5
    initial differences, 128
    null models, 74–5, 130
    reproductive character displacement. *See* speciation
    resource breadth, 128
    species-for-species matching, 131, 142
    symmetry, 128, 143, 146
    tests of, 143–52

character displacement (cont.)
  theory, 124–9
  trait over-dispersion, 130, 138–40, 147
  trophic level bias, 144–6
cichlid fishes. *See* fishes
coevolution
  arms race, 196
  competition. *See* character displacement
  escalation, 77
  escape and radiation, 77, 182–4
  predation. *See* predation
  self-augmenting diversity, 78
columbines. *See* plants
  comparative method, 11–12
competition
  aggressive interference, 157
  apparent competition. *See* predation
  ecological opportunity. *See* ecological opportunity
  for resources, 5, 68–9, 117, 163, 174
  for enemy free space. *See* predation

Daphnia, 91
developmental constraints. *See* genetics of divergence
divergent natural selection, 5, 66–8, 84–122
  definition, 66, 90
  direct measurement, 104–11
  tests of, 90–115
Drosophila. *See* insects

E. coli. *See* bacteria
ecological opportunity, 5, 69–70, 163–87
  key innovations. *See* key innovations
  morphological divergence, 164–75
  predation, 76–7
  speciation, 175–80
ecological speciation, 188–214
  definition, 5, 66
  speciation rate, 79
  tests of, 197–207
  theory, 70–1, 189–97
environment
  definition, 19
evolution
  genetic constraints. *See* genetic covariance
  neutral expectation. *See* genetic drift
  predictability of, 37, 55–9, 64, 204–6
  rate of, 61–3, 93
  trends, 4, 36–64, 173–4
evolvability, 186
experiments
  character displacement, 150–2
  divergent natural selection, 103, 108–10

laboratory, 8, 150, 175, 190, 193
reciprocal transplants, 95–104

fishes
  African cichlids, 4, 15, 59, 62, 80, 159–60, 180, 186, 198–200, 208–10
  Coregonus – whitefish, 57, 175
  Gasterosteus – stickleback, 57, 98, 130, 143, 146, 150, 151, 201–4, 211
  Lepomis – sunfish, 34, 105, 150, 172
  Poecilia – guppy, 108, 212
  postglacial, 57, 172, 178–9, 198–9
  Prosopium, 56
  Salmo, 57
  salmon, 204
  Salvelinus – charr, 57, 175
  Sebastes – rockfish, 15–16
fitness function. *See* natural selection
fitness surface. *See* natural selection
food webs, 78, 155, 156
  evolution of, 159–61
frogs. *See* amphibians
fungi, 43, 148

Galápagos finches. *See* birds
genetic covariance
  between environments, 99, 102–3
  between traits, 6, 119, 218–24, 229–30
genetic drift, 68, 119, 120
  along adaptive ridge, 72–3, 95, 104, 112, 121
  neutral expectation, 90–4
genetic variance, 218
  constraints on host shifts, 228–9
  heritability, 218
genetics of divergence, 6, 231–4, 215–35
  deleterious mutation model, 216
  genetic covariance. *See* genetic covariance
  genetic degrees of freedom, 220–2
  genetic line of least resistance, 219, 224–8
  genetic variance. *See* genetic variance
  infinitesimal model, 216–17
  mutation. *See* mutation
  quantitative trait loci, 224
  theory, 216–20

heritability, 218
hybrid fitness, 189, 198
  ecological mechanisms, 199–203
  genetic mechanisms, 199
  sexual selection, 190, 211
  sterility and inviability, 14, 189, 198, 209
hybrid zones, 202, 206–7

hybridization, 13, 70–1, 80, 109, 116, 157, 186, 188, 199, 201, 203, 206, 210

insects, 209
  aphids, 98, 103
  beetles, 55, 141, 150
  butterflies, 114
  Drosophila, 93, 176, 190, 194, 197, 204, 209
  Drosophila – Hawaiian, 18, 31, 34, 42, 47, 80, 160, 169, 176, 197, 208
  Enallagma – damselflies, 32, 106
  Eurosta, 202
  grasshoppers, 204
  Heliconius – butterflies, 154–5, 200
  Hymenoptera, 158
  Neochlamisus, 206
  Ophraella – leaf beetles, 228–9
  phytophages, 77, 160, 182–4
  Rhagoletis – apple maggot, 202–4
  Tegeticula – yucca moths, 157–8
isopods, 148, 165

key innovations, 5, 70, 163–4, 181–6
  examples, 182–5
  statistical tests, 181–8

Lepomis. *See* fishes
lizards
  Anolis, 23–7, 34, 43, 48, 55, 58, 62, 108, 130–1, 146, 150, 156, 169, 172, 176, 187
  Cnemidophorus, 59, 143, 149
  geckos, 55

mammals, 61–2, 69, 93, 173, 184
  cats, 130
  Dipodomys, 150
  horses, 163
  rodents, 70
mating systems, 52–3, 209
mimicry, 153
mites, 80
morphological diversity, 59–63
mutation, 41, 65, 73, 79, 90–5, 103, 119, 121, 129, 194, 196, 216–17, 224

natural selection
  adaptive landscape. *See* adaptive landscape
  convergent, 108
  density dependent, 124
  direct measurement, 104–11
  directional, 106
  disruptive, 104
  divergent. *See* divergent natural selection

experiments. *See* experiments
fitness function, 85–8, 105
fluctuating, 106, 117
frequency dependent, 124
parallel, 108
problem of correlated traits, 87, 108–9
retrospective analysis. *See* genetics of divergence
selection differential, 87
selection gradient, 87
stabilizing, 105
uniform, 73, 119
niche
  breadth. *See* specialization
  definition, 19
  divergence sequence, 49–52
nonadaptive radiation, 14, 18

parasitism. *See* predation
phenotype – environment correlation. *See* adaptive radiation – definition
phenotypic plasticity, 108, 117, 119, 136
phenotypic trait
  definition, 19
phylogeny
  ancestor reconstruction, 39
  common ancestry, 11
  comparative method, 12
  independent contrasts, 12
  monophyly, 11, 34
plants
  angiosperms, 4, 51, 52, 77, 183, 185, 209
  Aquilegia – columbines, 28–30, 42, 52, 82, 184–5
  Artemisia, 202
  Bidens, 35, 169, 171
  Brassica, 226
  Brocchinia, 160
  Cyanea, 31
  Dalechampia, 112–14
  Diodia, 109
  Dubautia. *See* plants – Hawaiian silversword alliance
  Encelia, 96
  Erodium, 150
  Gilia, 109, 110
  gymnosperms, 185
  Hawaiian silversword alliance, 3, 27–8, 43, 49, 52, 79, 199
  Hesperomannia, 176
  Impatiens, 108
  Ipomopsis, 202
  Iris, 203
  Lolium, 165

plants (*cont.*)
  Mimulus, 109, 203
  Passiflora, 154, 155
  Pinus, 180
  Populus, 202
  Raphanus, 218
  Schiedea, 52, 53
  Stylidium, 130, 157
  tarweeds, 3, 27, 28, 79, 171
  Tetramolopium, 55
  Trifolium, 165
pollinators, 51, 158, 184, 203
polyploidy. *See* speciation
postmating isolation. *See* hybrid fitness
predation, 78, 157–59
  apparent competition, 75–6, 152–6
  indirect mutualism, 75
  intraguild predation, 75
  parasitism, 158
premating isolation, 71, 80–1, 189–93, 197, 203–6, 209, 211, 235

reinforcement. *See* speciation
resources
  definition, 19

salamanders. *See* amphibians
selection landscape. *See* adaptive landscape
sexual selection, 71
  chase-away, 81, 194, 209
  divergent, 210–11
  ecological models, 193–6
  Fisherian, 81, 194
  rate of speciation, 208–9
  reinforcement, 80, 211
  sensory bias, 80, 194
  speciation theory, 79–82
shifting balance theory, 120
snails
  Cerion, 230

Littorina, 96–8
Mandarina, 59
Partula, 34, 46
snakes
  garter, 105
specialization, 37–49, 64, 83, 118
speciation, 188–214
  by-product mechanism, 71, 189–90
  coevolution, 196
  competitive, 190–3
  definition, 5, 13
  ecological. *See* ecological speciation
  ecological opportunity, 175–80
  genetic drift, 79
  habitat trends, 53–5
  non-ecological, 71, 78–81, 188, 198
  parallel, 204–6
  polyploidy, 71, 79, 198–9
  population persistence, 79, 189
  rate of, 13–18, 35, 176–9, 181, 184–5, 208–9
  reinforcement, 71, 80, 157, 191, 211
  reproductive character displacement, 80, 157
  role in adaptive radiation, 83
  sexual selection. *See* sexual selection
  sympatric, 190–3, 200–1
species
  definition, 13
stickleback. *See* fishes
sympatric speciation. *See* speciation

taxon cycle, 53–5
tetrapods, 164
trade-offs, 66, 97–9
transplant experiments. *See* experiments
trilobites, 59
trophic interactions. *See* food webs
trophic polymorphisms, 172

utility, 11